THE WORLD'S CLASSICS

ENDS AND BEGINNINGS

ALEXANDER HERZEN

ALEXANDER HERZEN (1812–70) was born in Moscow, an illegitimate child of an aristocrat. He early chose a socialist path, and became the greatest social thinker that Russia has ever produced. His views led him into exile within Russia in 1835, and in 1847 he left his country for good, living thereafter mainly in London. Herzen's major work is his memoirs, *My Past and Thoughts*, which Isaiah Berlin has called 'an autobiography of the first order of genius . . . a major classic, comparable in scope with *War and Peace*'.

AILEEN KELLY, who has edited this selection from *My Past and Thoughts* is a Lecturer in Russian at the University of Cambridge, and a fellow of King's College. Her recent work includes *Mikhail Bakunin: A Study in the Psychology and Politics of Utopianism* (Oxford, 1982; Oxford University Press), and an introduction to *Russian Thinkers* (London, 1979; Hogarth Press), a collection of essays by Isaiah Berlin.

THE WORLD'S CLASSICS

ALEXANDER HERZEN

Ends and Beginnings

Selected and edited with an introduction by
AILEEN KELLY
from
My Past and Thoughts

Translated by
CONSTANCE GARNETT

Revised by
HUMPHREY HIGGENS

Oxford New York
OXFORD UNIVERSITY PRESS
1985

Oxford University Press, Walton Street, Oxford OX2 6DP

London New York Toronto
Delhi Bombay Calcutta Madras Karachi
Kuala Lumpur Singapore Hong Kong Tokyo
Nairobi Dar es Salaam Cape Town
Melbourne Auckland

and associated companies in
Beirut Berlin Ibadan Mexico City Nicosia

Oxford is a trade mark of Oxford University Press

Introduction and selection © Aileen Kelly 1985
Revised translation © Chatto & Windus Ltd 1968

First published 1968 by Chatto & Windus
First issued as a World's Classics paperback 1985

British Library Cataloguing in Publication Data
Herzen, A. I.
Ends and beginnings.—(The World's classics)
1. Herzen, A. I. 2. Intellectuals—Russia
—Biography
I. Title   II. Kelly, Aileen   III. Higgens,
Humphrey   IV. Byloe i dumy. Part 3. English
Selections
891.78'308   DK209.6.H4
ISBN 0-19-281604-7

Set by Grove Graphics
Printed in Great Britain by
Hazell, Watson & Viney Ltd
Aylesbury, Bucks

CONTENTS

A Family Drama

Russian Shadows

THE FREE RUSSIAN PRESS AND 'THE BELL'

INTRODUCTION

ALEXANDER HERZEN was once described, in an essay by Isaiah Berlin, as one of the three moral preachers of genius born on Russian soil. Like Tolstoy and Dostoevsky, he placed his extraordinary creative powers in the service of an ideal of how men should live. But – and here he stands alone among Russia's great writers – there was no conflict in him between the artist and the preacher. On the contrary, his memoirs, the work which puts him in the first rank of Russian writers, owe much of their quality as art, as well as social commentary, to the ideal of freedom which pervades them. This ideal was the basis of an original and far-sighted political philosophy, whose relevance to the needs and problems of our own time is in itself sufficient justification for a further edition of *My Past and Thoughts* in English translation.

The problems faced by editors of Herzen's memoirs are compounded by the fact that the author left no authoritative final text of the work. Herzen began writing his memoirs in 1852, and over the next sixteen years published fragments, mainly in the periodicals which he edited in London, *The Pole Star* and *The Bell*. In the 1860s he embarked on a complete edition of the work. Only three volumes were published before his death in 1870, and it was another fifty years before the first complete version (its structure based on indications in Herzen's letters and papers) was published in M. K. Lemke's edition of Herzen's collected works (Moscow/Petrograd 1918–19). The first complete English translation of the memoirs, edited by Humphrey Higgens (4 vols, Chatto & Windus, 1968), was based on Constance Garnett's translation (Chatto & Windus, 1922–7). An abridged version, in one volume, of Higgens' edition, edited by Dwight Macdonald, appeared in 1973. In 1980 a translation by J. Duff of the first two parts of the memoirs was reprinted, together with an introduction by Isaiah Berlin, in the *World's Classics* series of the Oxford University Press, under the title *Childhood, Youth and Exile*. The present volume (its sequel) is an abridgement of the remaining parts, based on the Higgens edition.

As he confessed to a friend, Herzen had had deep hesitations over the form his memoirs should take: a political *apologia pro vita sua*, or a discursive account after the model of Goethe's *Dichtung und Wahrheit*. He finally opted for the second: hence the work's dazzling multiplicity of genres, styles and moods: comedy, satire, sober analysis, lyrical and philosophical digressions, and a subject-matter ranging in scale from the major upheavals of nineteenth-century Europe (the 1848 revolutions, the rise of the European bourgeoisie and the decay of the Russian serf-owning aristocracy) to the most intimate personal emotions and tragedies. The work also includes a number of polemical articles, responses to the crises of the age, which were first published in *The Bell*.

The present selection from the memoirs, unlike Macdonald's, does not attempt to convey its multidimensional nature.[1] One strand alone has been selected: the strictly autobiographical. Herzen's account of his life, both public and private, though sometimes remarkably frank, is very incomplete. But the frequent gaps and distortions in chronology do nothing to diminish its power as an account of a journey of the intellect and spirit. Though structurally interwoven with the multiple other strands, it can nevertheless stand alone. Like Tolstoy's (much shorter) *Confession*, it owes its dramatic power and unity to one central theme: the savagely destructive questioning, one by one, of contemporary faiths and authorities in the search for an ideal of self-fulfilment which would withstand this annihilating critique. But while the *Confession* is a linear progression to triumphant certainty, *My Past and Thoughts* is a much more complex account of experience and introspection. One of the generation of 'superfluous men' immortalised by Turgenev (Herzen himself wrote a novel on the subject, of which he was the thinly-disguised hero), he possessed to a very high degree the capacity for introspection and subtle dissection of motive characteristic of the type; and he acquired very early a profound scepticism about all political and moral formulae and systems which claimed to accommodate and satisfy men's complex and contradictory impulses. In this sense Herzen was far closer to Dostoevsky than to Tolstoy. The portrait of an individual and a generation presented in this volume sheds much light on the confusion of ideals and the unacknowledged motives behind the political disasters whose seeds were sown in the middle of the nineteenth century.

[1] See p. xvi for the marking of the omissions.

The first two parts of the memoirs, contained in *Childhood,
Youth and Exile*, relate the origins of Herzen's political protest in
the stifling atmosphere of Nicholas I's Russia. We see the
rôle of European romanticism in shaping an ideal of the
autonomous and many-sided personality to which the educated
élite of Herzen's generation aspired and which, vague though it
was, aroused the Russian government's suspicions: the volume
ends with Herzen's arrest and imprisonment on a nebulous
charge, followed by three years of exile in the remote provincial
town of Vyatka.

The present volume opens with the date of 1838, the year of
Herzen's marriage and of his transfer (for the remaining eighteen
months of his exile) to a town within reach of his Moscow friends.
Among the intelligentsia of the capital, the inward-looking
romanticism of the previous decade had been replaced by a
fascination with the historiosophical schemas of German Idealism,
and the burning question of Russia's historical destiny was beginning to separate Russia's intellectual élite into the camps of
Slavophiles and Westernisers. Back at the centre of Russian intellectual life (he was pardoned in 1839, although another spell of
banishment followed in 1841), Herzen entered into polemics with
both camps, in the course of which he elaborated that concept of
freedom on which he would build his political philosophy, and
which has been brilliantly analysed in the essay by Isaiah Berlin
which introduces *Childhood, Youth and Exile*. It was based on
the premiss (which owed much to the influence of the Left
Hegelians) that the chief obstacle to freedom was man's eternal
tendency to immolate himself and others in the name of doctrines,
teleologies and moral and political absolutes which were the
abstract constructions of his own mind. In his age's obsession
with the high-sounding concepts of progress, humanity and the
common good, Herzen saw the seeds of new forms of tyranny:
his own political ideal was a form of anarchism modelled on the
structure of the Russian peasant commune in which social relations would be regulated by the need to maintain a dynamic
balance between individual and general goals.

Herzen's onslaughts on political doctrinairism and philosophies
of progress were to alienate him from all the major political
groupings of his time in Russia and (after his emigration in 1847)
in Europe; the autobiographical narrative in *My Past and*

Thoughts is increasingly punctuated by polemical digressions and ironical sketches of the organisational confusion and ideological disarray of progressive parties and their leaders after the failure of the 1848 revolution. This material has been excluded from the present volume, with the exception of portraits of members of the Russian intelligentsia which are clearly pertinent to the autobiographical narrative.

Herzen believed that the particular historical experience of his generation contained lessons of great relevance to the problems which Europe was facing. The year before he began his memoirs he published a letter to the French historian Michelet, entitled *The Russian People and Socialism*, in which he declared that the alienation suffered by Russia's 'superfluous men' had given them an enormous compensating advantage in the battle to establish new foundations for human societies. Even the most progressive Europeans, he argued, were conservatives in their deepest being, doomed by their attachment to their cultural heritage to repeat old mistakes in new ways. The Russian intelligentsia had no such ties. Peter the Great's revolution had broken their links with their national traditions, but had replaced them with nothing that could command their allegiance. Nor, as the foundlings of the European family, could they share its commitment to values and institutions which could not be justified by rational argument. Having absorbed the finest fruits of European culture (the socialist vision of a just society), they had no scruples or commitments to hold them back from spelling out, with ruthless logic, all the consequences that followed from them. Herzen's generation of thinking Russians, he concluded, were, intellectually at least, the most emancipated creatures on earth: 'We are free agents, because we are self-made.' With their fearless and clearsighted consistency they could do much to ensure that the new social order to which all progressive Europeans were striving would not enshrine old forms of oppression under new names.

The bravado of this self-image was a challenge to the view then current among European radicals, that Russia was a barbarous land where even the educated elite were the despot's willing slaves. (The flamboyant Slav nationalism preached by Bakunin in 1848 had done much to foster the suspicions of Marxists in particular that Russian revolutionaries were the conscious or unconscious tools of Tsarist imperialism.) When, in 1852, Herzen began his memoirs, his desire to vindicate himself and his countrymen had been intensified by a scandal which reverberated throughout the European political emigration: his quarrel with the German radical poet Georg Herwegh, following Herwegh's affair with Herzen's wife and her tragic death. The account of the affair which Herwegh spread among the emigrés cast a shadow on Herzen's personal and political honour. As he wrote to Proudhon, it brought into question the integrity of his wife and himself as representatives of the society of the future; and he had hopes (which were unfulfilled) that a 'court of honour' composed of leading revolutionaries would be set up to judge his and Herwegh's conduct.

Although the long account of the affair contained in this volume was never published during Herzen's lifetime (it first appeared in Lemke's edition of his works), there is no doubt as to the importance that Herzen attached to it. As his Soviet editors suggest, had it not been for his desire to justify his conduct to posterity, the long-contemplated project of his memoirs, to which he proceeded in the same year, might never have been fulfilled. Unfortunately, Herzen's analysis of character and motivation in the chapters concerned with this drama does not show him at his best: even those who have not read E. H. Carr's account of the affair in *The Romantic Exiles* will find their credulity strained by a version of events in which the *beau rôle* is reserved for the author, who spares no effort to convince the reader of his enemy's total moral worthlessness. But the more he blackens Herwegh, the more evasive he is forced to be about the awkward fact that the man whom he represents as a villain and a coward, the object of ridicule in radical circles, was for over two years his only intimate friend in that milieu. However, if one takes account of the circumstances in which it was written, the imaginary 'court of honour' before which Herzen felt that not only he, but also his long-suffering and much-calumniated countrymen were on trial, one can excuse much of the rhetoric of A *Family Drama*.

The defects of these chapters rarely appear elsewhere in this volume, where there is no special pleading: Herzen's desire to explain and vindicate the Russian intelligentsia as a historical phenomenon leads him despite himself into an analysis of the psychology of Russia's superfluous men which results in a picture much darker, much more complex and contradictory, and altogether more credible than the propaganda image he presented to the West in 1851.

Herzen's account of his conflicts with his own and the next generation reveals that the iconoclasm which he had represented to Michelet as the outstanding characteristic of the Russian intelligentsia was counterbalanced by an equally marked thirst for new faiths and dogmas to fill the void created by their negation. Slavophiles who denounced Western influences on Russian culture drew their faith in a mythical national past from the dreams of European romanticism and German Idealist philosophy; Westernisers who preached the liberation of reason from the bonds of tradition believed uncritically in the universal validity and applicability of European values; while Bakunin, the most radical member of Herzen's generation, wasted his powers in the futile pursuit of his fantasy of universal destruction. The radicals of the 1860s denounced their 'fathers' for compromising with a society and culture that had to be extirpated root and branch; but they themselves were mirror-images of the regime which had warped and stunted them: their bullying intolerance, their neurotic amour-propre and their narrow dogmatism reeked of the barrack-room and government offices.

Herzen's acuteness of vision does not always extend to his own inconsistencies; beneath the self-justifying rhetoric of A Family Drama there is more than a hint that the sexual freedom he preached was not in practice extended to his own wife; while his dismissive contempt for the aesthetic deficiencies of the bourgeois culture of mid-nineteenth century France reveals this 'self-made' man to be steeped in the values of an aristocratic caste and a preindustrial society. But this unintentional self-revelation only adds weight to his depiction of that extraordinary mixture of clarity of vision and self-delusion which characterised all the best representatives of his generation, as it has characterised all great Russian apostles of negation, from Tolstoy to Solzhenitsyn. Herzen's exploration of the pathology of alienation in My Past and Thoughts was, and is, a powerful challenge to the sanguine belief that the dispossessed have only to throw off their chains for liberty and harmony to prevail. It magnificently illustrates his critique, in his most mature political writings, of all forms of political messianism which confer on a nation or class, by reason of its past history or present distance from power and privilege (or both), special virtues which give it a title to moral leadership or political power.

When he began writing his memoirs, Herzen himself had thought that the Russian intelligentsia, together with the peasants in their communes, possessed such virtues. Like other Russian romantic nationalists, he had believed that the 'historical freshness' and 'untapped forces' of the Russian people guaranteed it a brilliant rôle on a future historical stage. By the time he completed the memoirs, sixteen years later, he had come to believe that history conferred no special immunities from the prejudices, superstitions and persistent delusions which separated human consciousness from the goal of rational autonomy.

The years in which *My Past and Thoughts* was composed thus saw a fundamental shift in Herzen's political vision: from a form of nationalist utopianism characteristic of many of the Russian intelligentsia towards a critique of utopian thought in which he achieved greater consistency than any other European socialist, including Marx. But not only is the autobiographical narrative disappointingly thin and reticent on this crucial period: Herzen also misleads the reader on the date and nature of the crisis which led to the sober realism of his last years. His memoirs would have us believe that it came in 1848, when the events of that year destroyed a faith in the revolutionary regeneration of Europe that until then had been central to his existence. In reality, seduced by Slavophile dreams of renewal from the East, he had frequently suggested in the years before he left Russia that Europe might have reached the end of her development and might be on the verge of a cataclysmic collapse: he had been fond of comparing the existing state of Europe to the last years of the Roman Empire. Unlike the Slavophiles (whose wholesale contempt for the goals of western culture he did not share, any more than he shared their romantic conservatism and their Orthodox faith), he had not asserted that Russia's future as Europe's historical successor was automatically assured. For this to be possible the germ of a new order contained in the 'barbarian' East must first be nurtured with the aid of progressive ideas and technology from the West. But if this condition were fulfilled, he had suggested that it might fall to Russia's lot to put into practice the socialist ideals which were the most precious legacy of a decrepit Europe. The events of

1848 were thus not the unexpected blow that in retrospect he makes them out to be; nor was his despair so total. The darker the West, the brighter might be the saving dawn from the East: this message is conveyed by the contrast between 'old Europe' and 'young Russia' which runs through his propaganda writings of the 1850s. It was not 1848, but the Polish rising of 1863; not the blunders of French and German radicals, but the character of the emerging Russian revolutionary movement, which shattered Herzen's last illusions, and on this subject his memoirs are reticent.

One reason for this is plain : his portrayal of the Russian intelligentsia was directed largely to a European audience; and if it had turned out rather differently from what the readers of *The Russian People and Socialism* might have expected, Herzen had no intention of adding fuel to the anti-Russian feeling among the European Left (augmented in the late 1860s by the rift between Bakuninists and Marxists in the International) by dwelling on the causes of his own disillusionment. But its magnitude can be deduced by comparing his portraits of the Russian radicals towards the end of this volume with the picture of the typical Russian intellectual ('the most emancipated creature on earth') which he had painted for the edification of the West in 1851.

In the 1860s, writing in *The Bell* for a Russian readership, Herzen represented the emancipation of human consciousness as a slow and painful process, in which education and patient propaganda must prepare the way for the transformation of social institutions; and he attacked what he saw as the fatal inconsistency in the young Russian radicals' attempts to bring the new world into being with the methods of the old: force, terror, and the imposition of a single political orthodoxy. But his increasingly bitter polemics with the young radicals on the question of means did not destroy his faith in the revolutionary potential of the Russian peasant. Hence his decision to put the support of *The Bell* firmly behind the Polish rising, in the hope that it would spread to Russia, leading to a peasant insurrection and ultimately a revolution. In his account of this episode in the last chapter of the present volume, Herzen represents himself as having been opposed to the rising from the first, on the grounds that it was premature and ill-prepared, but forced to give it reluctant support largely out of loyalty to a heroic but pathetic figure from the past – Bakunin, who, having spent the preceding decade in prison and exile, had preserved intact the illusions that Herzen had lost in the aftermath of 1848. In reality, the distance between Herzen and Bakunin in 1863 was by no means as great as it seemed to Herzen in retrospect. The belief in the 'historical fresh-ness' and 'vital forces' of the Russian people, on which Bakunin based his hopes for a successful revolution, was one which Herzen had been proclaiming for over a decade, and from his cor-respondence in 1863 one may conclude that Herzen allied him-self to Bakunin and the Polish cause less from quixotic loyalty than because it offered him the last hope of witnessing that salvation from the East that he had long promised to a sceptical Europe. If the peasantry did not fulfil their hopes, he wrote to Ogarev, the co-editor of The Bell, they might as well both retire from the revolutionary struggle. As the rising gained momentum, Herzen became increasingly optimistic about the prospect of the 'dawn of our freedom'. In the event, the peasantry failed him, while the Polish revolutionaries and their Russian sympathisers, in their incapacity for organisation, lack of realism, and confusion as to their goals seemed to re-enact many of the fatal blunders of 1848. The Russian Minerva, as Turgenev had once remarked to Herzen à propos his faith in the Russian peasant, was not very different from her European sister – only somewhat broader in the beam.

For Herzen, 1863 was a personal as well as a political catas-trophe. His alliance with the Poles did not heal his rift with the extreme Left, and it lost him his only remaining audience in Russia – the moderate opposition, which shared the anti-Polish mood of the nationalist Right. The circulation of The Bell, which, when it was smuggled into Russia in the late 1850s, had been read by every section of educated society (including the Tsar's own officials), dropped dramatically and continued to dwindle until in 1867 it ceased publication (it was briefly revived in a French edition). The collapse of The Bell, which had shaped the political consciousness of a generation, is recorded laconically in the closing paragraph of this volume. Herzen left no autobiographical account of the years of political isolation and impotence which followed. It is likely that a combination of national pride and self-esteem (he would never directly admit to having shared those delusions which he describes with such olympian detachment) prevented Herzen from leaving a record of the most painful years of his public life. But *My Past and Thoughts* is a testimony to the wisdom he attained at such personal cost.

A few months before his death in 1870, Herzen published a cycle of letters to Bakunin in which he reflects on the divergence of their respective political paths since the time when, as young radicals fresh from the school of Left Hegelian negation, they had contemplated the ruins of 1848 and called on the forces of destruction to sweep away the whole rotten structure of Europe. Events had since revealed to Herzen the complexity and obliqueness of the paths of history, the obstinate attachment of human consciousness to the traditional and familiar, even when these were the source of its oppression. Their ultimate goal had remained the same; the question which divided them was whether, given the slowness of the pace of progress, they should attempt to force it:

You rush ahead as before, with your passion for destruction, which you take to be a creative passion, smashing every obstacle and respecting history only in the future. I no longer believe in the revolutionary paths of the past; I try to understand the pace of history in the past and the future, in order to know how to go in step with it, not falling behind and not running so far ahead that people can't follow me.

My Past and Thoughts is the living record of his achievement.

NOTE ON THE OMISSIONS

In making her selection, the editor has omitted passages of varying length. Some indication of the extent of each omission is given by the use of three different symbols: for part of a paragraph, an ellipsis (. . .) within the text, in square brackets; for one or more paragraphs, an ellipsis in a space; for one or more sections or chapters, an asterisk in a space. Ellipses within the text without square brackets are Herzen's own.

VLADIMIR

ON THE KLYAZMA

1838–1839

Do not expect from me long accounts of my inner life of that period ... Frightful events, woes of all sorts, are yet more easily put upon paper than quite bright and cloudless memories ... Can happiness be described?

Fill in for yourselves what is lacking, divine it with the heart – while I will tell of the external side, of the setting, only rarely, rarely touching by hint or by word on my ineffable secrets.

A. I. HERZEN: *My Past and Thoughts*

CHAPTER I

THE TWO PRINCESSES

WHEN I was five or six years old and was very naughty, Vera Artamonovna[1] used to say: 'Very well, very well, you wait a bit, I'll tell the princess everything as soon as she comes.' I was at once subdued by this threat and begged her not to complain.

Princess Marya Alexeyevna Khovansky, my father's sister, was a stern, forbidding old lady, stout and dignified, with a birthmark on her cheek and false curls under her cap; she used to screw up her eyes as she spoke, and to the end of her days, that is till she was eighty, used a little rouge and powder. Whenever she caught sight of me she persecuted me; there was no end to her lecturing and grumbling; she would scold me for anything, for a crumpled collar, or a stain on my jacket, would declare I had not gone up to kiss her hand properly, and make me go through the ceremony again. When she had finished lecturing me, she would sometimes say to my father, taking with her fingertips a pinch of snuff out of a tiny gold snuff-box: 'My dear, you should send your spoilt child to me to be corrected; he would be as soft as silk when he had been a month in my hands.' I knew that they would not give me up, but I shivered with horror at these words.

My fear of her passed off with the years, but I never liked the old princess's house; I could not breathe freely in it, I could not be myself, but like a trapped hare, looked uneasily from one side to the other to make my escape.

The old princess's household was not in the least like my father's or the Senator's.[2] It was an old-fashioned, orthodox Russian household in which they kept the fasts, went to the early service, put a cross on the doors on the eve of Epiphany, made marvellous pancakes at Shrove-tide, ate pickled pork with horse-radish, dined at exactly two o'clock and supped at nine. The Western contagion which had infected her brothers and thrown them somewhat out of their native rut had not touched the life of the old princess; on the contrary, she disapproved of the way in which 'Vanyusha and Levushka', as she called my father[3] and uncle, had been corrupted in 'that France'.

Princess Marya Alexeyevna lived in a wing of the house occupied by her aunt, Princess Anna Borisovna Meshchersky, a maiden lady of eighty.

This Princess Meshchersky was the living and almost the only link connecting all the seven ascending and descending branches of the family. At the chief holidays all the relations gathered about her. She reconciled those who were at variance and brought together those who had drifted apart. She was respected by all, and she deserved it. At her death family ties were loosened and lost their rallying-point, and the relations forgot each other.

She had seen the education of my father and his brothers through; after the death of their parents she looked after their property until they came of age. She put them into the Guards, and she made marriages for their sisters. I do not know how far she was satisfied with the fruits of her bringing up, which with the help of a French engineer, a kinsman of Voltaire, had turned them into landowners and *esprits forts*, but she knew how to inspire esteem for herself, and her nephews, though not greatly disposed to feelings of obedience and reverence, respected the old lady and often obeyed her to the end of her life.

Princess Anna Borisovna's house, preserved by some miracle at the time of the fire of 1812, had not been repaired nor redecorated for fifty years: the silk hangings that covered the walls were faded and blackened; the lustres on the chandeliers, discoloured by heat and turned into smoky topazes by time, shook and tinkled, glittering and shining dimly when anyone walked across the room. The heavy, solid mahogany furniture, ornamented with florid carvings that had lost all their gilt, stood gloomily along the walls; chests of drawers with Chinese incrustations, tables with little copper trellis-work, rococo porcelain dolls – all recalled another age and different manners.

Grey-headed flunkeys sat in the outer hall, occupied with quiet dignity in various trifling tasks, or sometimes reading half aloud from a prayer-book or a psalter, the pages of which were darker than its cover. Boys stood at the doors, but they were more like old dwarfs than children – they never laughed nor raised their voices.

A deathly silence reigned in the inner apartments; only, from time to time, there was the mournful cry of a cockatoo, its unhappy, faltering effort to repeat a human word, the bony tap of its beak against its perch, covered with tin, and the disgusting whimper of a little old monkey, shrunken and consumptive, that lived in the great hall, on a little shelf of the stove with its Dutch tiles. The monkey, dressed like a *débardeur*, in full, red trousers, gave the whole room a peculiar and extremely unpleasant smell. In another big room hung a multitude of family portraits of all sizes, shapes, periods, ages, and costumes. These portraits had a peculiar interest for me, especially from the contrast between the originals and their semblances. The young man of twenty with a powdered head, dressed in a light-green embroidered *caftan*, smiling courteously from the canvas, was my father. The little girl with dishevelled curls and a bouquet of roses, her face adorned with a patch, mercilessly tight-laced into the shape of a wine-glass, and thrust into enormous petticoats, was the formidable old Princess Marya Alexeyevna.

The stillness and the stiffness grew more marked as one approached the princess's room. Old maidservants in white caps with wide frills moved to and fro with little teapots, so softly that their footsteps were inaudible; from time to time a grey-headed manservant in a long coat of stout dark-blue cloth appeared in a doorway, but his footsteps too were as inaudible, and, when he gave some message to the head parlourmaid, his lips moved without making a sound.

The little, withered, wrinkled, but by no means ugly, old lady, Princess Anna Borisovna, was usually sitting or reclining on the big clumsy sofa, propped up with cushions. One could scarcely distinguish her; everything was white, her morning dress, her cap, the cushions, the covers on the sofa. Her waxen white face of lace-like fragility, together with her faint voice and white dress, gave her an air of something that had passed away, that was scarcely breathing.

The big English clock on the table with its loud, measured spondee — tick-tack, tick-tack — seemed marking off the last quarters of an hour of her life.

Between twelve and one Princess Marya Alexeyevna would enter and settle herself with dignity in a big easy-chair. She was dull in her empty apartments. She was a widow, and I still remember her husband, a little grey-headed old gentleman who, unknown to the princess, drank liqueurs and home-made beverages; he never played an important part in the house, and was accustomed to obey his wife implicitly – though he sometimes rebelled against her in word, especially after his secret potations, but never in deed. The princess would be surprised at the great effect produced on her spouse by the minute glass of vodka which he drank officially before dinner, and she used to leave him in peace to play the whole morning with his blackbirds, nightingales, and canaries, which trilled at the pitch of their throats in emulation of each other; he trained some of them with a little organ, others by whistling to them himself; he used to drive off very early to the bird-market to exchange, sell, and buy birds; he took an artistic delight in succeeding as he supposed, in cheating a dealer. . . . And so he continued his profitable life, until one morning, after whistling to his canaries, he fell forward on his face and two hours afterwards died.

His widow was left alone. She had had two daughters, both of whom married not for love but simply to escape from the maternal yoke. Both died in their first childbirth. The princess was really an unfortunate woman, but her troubles rather warped her character than softened it. The blows of fate made her not milder, not kinder, but harder and more forbidding.

Now she had no one left but her brothers and, most important, her aunt. The old lady, from whom she had scarcely parted all her life, drew her still closer to herself after her husband's death. The aunt had no say in the running of her household; the niece managed everything like an autocrat, and oppressed her aunt under the pretext of looking after her and caring for her wants.

There were always some old women of every sort, either *habituées* of the old princess's house, or encamped there temporarily, like nomads, ranged along the walls or sitting in various corners. Half saints and half vagrants, rather crazy and very devout, sickly and extraordinarily unclean, these old women trailed from one old-fashioned house to another: in one they were fed, in another presented with an old shawl; from one place they were sent groats and firewood, from another cloth and cabbage; and so they somehow made both ends meet. Everywhere they were a nuisance, everywhere they were passed over, everywhere put in the lowest seat, and everywhere received through boredom and emptiness and, most of all, through love of gossip. In the presence of other company these mournful figures were usually silent, looking with envious hatred at each other . . . . They sighed, shook their heads, made the sign of the cross, and muttered to themselves the number of their stitches, prayers, and perhaps even words of abuse. On the other hand, *tête à tête* with their *benefactresses* and *patronesses*, they rewarded themselves for their silence by the most treacherous gossip about all the other benefactresses who received them, fed them and made them presents.

They were continually begging from the old princess and, in return for her presents, often made without the knowledge of her niece, who did not like them to be indulged, brought her holy bread, hard as a stone, and unnecessary woollen and knitted articles of their own make, which the old lady afterwards sold for their benefit, regardless of whether the purchaser was willing or not.

Apart from birthdays, name-days, and other holidays the most solemn gathering of kinsmen and friends in Princess Anna Borisovna's house took place on New Year's Eve. On that day she 'elevated' the Iversky Madonna. The holy icon was carried through all the apartments by chanting monks and priests. The old princess walked under it in front, crossing herself, and after her all the visitors, men and maid servants, old people and children. After this they all congratulated her on the approaching New Year, and made her all sorts of trifling presents such as are given to children. She would play with them for a few days and then give them away.

My father used to take me every year to this heathen ceremony; everything was repeated in exactly in same order, except that some old men and women were missing, and their names were intentionally avoided; only the old lady would say:

'Our Ilya Vasilyevich is no longer here, the Kingdom of Heaven be his! . . . Whom will the Lord summon in the coming year?'

And she would shake her head dubiously.

And the spondees of the English clock would go on measuring out the days, the hours, the minutes, and at last it reached the fatal second. The old lady felt unwell on getting up one day; she walked about the rooms and was no better; her nose began bleeding, and very violently; she was feeble and tired, and lay down fully dressed on her sofa, fell quietly asleep . . . and did not wake up. She was over ninety.

She left her house and the greater part of her property to her niece, but the inner significance of her life she did not hand on to her. Princess Marya Alexeyevna could not continue the – in its own way – elegant rôle of head of the family, of the patriarchal link connecting many threads. With the death of Princess Anna Borisovna everything at once took on an aspect of gloom, as in mountainous places at sunset; long, black shadows on everything. Princess Marya Alexeyevna shut up her aunt's house entirely and remained living in a wing; the yard was overgrown with grass, the walls and picture-frames grew blacker and blacker; the front hall, in which ungainly, yellowish dogs were for ever asleep, fell out of the perpendicular.

Friends and relations came less frequently, her house was deserted; she was distressed at it, but did not know how to put things right.

The only survivor of the whole family, she began to fear for her own useless life, and mercilessly repulsed everything that could disturb her physical or moral equilibrium and cause her uneasiness or annoyance. Afraid of the past and of memories, she removed every object that had belonged to her daughters, even their portraits. It was the same with her aunt's belongings – the cockatoo and the monkey were exiled to the servants' hall, and then turned out of the house. The monkey lived out its days in the coachman's quarters at the Senator's, suffocated by the rank tobacco and amusing the postillions.

The egoism of self-preservation has a fearfully hardening effect on the heart of the old. When her last surviving daughter's illness became quite hopeless, they tried to persuade the mother to go home, *and she went*. At home she at once ordered spirits of various sorts and cabbage leaves (she used to tie them on her head) to be got ready, that she might have everything necessary at hand when the *terrible news* should come. She did not take leave of her dead husband nor of her daughter, she did not see them after their death and was not at their funerals. When later on the Senator, her favourite brother, died, she guessed what had happened from a few words dropped by a nephew, and *begged him* not to tell her the sad news nor any details of the end. With these precautions against one's own heart, and such a compliant heart, one may well live to eighty or ninety in perfect health and with undisturbed digestion.

...

To preserve her peace untroubled, the old princess established a special sort of police, and entrusted the command of it to skilled hands.

Besides the nomadic old women inherited from Princess Anna Borisovna, she had a permanent lady companion living with her. This post of honour was filled by the healthy, rosy-cheeked widow of a Zvenigorod government clerk, very proud of 'being a lady' and of her dead husband's rank of assessor; a bustling termagant who could never forgive Napoleon the premature death of her Zvenigorod cow, which had perished in the war of 1812. I remember how seriously troubled she was on the death of Alexander I about what width of crape weepers she should wear to conform with her *rank*.

This woman had played a very insignificant part in the household while Princess Anna Borisovna was alive, but afterwards she managed so adroitly to humour the younger princess's caprices and apprehensive anxiety about herself, that she obtained the same control over her as the princess herself had had over her aunt.

Draped in the weepers suited to her rank, this Marya Stepanovna Makashin bounced about the house like a ball from morning to night; she shouted and was noisy, gave the servants no peace, complained of them, held investigations into the doings of the maids, slapped the boys and pulled them by the ears, dealt with the accounts, ran into the kitchen, rushed out to the stables, brushed away the flies, rubbed the princess's feet, and made her take her medicine. The members of the household no longer had access to their mistress; the woman was an Arakcheyev,[4] a Biron,[5] in a word, a Prime Minister. The princess, a haughty and, although in the old-fashioned style, a well-bred woman, was often vexed, especially at first, by the Zvenigorod widow, by her scolding voice and market-woman's manners, but she gradually put more and more confidence in her, and saw with delight that Marya Stepanovna considerably diminished the household expenses, which had not been over-high before. For whom the princess was saving her money it is hard to say; she had no near relations except her brothers, who were twice as wealthy as she was.

For all that the princess was really bored after the death of her husband and daughters, and was glad when an old Frenchwoman, who had been her daughters' governess, came to spend a fortnight with her, or when her niece from Korcheva paid her a visit. But these were only passing, rare events, and the tedious *tête à tête* with her lady companion did not fill the intervals satisfactorily.

An occupation, a plaything, and a distraction had been provided for her in a quite natural way not long before her aunt's death.

CHAPTER II

THE FORLORN CHILD

In the middle of 1825 'The Chemist',[1] who found his father's affairs in great confusion when he took them over, sent his brothers and sisters from Petersburg to the Shatskoye estate; he assigned them the house there and their keep, proposing to arrange for their education and their future later on. My aunt, Princess Marya Alexeyevna, drove over to look at them. A child of eight[2] caught her attention by her mournfully pensive face; my aunt put her in the carriage, took her home and kept her.

The mother was glad, and went off with the other children to Tambov.

The Chemist gave his consent – it did not matter to him.

'Remember all your life,' Marya Stepanovna kept saying to the little girl when they had reached home, 'remember that the princess is your *benefactress* and pray that her days may be long. What would you be without her?'

And so into this house which had ceased to live, which was gloomily oppressed by two restless old women, one full of whims and caprices, the other her indefatigable spy, devoid of all trace of delicacy or tact, a child was brought, torn away from everything familiar to her, strange to everything surrounding her, and adopted out of boredom, as people take a puppy, or as my aunt's husband had kept canaries.

The little girl with a face so pale that it had a bluish tinge was sitting at the window in a long, woollen mourning frock when my father brought me a few days later to visit the princess. She was sitting in silence, scared and bewildered, gazing out of the window, afraid to look at anything else.

My aunt called her over and presented her to my father. Always frigid and ungracious, he patted her carelessly on the shoulder, observed that his late brother had not known what he was about, abused The Chemist, and began talking of something else.

The little girl had tears in her eyes; she sat down by the window again and again began to look out of it.

A hard life was beginning for her. Not one warm word, not one tender glance, not one caress; beside her, around her, were strangers, wrinkled faces, yellowed cheeks, sickly creatures whose life was smouldering out. Princess Marya Alexeyevna was always stern, exacting and impatient, and she kept the forlorn child at such a distance that it could never enter her head to seek refuge with her, to find warmth or comfort in being near her, or to shed tears. Visitors took no notice of her. Marya Stepanovna put up with her as one of the princess's whims, as something superfluous, but which could do her no harm; she even made a show of protecting the child and interceding for her with the princess, especially before outsiders.

The child did not grow used to her surroundings, and a year later was as little at home as on the day of her arrival, and even more sorrowful. Even the princess was surprised at her 'seriousness', and sometimes, seeing her sitting dejectedly for hours together at her little embroidery frame, would say to her:

'How is it you don't play and run about?'

The little girl would smile, flush, and thank her, but stay where she was.

And the old lady left her in peace, in reality caring nothing about the child's sadness and doing nothing to distract her. Holidays came, other children were given playthings, other children talked of treats, of new clothes. . . . No presents were given to the little orphan. The princess considered that she was doing enough for her in giving her shelter; she had shoes – what did she want with dolls? And in fact she did not need them, for she did not know how to play; besides, she had no one to play with.

. . .

Loneliness and harsh treatment at the tenderest age left a black trace on her soul, a wound which never fully closed.

'I do not remember,' she writes in 1837, 'any time when I could utter the word "mother" freely and spontaneously, any person on whose breast I could lay my head in security, forgetting everything. I have been a stranger to everyone since I was eight years old; I love my mother . . . but we do not know each other.'[3]

Looking at the pale face of the twelve-year-old girl, at her big eyes with dark rings round them, at her tired listlessness and ever-lasting sadness, many people thought she was one of the pre-destined, early victims of consumption, those victims marked from childhood by the finger of death with a special imprint of beauty and premature thoughtfulness. 'Perhaps,' she says, 'I should not have survived this struggle if I had not been saved by our meeting.'

And I was so slow to understand her and read her heart!

Till 1834 I failed to appreciate the richly gifted nature that was unfolding beside me, although nine years had passed since the old princess had presented her to my father in her long woollen dress. It is not hard to explain. She was *farouche*, I was preoccupied; I was sorry for the child who sat so solitary and sad in the window, but we saw each other very seldom. It was only rarely and always unwillingly that I went to the princess's; still more rarely did she bring her to see us. Besides, my aunt's visits almost always left unpleasant impressions. She and my father usually quarrelled over trifles and, when they had not seen each other for two months, they said barbed things to each other, hiding them in affectionate turns of phrase, just as nasty medicines are covered with a coating of sugar. 'My dear boy,' the princess would say; 'My dear girl,' my father would answer, and the quarrel would go its way. We were always glad when the princess drove away. Moreover, it must not be forgotten that at that time I was completely absorbed by my political dreams and my studies, and my life was in the university and my comrades.

But what had *she* to live on, besides her melancholy, during those long, dark *nine years*, surrounded by stupid bigots, haughty relations, tedious priests and fat priests' wives, hypocritically patronised by the 'lady companion', not allowed to go further from the house than the gloomy courtyard overgrown with grass and the little garden at the back?

From the foregoing lines it may be seen that the princess was not particularly lavish in her expenditure on the education of her adopted child. Her moral training she undertook herself; it consisted in external observances and in inoculation with a complete system of hypocrisy. The child had from early morning to be laced up, standing at attention, with her hair properly dressed: this might be admissible so far as it was not injurious to health; but the princess put her soul in stays as well as her waist, suppressing every open, spontaneous feeling; she insisted on a smile and an air of gaiety when the child was sad, on amiable phrases when she wanted to cry, on an appearance of participation in subjects of no interest – in short, on continual duplicity.

At first the poor girl was taught nothing, on the pretext that learning things early was useless; later on, that is *three or four years later*, wearied by the observations made by the Senator and even by outsiders, the princess made up her mind to arrange for her to be taught, keeping in view the strictest economy.

For this purpose she availed herself of an old governess who considered herself under an obligation to the princess and sometimes was in need of her assistance. In this way the French language was brought down to the lowest price; but in return for that, it was taught *à bâtons rompus*.

But the Russian language, too, was equally cheapened; to teach it and *all other subjects*, the princess engaged the son of a priest's widow, to whom she had been a benefactress – of course, at no special expense to herself: through her good offices with the metropolitan the widow's two sons had been made priests in the cathedral. The instructor was their elder brother, the deacon of a poor parish, burdened with a large family. He was in the lowest depths of poverty, was glad of any payment, and dared not haggle over terms with his brothers' benefactress.

Nothing could have been more pitiful, more insufficient than such an education; and yet all went well, it all brought forth marvellous fruits, so little is needed for development if only there is something to develop.

The poor deacon, a tall, thin, bald man, was one of those enthusiasts whom neither years nor misfortunes can cure of their dreams; on the contrary, their troubles tend to keep them in a state of mystic contemplation. His faith, which approached fanaticism, was sincere and not without a shade of poetry. Between these two, the father of a hungry family and the forlorn child fed on alien bread, a good understanding sprang up at once.

The deacon was received in the princess's household as a defenceless, and at the same time mild-tempered pauper usually is received, with barely a nod, barely a condescending word. Even the 'lady companion' thought it necessary to show her disdain; while he scarcely noticed either them or their manners, taught his subjects with love, was touched by his pupil's readiness of understanding, and could move her to tears. This the old princess could not understand; she scolded the child for being a cry-baby and was greatly displeased, declaring that the deacon was upsetting her nerves. 'It's . . . I don't quite know what,' she said: 'it's not at all like a child!'

Meanwhile the old man's words were opening before the young creature another world, attractive in a very different way from that in which religion itself was turned into an affair of diet, reduced to keeping the fasts, and going to church at night, where fanaticism, nourished by fear, walked side by side with imposture, where everything was limited, artificial, and conventional, and oppressed the soul with its narrowness. The deacon put the Gospel into his pupil's hands – and it was long before she let it go again. The Gospel was the *first book* she read, and she read it over and over again, with her one friend, Sasha,[4] her old nurse's niece, now a young maid of the princess's.

Later on I knew Sasha very well. Where and how she had managed to develop I never could understand, born, as she was, between the coachman's quarters and the kitchen, and never leaving the maids' room, but developed she was, unusually so. She was one of those innocent victims who perish unnoticed in the servants' quarters, and more often than we suppose, are crushed by the state of being a serf. They perish not only without compensation, without commiseration, without an hour of brightness, without a joyful memory, but without knowing, without themselves suspecting, what is perishing in them and how much is dying with them.

Their mistress says with vexation: 'The wretched girl was just beginning to be trained to her work when she took to her bed and died.' . . . The seventy-year-old housekeeper grumbles: 'What are servants coming to nowadays? They are worse than any young lady,' and goes to the funeral dinner. The mother weeps and weeps and begins to drink – and that is the end of the business.

And we pass hurriedly by, seeing the terrible dramas enacted at our feet, thinking we have more important things to fill our time, and feeling that we have done our part with a few roubles and a kindly word. And then suddenly, astounded, we hear the heart-rending moan with which the broken spirit lets itself be heard of for all time, and as though awakening from sleep we ask ourselves whence came that spirit, that strength.

Princess Marya Alexeyevna killed her maid, unintentionally and unconsciously, of course; she worried her to death over trifles, broke her heart, oppressed her whole life, wore her out with humiliations and with rough churlish treatment. For several years she would not let her marry, and allowed it only when she could see consumption in her suffering face.

Poor Sasha, poor victim of the loathsome, accursed Russian life, defiled by serfdom, by death you escaped to freedom! And yet you were incomparably happier than others: in the harsh bondage of the princess's house you met a friend, and the affection of her whom you loved so immeasurably accompanied you invisibly to the grave. You cost her many tears; not long before her own death she still thought of you, and blessed your memory as the one bright image that had appeared in her childhood!

The two young girls (Sasha was a little the elder) used to get up early in the mornings when all the household was still asleep, read the Gospel and pray, going out into the courtyard under the open sky. They prayed for the princess and her lady-companion, besought God to soften their hearts; they invented ordeals for themselves, ate no meat for weeks together, dreamed of a nunnery and of the life beyond the grave.

Such mysticism is in keeping with adolescence, with the age in which everything is still a secret, still a religious mystery, when the awakening thought is still shining dimly through the mist of early morning, and the mist is not yet dissipated by experience or passion.

At quiet and gentle moments I loved in after years to hear of these childish prayers, with which one full life began and one unhappy existence ended. The image of the forlorn child outraged by coarse patronage, and of the *slave girl* outraged by the hopelessness of her situation, praying for their oppressors in the neglected courtyard, filled the heart with tenderness, and breathed a rare peace upon the spirit.

The pure and gracious being, whom none of her equals appreciated in the princess's senseless household, won, besides the devotion of the deacon and Sasha, a warm response and adoration from all the servants. These simple people saw in her more than a kind and affable young lady: they divined in her something higher for which they felt reverence; they had faith in her. The girls of the princess's household, when they were going to their weddings, would beg her to pin a ribbon on them with her own hands. One young maidservant – I remember her name was Yelena – was suddenly stricken with a stitch in the side; it turned out to be acute pleurisy; there was no hope of saving her, and the priest was sent for. The frightened girl kept asking her mother whether it was all over with her; the mother, sobbing, told her that God would soon summon her. Then the sick girl clung to her mother and besought her with bitter tears to fetch her young lady that she might come herself to bless her with the holy icon for the other world. When she came the sick girl took her hand, laid it on her forehead, and repeated: Pray for me, pray for me!' The young girl, herself in tears, began praying in a low voice, and the sick girl died as she prayed. All in the room knelt round her, crossing themselves; Natalie closed the dead girl's eyes, kissed the cold forehead, and went out.

Only dry and ungifted natures know nothing of this romantic period; they are as much to be pitied as those frail and sickly beings in whom mysticism outlives youth and remains for ever. In our age this does not happen with realistic natures; but whence could the *secular* influence of the nineteenth century penetrate into the princess's house, where every crevice was so well caulked?

A crack was found, nevertheless.

My Korcheva cousin[5] used sometimes to come on a visit to the princess. She was fond of the 'little cousin', as one is fond of children, especially if they are unhappy, but she did not understand her. With amazement, almost with dread, she discovered later on her exceptional nature, and, impulsive in everything, at once determined to make up for her neglect. She begged from me Hugo, Balzac, or anything new I might have.

'The little cousin,' she said to me, 'is a genius, we ought to guide her onward!'

The 'big cousin' – and I cannot help smiling at this name for her, for she was a tiny creature – at once communicated to her protégée every stray thought in her own mind – Schiller's ideas and Rousseau's, revolutionary ideas picked up from me and the dreams of a girl in love picked up from herself. Then she secretly lent her French novels, verses, poems; they were for the most part books that had appeared since 1830. With all their defects, they stimulated thought, and baptised youthful hearts with the spirit and with fire. In the novels and stories, the poems and songs of that period, whether the author knew it or not, the vein of social feeling everywhere pulsed strongly: everywhere social sores were revealed and the groan of the hungry could be heard, the innocent galley-slaves of labour; even by that date their murmur and complaint was no longer feared as a crime.

I need hardly say that my cousin lent the books without any discrimination, without any explanations, and I think there was no harm in that; there are natures which never need help, support or guidance from others, who always walk most safely where there is no rail.

Another person who carried on the secular influence of my Korcheva cousin was soon added to the list. The princess at last made up her mind to engage a governess, and to avoid expense had taken on a young *Russian* girl[6] who had only just left a boarding-school.

Russian governesses are not thought much of with us; at any rate they were not in the 'thirties, yet for all their defects they are better than the majority of French girls from Switzerland, of courtesans on indefinite leave and retired actresses who catch at teaching in despair as a last means of earning their daily bread, a resource needing neither talent nor youth, nothing in fact but the ability to pronounce 'hrrra' and the manners *d'une dame de comptoir*, which in the provinces are often taken for 'good' manners. Russian governesses come from boarding-schools, or foundling hospitals, and so have had some sort of regular education, and are free from the *petit bourgeois* coating which the foreign women bring with them.

The French governesses of today must be distinguished from those who used to come to Russia before 1812. In those days France was less *bourgeois* and the women who came to Russia belonged to quite a different social stratum. To some extent they were the daughters of emigrants, of ruined noblemen, or widows of officers, often their deserted wives. Napoleon used to marry off his warriors in the way that our landowners marry their serfs, without much regard for love or inclination. He wanted, by these marriages, to unite his new gun-powder aristocracy with the old nobility; he wanted to knock his Skalozubs[7] into shape by means of their wives. Accustomed to blind obedience, they married without protest, but soon abandoned their wives, finding them too stiff for the festivities of the barracks and the bivouac. The poor women made their way to England, to Austria, to Russia. The old Frenchwoman who used to stay with the princess belonged to this earlier class of governess. She spoke with a smile in an exquisite style and never made use of a single strong expression. She was entirely made up of good manners and never forgot herself for a minute. I am convinced that even at night in her bed she was more preoccupied with teaching the proper way to sleep than with sleeping.

The young governess was an intelligent, alert, energetic girl with a good share of boarding-school enthusiasm and an innate feeling for what is fine. Active and high-spirited, she brought more life and movement into the existence of her pupil and friend.

There had been a shade of mourning, of melancholy in the sad and depressing friendship with the wasting Sasha. Her company, together with the words of the deacon and the absence of every kind of diversion, was drawing the young girl away from the world and from people. The arrival of this third person was very timely, for she was young, full of life and gaiety, and at the same time sympathetic with everything visionary and romantic, and came in the nick of time: she drew her back to earth, to the true, the real soil.

At first the pupil to some extent adopted her Emilia's outward ways; a smile was more often to be seen on her face, and her conversation grew livelier; but within a year the natures of the two girls had occupied places that corresponded to the specific gravity of each. The nice, dreamy Emilia gave way before the stronger nature and was completely dominated by her pupil, saw with her eyes, thought her thoughts, lived in her smile and in her affection.

Before I had finished my studies at the university, I took to going more frequently to the princess's house. The younger girl seemed pleased when I came, and sometimes her cheeks glowed and her talk grew more animated, but she quickly withdrew into her usual pensive stillness, recalling the cold beauty of sculpture or Schiller's 'Mädchen aus der Fremde' who put a stop to every approach.

It was not unsociability or coldness, but an active inner life; not understood by others, she did not as yet understand herself, and had rather a presentiment than a knowledge of what was in her. In her lovely features there was still something incomplete, not fully expressed; they lacked a spark, a touch of the sculptor's chisel which would decide whether she was destined to pine and fade away in sandy soil, knowing neither herself nor life, or to reflect the glow of passion, to be enfolded by it, and to live, perhaps to suffer – certainly, indeed, to suffer, but *to live abundantly.*

I first saw the impress of life coming out on her half-childish face on the eve of our long separation.

Well I remember her eyes with quite a different light in them, and all her features with their significance transformed, as though penetrated by a new thought, a new fire . . . as though the secret had been guessed and the inner mist dissipated. This was[8] when I was in prison. A dozen times we said goodbye, and still we could not bear to part. At last my mother, who had come with Natalie* to the Krutitsky Barracks, resolutely got up to go. The young girl trembled, turned pale, squeezed my hand with unexpected strength, and repeated, turning away to hide her tears, 'Alexander, don't forget your sister.'

The gendarme saw them out and started his walking up and down. I flung myself on my bed and gazed long at the door behind which that bright apparition had vanished. 'No,' I thought, 'your brother will not forget you.'

The next day I was taken to Perm, but before I speak of our time of separation I shall tell of something else that prevented me, before my prison days, from understanding Natalie better and from being more friendly with her. I was in love!

Yes, I was in love, and the memory of that pure, youthful love is as dear to me as the memory of a spring day spent by the sea amid flowers and singing. It was a dream that wafted over me much that was lovely, and vanished as dreams usually do vanish !

I have mentioned already that there were few women in our

---

* I know very well how affected the French translation of names sounds, but a name is a traditional thing and how is one to change it? Besides, all unSlavonic names are with us, as it were, truncated and less musical; we, educated to some extent, 'not in the law of our fathers', [from *Graf Nulin*, by A. S. Pushkin] in our youth 'romanticised' names, while the powers in authority 'Slavonicised' them. As a man is promoted and attains to influence at court, the letters in his name are changed – thus, for instance, Count Strogonov remained to the end of his days Sergey Grigorevich, but Prince Golitsyn was always called Sergiy Mikhaylovich. The last example of such a promotion we saw in General Rostovtsev, *so celebrated* in connection with the 14th of December; throughout the reign of Nicholas Pavlovich he was Yakov, as was Yakov Dolgoruky, but with the accession of Alexander II he became Iakov, the same as the brother of our Lord!

circle, especially of the sort with whom I could have been on intimate terms: my affection for my Korcheva cousin, ardent at first,
little by little became more even in tone. After her marriage we
saw each other less often, and then she went away. A desire for a
warmer, tenderer feeling than the affection of my men friends
hovered undefined about my heart. Everything was ready, all that
was lacking was 'she'. In one of the families of our acquaintance
there was a young girl[9] with whom I quickly made friends. It was
a strange chance that brought us together. She was engaged to be
married, but some dissension suddenly arose and her fiancé abandoned her and went off to the other end of Russia. She was in
despair, overcome with distress and mortification. With deep and
sincere sympathy I saw her being consumed by grief. Without
daring to hint at the cause, I tried to comfort her and distract her
mind, brought her novels, read them aloud to her, told her long
stories, and sometimes neglected to prepare for my lectures at the
university in order to stay longer with the distressed girl.

Little by little her tears fell less frequently, and from time to
time a smile glimmered through them; her despair passed into a
languid sadness. Soon she began to feel frightened for her past, she
struggled with herself and defended it against the present, from
a *point d'honneur* of the heart, as a soldier defends the flag,
though he knows that the battle is lost. I saw these last clouds
faintly lingering on the horizon and, myself carried away, with
a beating heart, softly, softly drew the flag out of her hands, and
when she ceased to cling to it I was in love. We believed in our
love. She wrote verses to me, I wrote whole essays to her in prose,
and then we dreamed together of the future, of exile, of prisons,
for she was ready for anything. The external side of life never took
a very clear shape in our imaginations; doomed to battle with a
monstrous power, we felt success almost impossible. 'Be my
Gaetana,' I said to her after reading Saintine's[10] 'The Mutilated
Poet', and I used to fancy how she would accompany me to the
Siberian mines.

'The Mutilated Poet' was the poet who wrote a lampoon upon
Sixtus V and gave himself up when the Pope gave his word not to
inflict the death penalty. Sixtus V ordered his tongue and hands
to be cut off. The figure of the luckless victim, choked by the mass
of ideas which swarmed in his brain and found no outlet, could
not help attracting us in those days. The martyr's sad and

exhausted eyes found peace when they rested with gratitude and some remnant of joy on the girl who had loved him before and did not abandon him in misfortune. Her name was Gaetana.

This first experience of love was soon over, but it was perfectly sincere. Perhaps, indeed, it was right for this love to pass, or it would have lost its finest, most fragrant worth, its innocent freshness, its nineteen-year-old charm. Lilies of the valley do not flower in winter.

And can it be, my Gaetana, that you do not recall our meeting with the same serene smile? Can it be that there is any bitterness mixed with your memory of me after twenty-two years? That would be very grievous to me. And where are you, and how have you spent your life?

I have lived my life and now am trudging downhill, broken, and morally 'mutilated'. I seek no Gaetana, I go over the memories of the past and gladly meet your image among them ... Do you remember the window in the corner facing the little side street into which I had to turn, and how you always came to it to see me off, and how disappointed I was if you did not come, or moved away before it was time for me to turn?

But I should not want to meet you in reality; in my imagination you have remained with your youthful face, your *blond cendré* curls: remain as you were. And you, too, if you think of me, will remember a well-knit lad with sparkling eyes and fiery speech, and may you think of him like that and never know that the eyes have lost their lustre, that I have grown heavy, that my brow is furrowed, that long ago my face lost its former radiant, eager look which Ogarëv [11] used to call 'the look of hope'. And, indeed, hope too is gone.

We ought to be for each other as we were then ... neither Achilles nor Diana grows old ... I do not want to meet you as Larina met Princess Alina: [12]

> 'Do you remember Grandison?
> Cousin, how is Grandison? –
> Oh, Grandison! In Moscow, living at Semeon's.
> He visited me on Christmas Eve;
> A son of his was married lately.'

The last, flickering flame of love for a moment lit up the prison vault, warmed the heart with its old dreams, and then we each went our way. She went away to the Ukraine while I was setting off into exile. Since then I have had no news of her.

CHAPTER III

SEPARATION

'Ah, people, wicked people,
You separated their . . .'

So my first letter to Natalie ended, and it is noteworthy that, frightened by the word 'hearts', I did not write it, and signed the letter 'your brother'.

How dear 'my sister' was then to me and how continually in my thoughts is clear from the fact that I wrote to her from Nizhny, from Kazan and from Perm[1] on the very day after my arrival there. The word 'sister' expressed all that was recognised in our affection; I liked it immensely and I like it now, used not as the limit of the feelings but, on the contrary, as the mingling of them all; in it are united friendship, love, the tie of kinship, a common devotion, surroundings, family, and habitual association. I had called no one by that name before, and it was so dear to me that in later years, too, I often used it to Natalie.

Before I fully understood our relationship, and perhaps just because I did not understand it fully, a different temptation awaited me which did not pass for me as such a luminous spell as was my encounter with Gaetana; a temptation that humbled me and cost me much regret and inner distress.

Having very little experience of life, and being flung into a world completely alien to me, after nine months of prison, I lived carelessly at first and, without looking round; the new country, the new surroundings dazzled me. My social position was transformed. In Perm and in Vyatka I was regarded quite otherwise than I had been in Moscow; there I had been a young man living in my father's house, here in this slough I stood on my own feet, and was accepted as a government official, although I was not quite one. It was not hard for me to perceive that without much effort I might play the part of a man of the world in the drawing-rooms beyond the Volga and the Kama, and be a lion in Vyatka society.

In Perm, before I had time to look about me, the landlady to whom I had gone to take lodgings asked me whether I wanted a kitchen garden and whether I was keeping a cow! It was a question by which I could, with horror, judge the depth of my descent from the academic heights of student life. But at Vyatka I made acquaintance with all the *monde*, especially with the younger people of the merchant class, which is much better educated in these remote provinces than in those nearer the centre, though they are no less given to dissipation. Turned aside from my usual pursuits by my work in the office, I led a restlessly idle life; owing to my peculiar impressionability, or perhaps mobility, of character and absence of experience, adventures of all sorts might well be expected.

From a coquettish passion *de l'approbativité* I tried to please right and left indiscriminately, forced my sympathies, made friends over a dozen words, became far more intimate than I need have, recognised my mistake a month or two later, said nothing, out of tact, and dragged a weary chain of false relationships until it was broken by an absurd quarrel in which I was blamed for capricious impatience, ingratitude, and inconstancy.

At Vyatka I did not live alone at first. A strange, comic figure, which from time to time appears at all the turning points of my life, at all its important events, the person who nearly gets drowned to make me acquainted with Ogarëv, and waves a hand-kerchief from Russia when I cross the frontier at Taurogen – K. I. Sonnenberg[2] – lived with me at Vyatka; I forgot to mention this when I described my exile.

This was how it happened: at the moment when I was being sent to Perm, Sonnenberg was preparing to go to the Fair at Irbit. My father, who always liked to complicate anything simple, suggested to Sonnenberg that he should go to Perm and there *furnish my house*, undertaking in return to pay his travelling expenses.

At Perm Sonnenberg zealously set to work, that is, to purchasing unnecessary articles, all sorts of crockery, saucepans, cups, glass, and provisions. He went himself to Obva to procure a Vyatka horse *ex ipso fonte*. When everything was complete I was transferred to Vyatka. We sold, at half-price, the goods he had purchased and left Perm. Sonnenberg, conscientiously carrying out my father's wishes, thought it his duty to go to Vyatka too to furnish my house. My father was so well pleased with his devotion and self-sacrifice that he offered him a salary of a hundred roubles a month so long as he would stay with me. This was more profitable and more secure than Irbit – and he was in no hurry to leave me.

At Vyatka he bought not one but three horses, one of which belonged to himself, though it too was bought at my father's expense. These horses elevated us extraordinarily in the eyes of Vyatka society. Karl Ivanovich, as I have already said, was, in spite of his fifty years and of considerable defects in his looks, a great dangler after women, and was agreeably convinced that any girl or woman who came near him risked the fate of a moth flying near a lighted candle. Karl Ivanovich had no intention of wasting the effect produced by the horses, but tried to turn them to his advantage in the erotic department. Moreover, all our circumstances favoured his designs; we had a verandah looking out into a courtyard beyond which there was a garden. From ten o'clock in the morning Sonnenberg, arrayed in Kazan *ichigi*, a gold embroidered *tyubiteyka*, and a Caucasian *beshmet*,[3] with an immense amber mouthpiece between his lips, would sit on watch, pretending to be reading. The *tyubiteyka* and the amber mouthpiece were all aimed at three young ladies who lived in the next house. The young ladies for their part were interested in the new arrivals and gazed with curiosity at the oriental-looking doll smoking on the verandah. Karl Ivanovich knew when and how they secretly lifted their blind, thought that things were going successfully – and tenderly blew a gentle stream of smoke in the direction of the objects of his devotion.

Soon the garden gave us the opportunity of making our neighbours' acquaintance. Our landlord had three houses, and the garden was common to them all. Two of the houses were occupied: we lived in one of them, together with the landlord and his stepmother, a fat, flabby widow who looked after him so maternally and with such jealousy that it was only on the sly that he ventured to speak to the ladies of the garden. In the second house lived the young ladies and their parents, and the third house stood empty. Within a week Karl Ivanovich was quite at home with the female society of our garden. He constantly spent several hours a day swinging the young ladies in the swing and running to fetch their capes and sunshades, in a word he was *aux petits soins*. The young ladies behaved more foolishly with him than with anybody else, just because he was more beyond suspicion than Caesar's wife: a mere glance at him was enough to check the most audacious piece of scandal.

In the evenings I used to walk into the garden, too, from that herd instinct which makes people do what others are doing, apart from any inclination. To the garden came, besides the lodgers, their acquaintances; the chief subject of talk and interest was flirtation and spying on one another. Karl Ivanovich devoted himself to sentimental espionage with the vigilance of a Vidocq,[4] and always knew who walked with whom most often, and who looked meaningly at whom. I was a terrible stumbling-block for all the secret police in our garden; the ladies and the men wondered at my reserve, and for all their efforts could not discover whom I was making up to, and who particularly attracted me; and indeed it was not easy to do so, for I was not making up to anyone and I did not find any of the young ladies particularly attractive. In the end they were annoyed and offended by this, they began to consider me haughty and sarcastic, and the friendliness of the young ladies grew noticeably cooler – though everyone of them tried her most deadly glances upon me when we were alone.

While things were like this, Karl Ivanovich informed me one morning that the landlady's cook had opened the shutters of the third house that morning and was cleaning the windows. The house had been taken by a family who had arrived in the town.

The garden was absorbed exclusively in details concerning the new arrivals. The unknown lady, who was either tired from the journey or had not yet had time to unpack, refused, as though to spite us, to show herself outside. Everyone tried to see her at a window or in the front hall, some succeeded, while others watched for days together in vain; those who saw her reported her to be pale and languid, interesting, in short, and good-looking. The young ladies said that she looked melancholy and sickly. A young clerk in the governor's office, a scamp and a quite intelligent fellow, was the only one who knew the strangers. He had once served in the same province with them, and everyone besieged him with questions.

This sprightly clerk, pleased at knowing what other people did not, held forth endlessly upon the charms of their new neighbour. He praised her to the skies, and declared that you could see she was a lady from Petersburg or Moscow.

'She is intelligent,' he repeated, 'nice, cultured, but she won't look at fellows like us. Ah, upon my soul,' he added, suddenly turning to me, 'there's a happy thought; you must keep up the honour of Vyatka society and get up a flirtation with her ... Why, you are from Moscow, you know, and in exile; no doubt you write verses. She's a godsend for you.'

'What rot you're talking,' I said, laughing, but I flushed crimson: I longed to see her.

A few days later I did meet her in the garden and found that she really was a very attractive blonde. The gentleman who had talked about her introduced me. I was agitated and was as little able to hide it as my sponsor could his smile.

The shyness of vanity passed and I got to know her; she was very unhappy and, deceiving herself by an assumed composure, was pining away and languishing in a kind of inertia of the heart.

Mme R—[5] was one of those secretly passionate natures only to be met among women of a fair complexion. The ardour of their hearts is masked by the mildness and gentleness of their features; they turn pale with emotion, and their eyes do not flash but rather grow dim when feeling brims over. Her languid eyes looked exhausted with a vague craving, her unsatisfied breast heaved unevenly. There was something restless and electric in her whole being. Often when walking in the garden she would suddenly turn pale and, inwardly troubled or alarmed, would answer absent-mindedly and hurry into the house. It was just at those moments that I liked to look at her.

I soon saw what was passing within her. She did not love her husband and could not love him; she was twenty-five, he was over fifty, yet that disparity she might have got over, but the difference of education, of interests, of temperament, was too great.

Her husband scarcely ever came out of his room; he was a dry, harsh, elderly man, an official with pretensions to being a land-owner, irritable like all invalids and like most people who have lost their fortune. She had been sixteen when she married him and then he had some property, but afterwards he had lost every-thing at cards and was forced to go into the service for a living. Two years before he was transferred to Vyatka he began to fall into ill-health; a sore on his leg developed into caries. The old man became surly and ill-humoured, was afraid of his illness, and looked with alarmed, helpless suspicion at his wife. She waited upon him with mournful self-sacrifice, but she did this as her duty. Her children could not make amends for everything; her unoccupied heart longed for something more.

One evening, speaking of one thing and another, I said that I should very much like to send my cousin my portrait, but that I could not find a man in Vyatka who could hold a pencil.

'Let me try,' said the lady. 'I used to draw rather successful portraits in pencil.'

'I shall be delighted. When?'

'Tomorrow before dinner, if you like.'

'Of course. I will come tomorrow at one o'clock.'

All this was in her husband's presence; he said not a word.

The next morning I got a note from Mme R—. It was the first I ever received from her. She very courteously and circumspectly informed me that her husband was displeased at her having offered to draw my portrait, begged me not to judge harshly of the whims of an invalid, said that he must not be worried, and, in conclusion, offered to make the sketch some other day, saying nothing about it to her husband, that he might not be upset by it.

I thanked her warmly, perhaps excessively warmly. I did not accept her offer to draw the portrait in secret, but nevertheless these two notes drew us much closer together. Her attitude to her husband, upon which I should never have touched, was openly expressed; a secret understanding, a league against him, was involuntarily formed between us.

In the evening I went to see them – not a word was said about the portrait. If her husband had been cleverer he must have guessed what had happened; but he was not clever. I thanked her with my eyes, and she answered with a smile.

Soon they moved into another part of the town. The first time I went to see them I found her alone in a barely furnished drawing-room; she was sitting at the piano, and her eyes were tear-stained. I begged her to go on; but the music dragged, she played false notes, her hands trembled, the colour left her face.

'How stifling it is!' she said, getting up quickly from the piano.

In silence I took her hand, a weak, feverish hand; her head, like a flower grown too heavy, as though passively obeying some external force, sank on my breast, she pressed her forehead against me and instantly fled.

The next day I received a rather frightened note from her, trying to gloss over the evening before; she wrote of the fearfully nervous state in which she had been when I came in and of scarcely remembering what had happened. She apologised for her behaviour – but the thin veil of her words could not conceal the passion that blazed between the lines.

I went to see them; that day her husband was better, though he had not got up from his bed since they had been in their new quarters. I was excited, played the fool, fired off witty jokes, talked all sorts of nonsense, made the invalid almost die with laughter, and of course all this was to smother her embarrassment and my own. Moreover, I felt that the laughter was intoxicating and captivating her.

Two weeks went by. Her husband was less and less well, and at half past nine he would ask visitors to leave. His weakness, emaciation and pain increased. One evening, about nine, I said goodnight to the sick man, and Mme R— came to see me out. In the drawing-room a full moon laid three oblique bands of pale mauve across the floor. I opened the window: the air was pure and fresh, and I was bathed in it.

'What an evening!' I said. 'And how I wish I weren't going.'

She came to the window.

'Stay here for a bit.'

'I can't. This is when I change his bandage.'

'Come back afterwards: I'll wait for you.'

She was silent, and I took her hand.

'Well, come, then. I beg you . . . you'll come?'

'I really can't. I put an overall on first.'

'Come in an overall. I've found you in an overall in the mornings, many times.'

'But if someone sees you?'

'Who? Your manservant is drunk: let him go to bed; and your Darya loves you more than she does your husband, truly. And she's a friend of mine. Anyhow, where's the harm? Goodness, it's only just after nine: you wanted to ask me to do something for you, and asked me to wait . . .'

'In the dark . . . ?'

'Have some candles brought. Though this night is as good as day.'

She still hesitated.

'Come, darling, come!' I whispered in her ear, calling her that for the first time.

She shivered.

'I'll come – but only for a minute.'

I waited for more than half an hour. All was quiet in the house: I could hear the old man sighing and coughing, his slow talk, a table being moved. The drunk servant was whistling as he prepared his bed on a chest in the hall; he swore, and in a minute began to snore. The heavy tread of the house-maid, as she left the bedroom, was the last sound I heard. Then silence, a groan from the sick man; and silence again . . . Then a rustle, the floor creaked, light footsteps, and a white overall glimmered in the doorway.

She was so violently agitated that at first she could not pronounce a single word; her lips were cold and her hands like ice. I could feel how hard her heart was beating.

'I've carried out your[6] wish,' she said at last. 'Now let me go. . . . Goodbye . . . goodbye, for God's sake – and you go home,' she added in a voice of sad entreaty.

I embraced her and pressed her firmly to my breast.

'My dear . . . but go!'

That was impossible. *Troppo tardi*. . . . To leave her at a moment when her heart and mine were beating so – that would have been beyond human power, and very foolish. . . . I did not go: she remained. . . . The moon traced out its bands in a different direction. She sat in the window and wept bitterly. . . . I kissed her wet eyes and wiped them with the locks of her hair that fell on her dull, pale shoulder, into which the moonlight was absorbed and lost without reflection in the dim, tender lustre.

I was sorry to leave her in tears, and I gabbled some nonsense to her in a half whisper. She looked up at me, and so much happiness gleamed in her eyes through the tears that I smiled. She seemed to follow my thought, and covered her face with both hands and stood up. . . . Now it really was time for me to go: I seized her hands, kissed them and her face over and over again – and went.

The maid let me out quietly, and I went by her without daring to look her in the face. The moon, grown heavy, was setting in a great, red ball. The dawn was beginning. It was cool and fresh, and the wind blew right in my face: I breathed it in deeper and deeper, for I needed to cool myself. As I approached my house the sun came up, and the good people I met were surprised that I had got up so early 'to take advantage of the good weather'.

This orgy of love lasted for a month; then my heart seemed to become tired, exhausted; I began to have moments of depression; I studiously concealed them, tried not to believe in them, wondered what was passing within me – while still love was cooling, cooling.

I began to feel constrained by the presence of the old man. I found his company uncomfortable and repellent. Not that I felt myself in the wrong as regards the man who had the civil and ecclesiastical rights of property in a woman who could not love him and whom he was incapable of loving, but my double part struck me as degrading; hypocrisy and duplicity are the vices most foreign to my nature. While burgeoning passion was in the ascendant I thought of nothing, but as soon as it was somewhat cooler I began to falter.

One morning Matvey[7] came into my bedroom with the news that old R— 'had passed away'. I was overcome by a strange feeling at this news; I turned on the other side and was in no hurry to dress. I did not want to see the dead man. Vitberg[8] came in, quite ready to go out. 'What?' he said, 'you're still in bed! Haven't you heard what's happened? I expect poor Mme R— is all alone, so let us go and make inquiries; hurry up and dress.' I dressed – and we went.

We found Mme R— in a swoon or in a sort of nervous lethargy. There was no pretence about it: her husband's death had recalled her helpless situation; she was left alone with her children in a strange town, without money, without friends or relations. Besides, she had on previous occasions fallen into this cataleptic condition, which was brought on by some violent shock and lasted several hours. On these occasions she would lie pale as death, with her face cold and her eyes closed, from time to time giving a gasp, not breathing at all in the intervals.

Not one woman came to help her, to show her sympathy, or to look after the children or the house. Vitberg remained with her; the prophetic clerk and I set about seeing to things.

The old man, looking black and shrunken, lay in his uniform on the drawing-room table, frowning as though he was angry with me. We laid him in the coffin, and two days later lowered him into the grave. After the funeral we went back to the dead man's house; the children in their black frocks with crape weepers huddled in the corner, more surprised and frightened than grieved:

they whispered to each other and walked on tiptoe. Mme R— sat with her head leaning on her hands, as though pondering, and did not say a single word.

In that drawing-room, on that sofa I had waited for her, listened to the sick man's groans and the drunken servant's swearing. Now everything was so black. . . . In the midst of the funereal surroundings and the smell of incense, I was haunted by confused and gloomy recollections of words and minutes of which I still could not think without tenderness.

Little by little her grief subsided and she looked more resolutely at her situation; then, little by little, other thoughts, too, began to light up her careworn and despondent face. Her eyes rested upon me with a sort of agitated inquiry, as though she were waiting for something . . . a question . . . an answer. . . .

I said nothing – and she, frightened, alarmed, began to feel doubts.

Then I saw that her husband had in reality been an excuse for me in my own eyes – love had burnt itself out in me. It was not that I had no feeling for her, far from it, but the feeling was not what she needed. My thoughts were now occupied by ideas of a different order, and that outburst of passion seemed to have possessed me simply to make another feeling clear to myself. Only one thing can I say in my defence – I was perfectly sincere in my infatuation.

While I had lost my head and did not know what to do, while with cowardly weakness I was waiting for a change to be brought about by time or circumstance, time and circumstance complicated my situation still further.

Tyufyayev,[9] seeing the helpless condition of a young, beautiful widow cast away without any support in a remote town in which she was a stranger, like the true 'father of the province', showed her the tenderest solicitude. At first we all thought that he felt real sympathy for her. But soon Mme R— noticed with horror that his attentions were by no means so simple. Two or three dissolute governors before him had kept Vyatka ladies as mistresses, and Tyufyayev, being used to such women, lost no time but at once began making declarations of love to her. Mme R— of course responded with cold disdain and mockery to his servile blandishments. Tyufyayev would not consider himself rebuffed, but persisted in his insolent attentions. Seeing, however, that he was making little progress, he gave her to understand that her children's future lay in his hands, that without his assistance she could not place them in schools at government expense, and that he on his side would not exert himself in her favour if she did not adopt a less chilly attitude to him. The insulted woman sprang up like a wounded wild beast. 'Kindly leave my house and don't dare to set foot in it again,' she said, pointing to the door.

'Ough, how angry you are!' said Tyufyayev, trying to turn things off with a jest.

'Pëtr, Pëtr,' she shouted into the hall, and the terrified Tyufyayev, fearing a public scandal, fled to his carriage, abashed and humiliated, gasping with fury.

In the evening Mme R— told Vitberg and me all that had happened. Vitberg realised at once that her gallant, put to flight and insulted, would not leave the poor woman in peace; Tyufyayev's character was pretty well known to us all. Vitberg resolved to save her at all costs.

Persecutions soon began. The petition with regard to the children was written in such a way that refusal was inevitable. The landlord and the shopkeepers demanded payment with particular insistence. God knows what might not be expected; the man who had done Petrovsky to death in a mad-house[10] was not to be trifled with.

Though burdened with an immense family and weighed down by poverty, Vitberg did not hesitate for one minute, but invited Mme R— to move with her children into his house two or three days after his wife's arrival at Vyatka. In his house Mme R— was safe, so great was the moral power of this exile. His inflexible will, his noble appearance, his fearless words, his scornful smile were dreaded even by the Vyatka Shemyaka.[11]

I lived in a separate apartment in the same house and dined at Vitberg's table, so we two found ourselves under the same roof, just when we ought to have been oceans apart.

In this close proximity she saw that there was no bringing back the past.

Why had it been precisely I whom she met, when at that time I was so unstable? She might have been happy: she deserved to be happy. The sorrowful past was over; a new life of love and harmony was so possible for her! Poor woman! Was it my fault that this storm-cloud of love, which had swooped down upon me so irresistibly, had blown upon me with its fiery breath, had intoxicated me and carried me away – and had then dispersed?

I lived in a state of alarm. Perplexed, foreseeing misfortune, and dissatisfied with myself, again I turned to dissipation and sought distraction in noise, was vexed at finding it and vexed at not finding it, and waited for a few lines from Natalie in Moscow as for a breath of pure air in the midst of sultry heat. The gentle image of the child on the verge of womanhood rose brighter and brighter above all this ferment of passion. My outburst of passion for Mme R— made my own heart clear to me and revealed its secret.

More and more absorbed by my feeling for my far-away cousin, I had not clearly analysed the sentiment that bound me to her. I was used to the feeling and did not watch closely to see whether it had changed or not.

My letters became more and more troubled; on the one hand I felt deeply not only the wrong I had done Mme R—, but the fresh wrong I did in the lying of which I was guilty by my silence. It seemed to me that I had fallen, that I was unworthy of any other love . . . while my love was growing and growing. .

The name of *sister* began to fret me; affection now was not enough for me: that calm emotion seemed cold. Her love was apparent in every line of her letters, but now that did not satisfy me. I needed not only the love but the very word itself, and I wrote: 'I am going to put a strange question to you. Do you believe that the feeling you have for me is only affection? Do you believe that the feeling I have for you is only affection? *I don't believe it.*'

'You seem somewhat agitated,' she answered. 'I knew your letter frightened you more than it frightened me. Set your mind at rest, my dear, it has changed absolutely nothing in me; it could not make me love you either more or less.'

But the word had been uttered: 'The mist has vanished,' she writes; 'all is clear and bright again.'

With unclouded joy she gave herself up to the feeling that had been given its name; her letters are one youthful song of love rising from a childish lisp to lyrical heights.

'Perhaps at this moment,' she writes, 'you are sitting in your study, not writing, not reading, but pensively smoking a cigar, and your eyes are fixed on the vague distance and you have no answer for the greeting of anyone who comes in. Where are your thoughts? What are you seeking to see? Do not answer, let them come to me . . .'

'Let us be children, let us fix an hour for both of us to be in the open air, an hour in which we can both be sure that nothing separates us but distance. At eight o'clock in the evening you, too, are surely free? Or else – what happened once before: I stepped out into the porch – and came back at once thinking that you were in the room.'

'Looking at your letters, at your portrait,[12] thinking of my letters, of my bracelet,[13] I have wished I could skip a hundred years and see what their fate would be. The things which have been for us holy relics, which have healed us, body and soul, with which we have talked and which have to some extent deputised for us to each other in absence; all these weapons with which we have defended ourselves from others, from the blows of fate, from ourselves, what will they be when we are gone? Will their virtue, their soul remain in them? Will they awaken, will they warm some other heart, will they tell the story of us, of our sufferings, of our love, will they win the reward of a single tear? How sad I feel when I imagine that your portrait will one day hang unknown in someone's study, or that a child perhaps will play with it, and break the glass and efface the features.'

My letters were not like this;* in the midst of complete, enraptured love there is a note of bitter vexation with myself and repentance; the dumb reproach of Mme R— was gnawing at my heart and troubling the clear radiance of my feeling; I seemed to myself a liar, and yet I had not been lying.

---

* The difference between the style of Natalie's letters and mine is very great, especially in the early part of our correspondence; afterwards it was less unequal and in the end becomes similar. In my letters, together with genuine feeling there are affected expressions, far-fetched high-flown phrases; the influence of the school of Hugo and the new French novelists is apparent. There is nothing of the sort in her letters: her language is simple, poetic, and sincere, the only influence that can be discerned in it is the influence of the Gospel. At that time I was still trying to write in the grand style and wrote badly, because it was not my own language. A life in spheres cut off from practical experience, and too much reading, prevent a young man for years from speaking and writing naturally and simply. Intellectual maturity only begins when the style is established and has taken its final form.

How could I make my avowal? How was I to tell Mme R— in January that in August I had been mistaken in telling her of my love? How could she believe in the truth of my story – a new love would have been easier to understand, betrayal would have been simpler. How the far-away image of her who was absent could enter into conflict with the present, how the stream of another love could have flowed through that furnace and have emerged stronger and more recognisable – all that I did not understand myself, but I felt that it was all true.

Finally Mme R— herself with the elusive agility of a lizard swerved away from any serious explanation; she had an inkling of danger, was seeking an answer to the enigma, and at the same time was fending off the truth. It was as though she foresaw that my words would reveal terrible truths, after which everything would be finished, and she cut short all talk at the point where it became dangerous.

At first she looked about her, and for a few days thought she had found her rival in a nice, lively young German girl whom I liked as one likes a child, and with whom I was at ease just because it had never entered her head to flirt with me, nor mine to flirt with her. In a week she saw that Paulina was not at all dangerous. But I cannot go further without saying a word about the latter.

In the government dispensary at Vyatka there was a German chemist, and there was nothing surprising about that, but what was surprising is that his assistant was Russian and was called Bolman. With this man I became acquainted; he was married to the daughter of a Vyatka government clerk, a lady who had the longest, thickest, and most beautiful hair I have ever seen. The dispenser himself, Ferdinand Rulkovius, was absent at first, and Bolman and I used to drink together various 'fizzing drinks' and artistic cordials compounded by the pharmacist. The dispenser was away in Reval; there he made the acquaintance of a young girl and offered her his hand; the girl, who hardly knew him, rushed into marriage with him, as generally happens to girls, and to German girls in particular; she had no notion even into what wilds he was taking her. But when after the wedding they had to start off, she was overcome by fear and despair. To comfort his bride, the dispenser invited a young girl of seventeen, a distant relation of his wife, to go with them to Vyatka. She, even more precipitately, and with no more idea of what was meant by

Vyatka, consented. Neither of the German girls spoke a word of Russian, and in Vyatka there were not four men who spoke German. Even the teacher of that language in the high school did not know it, a fact which surprised me so much that I actually ventured to ask him how he managed to teach it.

'With the grammar,' he answered, 'and with dialogues.'

He further explained that he was really a teacher of mathematics, but that, since there was no post vacant, he was teaching German meanwhile, and that he received, however, only half the salary. The German girls were dying of *ennui*, and seeing a man who, if he could not speak German well, could at least express himself intelligibly, were greatly delighted, regaled me with coffee and some sort of *kalte Schale*, told me all their secrets, hopes and desires, and within two days called me their friend and treated me still more hospitably to sweet, floury, cinnamon-cakes. Both were fairly well educated, that is, they knew Schiller by heart, played the piano, and sang German songs. There the likeness between them ended. The dispenser's wife was a tall, fair, lymphatic woman, very good-looking but sleepy and listless; she was extremely good-natured and, indeed, with her physique it would have been hard to be anything else. Being once convinced that her husband was her husband, she loved him quietly and steadily, looked after the kitchen and the linen, read novels in her spare moments, and in due time successfully bore the chemist a daughter with white eyebrows and eyelashes, who was scrofulous.

Her friend, a short, swarthy brunette, vigorously healthy, with big, black eyes and an air of independence, was a beauty of the sturdy peasant type; a great deal of energy was apparent in her words and movements, and when at times the dispenser, a dull, close-fisted fellow, made somewhat discourteous observations to his wife, and she listened to them with a smile on her lips and a tear on her eyelash, Paulina would flush crimson and give the irate husband such a look that he would instantly subside, pretend to be very busy, and go off to his laboratory to pound and mix all sorts of nasty things for the restoration of the health of the officials of Vyatka.

I liked the simple-hearted girl who knew how to stand up for herself, and I do not know how it happened, but it was to her I first talked of my love and translated some of Natalie's letters. Only one who has lived for long years with people who are complete strangers know how precious are these heart-to-heart talks. I rarely talk of my feelings, but there are moments, even now, when the longing to speak out becomes unbearable, and at that time I was four-and-twenty, and I had only just realised my love. I could bear separation, I could have borne silence too, but, meeting with another child on the threshold of womanhood, in whom everything was so unaffectedly simple, I could not restrain myself from blurting out my secret to her. And how grateful to me she was for this, and how much good she did me!

I was sometimes wearied by Vitberg's conversation, which was always serious, and, since I was fretted by my difficult relationship with Mme R—, I could not be at my ease with her. In the evening I often used to go off to Paulina, read foolish stories aloud to her, listen to her ringing laugh and to her singing, especially for my benefit, 'Das Mädchen aus der Fremde' – by which she and I understood another 'maiden from a strange land'; and the clouds were dissipated, there was an unfeigned gaiety, an untroubled serenity in my heart, and I would go home at peace. Then the dispenser, after stirring his last mixture and smearing his last plaster with ointment, would arrive to bore me with absurd political inquiries – not, however, before he had drunk off his 'medicine' and eaten the herring salad prepared by the little white hands *der Frau Apothekerin*.

Mme R— suffered, while I with pitiful weakness waited for time to bring some chance solution, and prolonged the half-deception. A thousand times I wanted to go to Mme R—, to throw myself at her feet, to tell her everything, to face her wrath, her contempt ... but it was not indignation that I feared – I should have been glad of that – I feared her tears. One must have experienced much evil to be able to endure a woman's tears, to be able to doubt while they trickle still warm over the flushed cheek. Besides, her tears would have been genuine.

A good deal of time passed like this. It began to be rumoured that my exile might soon come to an end. The day no longer seemed so remote on which I should fling myself into a chaise and dash off to Moscow; familiar faces hovered before my eyes and among them, foremost of them, the cherished features; but scarcely had I abandoned myself to these dreams when the pale, sad form of Mme R— would appear at the other side of the chaise with tear-stained eyes, full of pain and reproach, and my joy was dimmed: I felt sorry for her, mortally sorry.

I could no longer remain in a false position, and plucking up all my courage I resolved to break out of it. I wrote her a full confession. Warmly, frankly, I told her the whole truth. On the next day she said she was ill and did not leave her room. All the sufferings of a criminal, all the fear that he will be detected, I went through on that day. She had another attack of her nervous stupor – I dared not visit her.

I needed a greater penance. I shut myself up with Vitberg in my study and told him my whole story. At first he was astonished, then he listened to me not as a judge but as a friend, did not torment me with questions, did not preach to me with stale morality, but devoted himself to helping me find means to soften the blow – he alone could do that. His affection was very warm for those of whom he was fond. I had been afraid of his rigorous morals, but his affection for me and for Mme R— completely outweighed that. Yes, in his hands I could leave the unhappy woman to whose disconsolate existence I had given the finishing blow; in him she found strong moral support and authority. She respected him like a father.

In the morning Matvey gave me a note. I had scarcely slept all night. With a trembling hand I broke the seal. She wrote gently, in a noble and deeply mournful spirit; the flowers of my eloquence had not concealed the asp beneath them; in her words of resignation could be heard the stifled moan of a wounded heart, the cry of pain, repressed by a supreme effort. She blessed me on my way to my new life, wished me happiness, called Natalie a sister, and held out a pleading hand to us for forgetfulness of the past and friendship for the future – as though she had been to blame!

Sobbing, I read her letter over and over again. *Qual cuor tradisti!*

Later on I met her. She gave me the hand of friendship, but we felt uncomfortable; each of us had left something unsaid: there was something that each of us tried to avoid alluding to.

A year ago I heard of her death.

When I left Vyatka I was tormented for a long time by the memory of Mme R—. When I regained my composure I set to work to write a story of which she was the heroine. I described a young nobleman of the period of Catherine who has abandoned the woman who loves him and married another. She pines away and dies. The news of her death is a heavy blow to him, he becomes sombre and melancholy, and eventually goes out of his mind. His wife, who is the ideal of gentleness and self-sacrifice, after trying everything, leads him in one of his quieter moments to the Devichy Convent and kneels down with him at the unhappy woman's grave, asking for her forgiveness and intercession. From the windows of the convent the words of a prayer reach them, soft feminine voices sing of forgiveness – and the young man recovers. The story was a failure. At the time when I wrote it Mme R— had no thought of coming to Moscow, and the only man who guessed that there had been anything between us was the 'wandering German', K. I. Sonnenberg. After my mother's death in 1851 we heard no word of him. In 1860 a tourist, describing his acquaintance with Karl Ivanovich, now a man of eighty, showed me a letter from him. In a postscript the old man told him of the death of Mme R— and said that *my brother* had had her buried in the Devichy Convent!

I need hardly say that neither of them knew of the story I wrote.

CHAPTER IV

IN MOSCOW WHILE I WAS AWAY

My peaceful life at Vladimir was soon troubled by news from Moscow which now reached me from all sides and distressed me deeply. To make this intelligible I must go back to 1834.

The day after I was arrested in 1834 was the name-day of my aunt, the princess, and so when Natalie had parted from me in the graveyard she had said: 'Until tomorrow'. She was expecting me, and several members of the family had arrived, when suddenly my cousin appeared and told them the full details of my arrest. She was shocked by this news, so utterly unexpected; she got up to go into another room, and after taking two steps fell unconscious on the floor. The princess saw it all and understood it all; she determined to oppose this dawning love by every means in her power.

What for?

I do not know: she had recently, that is since I had finished my studies, been very well disposed to me; but my arrest and rumours of our *free-thinking* attitude, of our betrayal of the Orthodox Church by entering the Saint-Simon 'sect'[1] infuriated her; from that time forward she never spoke of me except as 'the political criminal' or 'that unfortunate son of my brother Ivan'. The Senator had to use all his authority to induce her to allow Natalie to come to the Krutitsky Barracks to say goodbye to me.

Fortunately I was in exile and the princess had plenty of time before her.

'And where is this Perm or Vyatka? He'll be sure to break his neck there, or have it broken for him; and he'll forget her there: that's the main thing.'

But as though to spite the princess, I had an excellent memory. Natalie's correspondence with me, long concealed from the old lady, was at last discovered, and she strictly forbade the maids and menservants to transmit letters to the young girl, or to take her letters to the post. In two years there began to be talk of my return. 'So I dare say some fine morning that unfortunate son of my brother's will open the door and walk in; it's no use wasting time thinking about it, and putting things off – we'll make a match for her and save her from the political criminal who has no religion or principles.'

Formerly the princess had used to sigh and say of the poor, forlorn girl that she had scarcely anything, that it would not do for her to pick and choose, and that she would like to see her settled *somehow* in her own lifetime. She had indeed, with the help of her hangers-on, *somehow* settled the fate of one distant cousin who had no dowry by marrying her off to an attorney of some sort. A nice, good-natured, well-educated girl, she married to satisfy her mother; two years later she died, but the attorney was still alive, and from gratitude was still looking after her Excellency's affairs. In this case, quite the contrary, the orphan was by no means a penniless bride; the princess was prepared to give her in marriage like a daughter of her own, to give her a dowry of a hundred thousand roubles in cash alone and to leave her something in her will besides. On such terms suitors are always to be found, not only in Moscow but anywhere you like, especially when there is the title of princess as well as a 'lady companion' and the nomadic old ladies.

The whispering, negotiations, rumours and the maid-servants brought Princess Marya Alexeyevna's intention to the ears of the unhappy victim of so much solicitude. She told the 'lady companion' that she would certainly not accept any proposal. Then began an incessant, outrageous, and ruthless persecution without one trace of delicacy, a petty persecution that attacked her every minute and clutched her at every step, at every word.

'Imagine bad weather, fearful cold, wind, rain, an overcast sky without expression, a nasty little room which looks as though a corpse had just been carried out of it, and these *children*, without object or even enjoyment, make a din, shout, break and defile everything near them; and it would be bad enough if one had simply to look at them, but when one is forced to live among them ...' So she writes in a letter from the country where the princess had gone in the summer; and she goes on: 'there are three old women sitting here with us, and they are all three describing how their late husbands were paralysed and how they looked after them; and it is chilly enough without that.'

Now to this environment was added systematic persecution, and it was carried on not by the princess alone but also by the wretched old women, who constantly tormented Natalie, exhorting her to marry and abusing me; as a rule she said nothing in her letters of the continual annoyances she had to endure, but sometimes bitterness, humiliation and boredom were too much for her. 'I don't know,' she writes, 'whether they can invent anything more to oppress me with. Will they possibly have wit enough for that? Do you know that I am actually forbidden to go into another room, or even to move to another seat in the same room? It is a long while since I played the piano; lights were brought and I went into the drawing-room, thinking they might be merciful, but no, they brought me back and made me knit; perhaps, at least I might sit at another table – I can't endure being beside them – might I do even that? No, I must sit just here beside the priest's wife, listen, look, and talk, while they do nothing but talk about Filaret[2] or criticise you. For a moment I was angry and grew red, but suddenly my heart was weighed down by a feeling of bitter sadness, not because I had to be their slave, no ... I felt mortally sorry for them.'

The matchmaking was officially beginning.

'A lady has been here who is fond of me, and whom for that reason I do not like ... She is doing her very utmost to get me settled, and she made me so angry that as she left I sang –

> *"I had rather be dressed in my winding-sheet*
> *Than the wedding veil without my sweet."'*

A few days later, 26 October 1837, she writes: 'What I have been through today, my dear, you can't imagine. They dressed me up and took me off to Mme S—, who has been extremely gracious to me ever since I was a child; Colonel Z— goes there every Tuesday to play cards. Imagine my situation: on the one side the old ladies at the card-table, on the other various ugly figures, and he. The conversation, the company – everything was so alien to me, so strange and repugnant, so lifeless and vulgar, that I was more like a statue than a living creature. Everything that was going on seemed like an oppressive, suffocating nightmare. I kept asking to go home, like a child: they would not heed me. The attention of the host and of *the visitor* overwhelmed me; he got as far as writing half my monogram in chalk. Oh heavens, I am not strong enough and I can look for support to no one among those who might be a help; I am all alone on the edge of a precipice, and a whole crowd of them are doing everything they can to push me over; sometimes I am weary, my strength fails me, and you are not near and in the distance I cannot see you; but the mere thought of you – and my soul is stirred and ready to do battle again in the armour of love.'

Meanwhile everyone liked the Colonel: the Senator was friendly to him, and my father gave it as his opinion that 'a better match could not be expected and should not be desired'. 'Even his Excellency D. P. (Golokhvastov)[3] is pleased with him,' wrote Natalie. The princess said nothing directly to Natalie, but restricted her freedom even more severely and tried to hasten things up. Natalie tried to pretend in his presence to be a complete imbecile, hoping to frighten him off, but not at all; he went on coming more and more frequently.

'Yesterday,' she writes, 'Emilia was here and this is what she said: "If I heard that you were dead I should cross myself with joy and thank God." She is right in a great deal but not altogether; her soul living by grief alone[4] could fully grasp the suffering of my heart, but the bliss with which love fills it she can hardly understand.'

But the princess was not losing heart. 'Wishing to have a clear conscience, the princess summoned a priest who is acquainted with Z— and asked him whether it would not be a sin to marry me by force. The priest said it would be actually a godly work to make provision for an orphan. I am sending for my own priest,' Natalie adds, 'and shall tell him the whole story.'

'30 October – My clothes are here, my attire for tomorrow, and the icon, the rings; there is fuss and preparations, and not a word to me. The Nasakins and others have been invited. They are preparing a surprise for me and I am preparing a surprise for them.

'Evening. – Now a family council is going on. Lev Alexeyevich (the Senator) is here. You urge me to be strong – there is no need, my dear. I am equal to extricating myself from the awful, loathsome scenes into which they are dragging me on the chain. Your image is bright above me, there is no need to fear for me, and my very distress and sadness are so sacred and have taken so firm a hold on my soul that tearing them away would hurt even more, the wounds would re-open.'

However, though they did their best to mask and cover up the affair, the Colonel could not avoid seeing the determined aversion of his proposed bride; he began to be less frequent in his visits, declared himself ill, and even hinted at some addition to the dowry; this greatly incensed the princess, but she got over even that humiliation and was ready to give her estate near Moscow as well. This concession he had not apparently expected, for after it he disappeared altogether.

Two months passed quietly. All at once the news came that I had been transferred to Vladimir. Then the princess made her last desperate effort to marry off her *protégée*. One of her acquaintances had a son, an officer, who had just returned from the Caucasus; he was young, cultivated, and a very decent fellow. The princess cast away her pride and herself suggested to his sister that she should 'sound' her brother and see whether he cared for the match. He yielded to his sister's suggestion. My young cousin did not care to play the same disgusting and tedious part a second time, so, seeing that the affair was taking a serious turn, she wrote to the young man a letter, told him directly, openly, and simply that she loved another man, trusted herself to his honour and begged him not to add to her sufferings.

The officer with great delicacy drew back. The princess was amazed and affronted and made up her mind to find out what had happened. The officer's sister, to whom Natalie had spoken herself, and who had given her word to her brother to pass nothing on to the princess, told the whole story to the 'lady companion'; the latter of course at once reported it to her mistress.

The princess almost choked with indignation. Not knowing what to do, she ordered the young girl to go upstairs to her room and not to let her set eyes on her; not content with that, she ordered her door to be locked and put two maids on guard; then she wrote notes to her two brothers and one of her nephews and asked them to come and give her advice, saying that 'she was so distressed and upset that she could not think what to do in the misfortune that had befallen her'. My father refused, saying that he had plenty of worries of his own, that there was no need to attach such importance to what had happened, and that he was a poor judge in affairs of the heart. The Senator and D. P. Golokhvastov appeared the next evening in answer to her summons.

They talked for a long time without reaching any agreement and at last asked to see the prisoner. The young girl came in, but she was no longer the shy, silent orphan they had known. Unshakable firmness and irreversible determination were apparent in the calm, proud expression of her face; this was no child, but a woman who had come to defend her love – my love.

The sight of the 'accused' confounded the Areopagus. They felt uncomfortable; at last Dmitry Pavlovich, *l'orateur de la famille*, expatiated at length on the cause of their coming together, the distress of the princess, her heartfelt desire to settle her *protégée's* future, and the strange opposition on the part of the one for whose benefit it was all being done. The Senator with a nod and a movement of his forefinger expressed his assent to his nephew's words. The princess said nothing but sat with her head turned away, sniffing salts.

The 'accused' heard all they had to say and asked with straightforward simplicity what they required of her.

'We have no thought of requiring anything from you,' observed the nephew. 'We are here at Aunt's desire to give you sincere advice. A match is being offered to you that is excellent in all respects.'

'I cannot accept it.'

'What is your reason for that?'

'You know it.'

The orator of the family coloured a little, took a pinch of snuff, and screwing up his eyes went on:

'There is a great deal here to which objection might be urged. I would call your attention to the precariousness of your hopes. It is so long since you have seen our unfortunate Alexander; he is so young and impetuous – are you certain of him?'

'Yes, and whatever his intentions may be, I cannot change mine.'

The nephew was *au bout de son Latin*; he got up saying: 'God grant that you may not regret it! I feel very anxious about your future.'

The Senator frowned; the luckless girl now appealed to him.

'You have always shown sympathy for me,' she said to him. 'I implore you, save me: do what you like, but take me out of this life. I have done no harm to anyone, I ask for nothing, I am not trying to do anything, I am only refusing to deceive a man and ruin myself by marrying him. What I have to endure on account of it you cannot imagine; it pains me to have to say this in the presence of the princess, but to put up with the slights, the insulting words, the hints of her friend is too much for me. I cannot, I ought not to allow it, for insulting me is insulting. . . .'

Her nerves gave way and the tears gushed from her eyes; the Senator leapt up and walked about the room in agitation.

Meanwhile the 'lady companion', boiling over with fury, could not restrain herself and said, addressing the princess:

'So that's our nice, modest girl: there's gratitude for you.'

'Of whom is she speaking?' shouted the Senator, 'Eh? How is it, sister, you allow that woman, devil knows who she is, to speak like that of your brother's daughter in your presence? And if it comes to that, why is this creature here at all? Did you invite her to the family council too? Is she a relation or what?'

'My dear,' answered the terrified princess, 'you know what she is to me and how she looks after me.'

'Yes, yes, that's all very nice, let her give you your medicine and what's necessary; that's not what I am talking about. I ask you, ma sœur, why is she here when family affairs are being discussed, and how dare she open her mouth? One might think, after this, that she runs the whole show, and then you complain – Hey, my carriage!'

The 'lady companion' flushed, and ran out of the room in tears.

'Why do you spoil her like this?' the Senator went on, carried away; 'she fancies she is sitting in a pot-house at Zvenigorod; how is it you aren't disgusted by it?'

'Leave off, my dear, please,' the poor princess groaned, 'my nerves are so upset – oh! You can go upstairs and stay there,' she added, addressing her niece.

'It's time to be done with all this Bastille business. It's all nonsense and leads to nothing,' observed the Senator and took his hat.

Before driving away he went upstairs; Natalie, overcome by all that had passed, was sitting in an armchair with her face hidden, weeping bitterly. The old man patted her on the shoulder and said:

'Calm yourself, calm yourself, it'll all be thrashed out. You must just try to stop my sister being angry with you; she is an invalid; she must be humoured; after all, she only wishes for your good, you know; but, there, you shan't be married against your will, I'll answer for that.'

'Better a nunnery, a boarding-school, to go to my brother at Tambov, or to Petersburg, than endure this life any longer,' she answered.

'Come, that'll do, try and soothe my sister, and as for that fool of a woman I'll teach her not to be rude.'

The Senator, as he crossed the great hall, met the 'lady companion': 'I'll ask you not to forget yourself,' he shouted at her, holding up a menacing finger; she went sobbing into the bedroom where the princess was already lying on the bed while four maids rubbed her hands and feet, moistened her temples with vinegar, and poured Hoffman's drops on lumps of sugar.

So ended the family council.

It is clear that the girl's situation was hardly likely to be improved by what had happened; the 'lady companion' was more on her guard, but, now cherishing a personal hatred for Natalie, and desirous of avenging her own injury and humiliation, she poisoned her existence by petty, indirect means. I need hardly say that the princess participated in this ignoble persecution of a defenceless girl.

This had to be ended. I made up my mind to come forward, and wrote a long, calm, frank letter to my father. I told him of my love and, foreseeing his reply, added that I did not want to hurry him, that I should give him time to see whether it was a passing feeling or not, and that all that I begged of him was that the Senator and he would put themselves in the poor girl's place and would remember that they had the same rights over her as the princess herself.

My father answered that he could not endure meddling in other people's affairs, that what the princess did in her own house was not his business; he advised me to abandon foolish ideas 'engendered by the idleness and *ennui* of exile', and added that I would do better to prepare myself for travel in foreign lands. We had often talked in past years of a tour abroad; he knew how passionately I wished for it, but he had found endless obstacles and had always ended by saying: 'You must first close my eyes, then you'll be free to go to the ends of the earth.' In exile I had lost all hope of going abroad soon. I knew how hard it would be to get his permission and, besides, it would have seemed a lack of delicacy to insist on a voluntary separation after the enforced one. I remembered the tears quivering on his old eyelids when I was setting off to Perm ... and now here was my father taking the initiative and suggesting I should go!

I had been open, I had spared the old man in my letter, had asked so little – and he had answered with irony and artifice.

'He doesn't want to do anything for me,' I said to myself; 'like Guizot he advocates *la non-intervention*. Very well then, I'll act myself, and now amen to concessions.' I had not once before thought about the ordering of the future; I believed, I knew that it was mine, that it was ours, and I left the details to chance; the consciousness of love was enough for us, and our desires did not go beyond a momentary interview. My father's letter forced me to take the future into my own hands. It was nothing to expect – *cosa fatta capo ha!* My father was not very sentimental, while as for the princess –

> 'Let her weep,
> Her tears mean nought to her!'[5]

Just at that time my brother[6] and Ketscher[7] came to stay at Vladimir. Ketscher and I spent whole nights together, talking, recalling the past, laughing through our tears, and laughing till we cried. He was the first of our set whom I had seen since I left Moscow. From him I heard the chronicles of our circle, what changes had taken place and what questions were absorbing it, what fresh people had arrived, where those were who had left Moscow, and so on. When we had talked everything over I told him of my plans. After considering how I ought to act, Ketscher concluded with a proposal the absurdity of which I appreciated afterwards. Desirous of exhausting all peaceful means, he offered to go to my father, whom he scarcely knew, and to talk to him *seriously.* I agreed.

Ketscher, of course, was better fitted for anything good or for anything bad, than for diplomatic negotiations, particularly with my father. He was endowed in the highest degree with everything that was bound to ruin any chance of success. His very appearance was enough to depress and alarm any conservative. A tall figure, with hair strangely dishevelled and arranged on no fixed principle, with a harsh countenance reminiscent of a number of the members of the Convention of 1793, especially of Marat, with the same big mouth, the same hard, disdainful lines about the lips, and the same expression of mournful and exasperated gloom; to this must be added spectacles, a wide-brimmed hat, extreme irritability, a loud voice, lack of all habit of self-control, and the power of arching his eyebrows higher and higher as he grew more indignant. Ketscher was like Laravigny in George Sand's excellent novel, *Horace*, with an admixture of something of the Pathfinder and Robinson Crusoe, as well as an element purely Muscovite. His open, generous temperament had set him from childhood in direct conflict with the world surrounding him; he did not conceal this antagonism and was accustomed to it. A few years older than we were, he was continually scolding us and was dissatisfied with everything. He used to quarrel and bring accusations against us, and he covered it all with the simple good nature of a child. His words were rude, but his feelings were tender and we forgave him much.

Imagine him, this last of the Mohicans with the face of a Marat, the 'friend of the people', setting off to admonish my father! Many times afterwards I made Ketscher describe their interview; my imagination was unequal to picturing all the oddity of this diplomatic intervention. It happened so unexpectedly that for a moment my old father lost his bearings and began expl... ing all the weighty considerations which led him to oppose my marriage; then, recovering himself, he changed his tone and asked Ketscher on what grounds he had come to discuss a matter which was none of his business. The conversation became more embittered. The diplomatist, seeing that the situation was deteriorating, tried to frighten the old man about my health, but it was already too late for that, and the interview ended, as might have been expected, in a series of malignant sarcasms from my father and rude rejoinders from Ketscher.

He wrote to me: 'Expect nothing from the old man.' That was all I needed to know. But what was I to do? How was I to begin? While I was thinking over a dozen different plans a day and unable to decide among them, my brother prepared to go to Moscow.

That was on the 1st of March, 1838.

CHAPTER V

THE 3RD OF MARCH
AND THE 9TH OF MAY 1838[1]

IN the morning I wrote letters; when I had finished we sat down to dinner. I could not eat, and we said nothing; I felt unbearably oppressed – it was between four and five, and at seven the horses were to come round. At the same time next day he would be in Moscow while I – and every minute my pulse beat faster.

'I say,' I said at last to my brother, looking at my plate, 'will you take me with you to Moscow?'

He put down his fork and looked at me, uncertain whether he had heard me aright.

---

* A fragment of this chapter was published in *The Pole Star*, Vol. I, page 79, together with the following note:

Who is entitled to write his reminiscences?

Everyone.

Because no one is obliged to read them.

In order to write one's reminiscences it is not at all necessary to be a great man, nor a notorious criminal, nor a celebrated artist, nor a statesman – it is quite enough to be simply a human being, to have something to tell, and not merely to desire to tell it but at least have some little ability to do so.

Every life is interesting; if not the personality, then the environment, the country are interesting, the life itself is interesting. Man likes to enter into another existence, he likes to touch the subtlest fibres of another's heart, and to listen to its beating ... he compares, he checks it by his own, he seeks for himself confirmation, sympathy, justification....

But may not memoirs be tedious, may not the life described be colourless and commonplace?

Then we shall not read it – there is no worse punishment for a book than that. And as to that, no special right to write one's memoirs can avail. Benvenuto Cellini's *Diary* is not interesting because he was an excellent worker in gold but because it is in itself as interesting as any novel.

The fact is that the very word 'entitled' to this or that form of composition does not belong to our epoch, but dates from an era of

'Take me through the town gate as your servant; I don't need anything more: do you agree?'

'Yes if you like, only, you know, afterwards you'll. . . .'

It was too late: his 'if you like' was already in my blood, in my brain. The idea that had only flashed upon me a minute before had now taken deep root.

'What is there to discuss? Anything may happen – so you'll take me?'

'Of course – I'm quite ready – only. . . .'

I jumped up from the table.

---

intellectual immaturity, from an era of poet-laureates, doctors' caps, corporations of savants, certificated philosophers, diploma'ed meta-physicians and other Pharisees of the Christian world. Then the act of writing was regarded as something sacred, a man writing for the public used a high-flown, unnatural, choice language; he 'expounded' or 'sang'.

We simply talk; for us writing is the same sort of secular pursuit, the same sort of work or amusement as any other. In this connection it is difficult to dispute 'the right to work'. Whether the work will find recognition and approval is quite a different matter.

A year ago I published in Russian part of my memoirs under the title of *Prison and Exile*. I published it in London at the beginning of the war. I did not reckon upon readers nor upon any attention outside Russia. The success of that book exceeded all expectations: the *Revue des Deux Mondes*, the most chaste and conceited of journals, published half the book in a French translation; the clever and learned *Athenaeum* printed extracts in English; the whole book has appeared in German and is being published in English.

That is why I have decided to print extracts from other parts.

In another place I speak of the immense importance my memoirs have for me personally, and the object with which I began writing them. I confine myself now to the general remark that the publication of contemporary memoirs is particularly useful for us Russians. Thanks to the censorship we are not accustomed to anything being made public, and the slightest publicity frightens, checks, and surprises us. In England any man who appears on any public stage, whether as a huckster of letters or a guardian of the press, is liable to the same critical examination, to the same hisses and applause as the actor in the lowest theatre in Islington or Paddington. Neither the Queen nor her husband are excluded. It is a mighty curb!

Let our imperial *actors* of the secret and open police, who have been so well protected from publicity by the censorship and paternal punishments, know that sooner or later their deeds will come into the light of day.

'Are you going?' asked Matvey, wanting to say something.

'I am,' I answered in such a tone that he said no more. 'I'll be back the day after tomorrow. If anyone comes, tell them I have a headache and am asleep. In the evening light the candles, and now get me my linen and my bag.'

The bells were tinkling in the yard.

'Are you ready?'

'Yes, and so good luck to us.'

By dinner-time next day the bells ceased tinkling, and we were at Ketscher's door. I bade them call him out. A week before, when he had left me at Vladimir, there had been no idea of my coming, and hence he was so surprised on seeing me that at first he did not say a word and then went off into a peal of laughter: but soon he looked anxious and led me indoors. When we were in his room he carefully locked the door and asked me: 'What has happened?'

'Nothing.'

'Then what have you come here for?'

'I couldn't stay at Vladimir; I want to see Natalie – that's all, and you must arrange it, and this very minute, because I must be back there tomorrow.'

Ketscher looked into my face and raised his eyebrows.

'What folly! The devil knows what to call it, to come like this with no need and nothing prepared! Have you written? Have you fixed a time?'

'I have written nothing.'

'Upon my word, my boy, but what are we to do with you? It's beyond anything, it's raving madness!'

'That's just the point – that we must think what to do without losing a minute.'

'You're a fool,' said Ketscher with conviction, raising his eyebrows higher than ever. 'I should be glad, extremely glad, if it didn't come off: it would be a lesson to you.'

'And rather a long lesson if I am caught. Listen: as soon as it is dark we'll drive to the princess's house; you shall call someone out into the street, one of the servants, I'll tell you which – and then we'll see what to do. That's all right, eh?'

'Well, there's no help for it, let's go; but how I should like it not to be a failure! Why didn't you write yesterday?' – and Ketscher, pulling his broad-brimmed hat over his brows with an air of dignity, threw on a black cloak lined with red.

'Oh, you damned grumbler!' I said to him as we went out, and Ketscher, laughing heartily, repeated: 'But really it's enough to make a hen laugh, to come like this without sending a word; it's beyond anything.'

I could not stay at Ketscher's – he lived terribly far away, and his mother had visitors that day. He took me to an officer of hussars whom he knew to be an honourable man, and who had never been mixed up with political affairs, and so was not under police supervision. The officer, a man with long moustaches, was sitting at dinner when we went in; Ketscher told him what we had come about. In reply the officer poured me out a glass of red wine and thanked us for the confidence we put in him; then he took me into his bedroom, which was adorned with saddles and saddle cloths, so that one might have supposed that he slept on horseback.

'Here is a room for you,' he said; 'no one will disturb you here.'

Then he called his orderly, a hussar like himself, and ordered him not to let anyone into that room on any pretext. I found myself once more under the guardianship of a soldier, with this difference, that at the Krutitsky Barracks[2] the gendarme had been keeping me from all the world, while here the hussar was keeping all the world from me.

When it was quite dark Ketscher and I set off. My heart beat violently when I saw the familiar streets and houses again which I had not seen for nearly four years ... Kuznetsky Bridge, Tverskoy Boulevard ... and here was Ogarëv's house; they had clapped a huge heraldic crest on it and it looked strange. On the ground floor, where we spent such happy youthful days, a tailor was living ... Here was Povarskaya Street – I held my breath: in the corner window of the attic room there was a candle burning: that was her room, she was writing to me, she was thinking of me, the candle twinkled so gaily, it seemed twinkling to *me*.

While we were considering how best to call someone out into the street, one of the princess's young footmen ran towards us.

'Arkady,' I said, going up to him. He did not recognise me. 'What's the matter with you?' I said, 'Don't you know your own people?'

'Oh, is it you, sir?' he cried.

I put my finger on my lips and said : 'If you would like to do me a friendly service, deliver this little note at once, as quickly as you can, through Sasha or Kostinka, do you understand? We will wait for the answer round the corner in the alley, and don't breathe a word to anyone about having seen me in Moscow.'

'Don't be uneasy, we'll pull it off in a flash,' answered Arkady, and he trotted into the house.

We walked up and down the alley for about half an hour before a little, thin, old woman came hurrying out, looking about her; this was that same spirited servant girl who in 1812 had asked the French soldiers for *'manger'*[3] for me; we had called her Kostinka ever since I was a child. The old woman took my face in both hands and showered kisses upon it.

'So you've flown to see us,' she said. 'Ah, you headstrong boy, when will you calm down, you rogue, you! – and you've given our young lady such a fright that she almost fainted.'

'And have you a note for me?'

'Yes, yes, see how impatient he is,' and she gave me a scrap of paper.

A few words had been scribbled in pencil with a trembling hand: 'My God, can it be true – you, here! Tomorrow between five and six in the morning I shall expect you. I can't believe it, I can't believe it! Surely it must be a dream!'

The hussar again put me into his orderly's keeping. At half-past five next morning I stood leaning against a lamp-post, waiting for Ketscher, who had gone in at the wicket-gate of the princess's house. I shall not attempt to describe what was passing in me while I waited at the lamp-post; such moments remain one's own secret because there are no words for them.

Ketscher beckoned to me. I went in at the little gate. A boy who had grown up since I left showed me in with a friendly smile, and here I was in the hall which at one time I used to enter with a yawn, and now I was ready to fall on my knees and kiss every plank in the floor. Arkady led me into the drawing-room and went out. I sank exhausted on the sofa, my heart throbbed so violently that it was painful; and besides I was frightened. I linger over my story for the sake of spending longer with these memories, though I see that my words give a poor idea of them.

She came in all in white, dazzlingly lovely; three years of

separation and the struggles she had been through had given the finishing touches to her features and her expression.

'It is you,' she said in her soft, gentle voice.

We sat down on the sofa and remained silent.

The expression of joy in her eyes resembled one of suffering. I suppose that when the feeling of gladness reaches its highest pitch it is mingled with an expression of pain, for she said to me: 'How tormented you look!'

I held her hand, she leaned her head on the other, and we had nothing to say to each other . . . a few brief phrases, two or three reminiscences, words from our letters, some idle remarks about Arkady, about the hussar, about Kostinka, that was all.

Then Kostinka came in, saying that it was time for me to go, and I got up without protesting, and she did not try to keep me . . . our hearts were so full, all thoughts of more or less, of shorter or longer, all vanished before the fullness of the present. . . .

When we had passed the town gate, Ketscher asked: 'Well, have you settled anything?'

'Nothing.'

'But you talked to her?'

'Not a word about that.'

'Does she consent?'

'I didn't ask. Of course she consents.'

'Well, upon my soul, you behave like a child, or a lunatic,' observed Ketscher, raising his eyebrows and shrugging his shoulders with indignation.

'I'll write to her and then to you, and now, goodbye. Now get along, all three of you!'[4]

It was thawing, the spongy snow was black in places, the endless white plain lay on both sides, little villages flashed by with their smoke, and then the moon rose and shed a different light on everything; I was alone with the driver and kept looking out, yet all the while was there with her, and the road and the moon and the fields were somehow mingled with the princess's drawing-room. And, strange to say, I remembered every word uttered by Kostinka, by Arkady, even by the maid who had led me out to the gate, but what I had said to her and what she had said to me I could not remember!

Two months were spent in an incessant bustle. I had to borrow money, and to get her baptismal certificate; it appeared that the princess had taken it. One of my friends – crawling, bribing, treating policemen and clerks – succeeded by all sorts of false statements in getting another from the Consistory.

When everything was ready, we, that is Matvey and I, set off.

At dawn on the 8th of May we were at the last posting-station before Moscow. The drivers had gone to get horses. The air was heavy, there were drops of rain, and it seemed as though a storm was coming on; I remained in the *kibitka* and hurried up my driver. Someone spoke near me in a strange, high, whining, sing-song voice. I turned round and saw a pale, thin girl of about sixteen, begging. She was in rags, with her hair hanging about her; I gave her some small silver coin; she laughed when she saw it, but instead of going away she clambered on to the driver's seat, turned towards me and began muttering half-coherent sentences, looking straight into my face; her eyes were clouded and pitiful, wisps of hair fell over her face. Her sickly face, her unintelligible mutterings, together with the light of early morning, aroused a sort of nervous timidity in me.

'She's crazy, you know, that is, she is simple,' observed the driver. 'And where are you poking yourself? I'll give you a lash with the whip and then you'll know! Upon my soul, I will, you saucy hussy!'

'Why are you scolding, what have I done to you – here your master's given me a silver bit, and what have I done to you?'

'Well, he's given it to you, and so be off to your devils in the forest.'

'Take me with you,' added the girl, looking piteously at me, 'do, really, take me. . . .'

'To put you in a show in Moscow as a freak, some sea monster,' observed the driver. 'Come, get down, we're just off.'

The girl made no attempt to move, but kept looking pitifully at me. I begged the driver not to hurt her, and he lifted her gently in his arms and set her on the ground. She burst out crying and I was ready to cry with her.

Why had this creature crossed my path just on that day, just as I was driving into Moscow? I thought of Kozlov's 'Mad Girl';[5] he had met her near Moscow, too.

We drove on: the air was full of electricity, unpleasantly heavy and warm. A dark blue storm-cloud with grey streamers reaching to the earth was slowly trailing over the fields, and all at once a zig-zag of lightning ran slanting through it, there was a clap of thunder and the rain came down in torrents. We were nearly seven miles from the Rogozhsky Gate and after reaching Moscow had an hour's drive to the Devichy field. We reached the Astra-kovs',[6] where Ketscher was to wait for me, literally without a dry stitch on us.

Ketscher was not there. He was at the bedside of a dying woman, Ye. G. Levashev. This woman was one of those marvellous phenomena of Russian life which reconcile one to it, one of those whose whole existence is a heroic feat, unknown to any but a small circle of friends. How many tears she had wiped away, how much comfort she had brought to more than one broken heart, of how many young lives she had been the support, and how much she had suffered herself! 'She spent herself in love,' I was told by Chaadayev, one of her closest friends, who dedicated to her his celebrated letter about Russia.[7]

Ketscher could not leave her, and wrote that he would come about nine o'clock. I was alarmed by this news. A man absorbed by a great passion is a dreadful egoist; in Ketscher's absence I could see nothing but a delay. . . . When it struck nine, when the bells began ringing for evening service and then another quarter of an hour passed, I was overcome by feverish anxiety and cowardly despair. . . . Half-past nine – no, he would not come; the sick woman was probably worse, and what was I to do? I could not remain in Moscow: one incautious word from the maid or from Kostinka in the princess's house could give everything away. It was possible to go back, but I felt I had not the strength to go back.

At a quarter to ten Ketscher appeared, in a straw hat and with the crumpled face of a man who has not slept all night. I rushed up to him and as I embraced him I showered him with reproaches. Ketscher looked at me with a frown and asked: 'Why, isn't half an hour enough to get from the Astrakovs' to the Povarskaya? I might have been gossiping with you here for an hour, and I dare say it would have been very nice, but I could not bring myself to leave a dying woman sooner than I needed for the sake of that. She sends you her greetings,' he added; 'she blessed me with her dying hand, and wished us success and gave me a warm shawl in case of need.' The dying woman's greeting was particularly dear to me. The warm shawl was very useful in the night, and I had no time to thank her nor to press her hand . . . soon afterwards she died.

Ketscher and Astrakov set off. Ketscher was to drive out of the town with Natalie, and Astrakov was to come back and tell me whether everything had gone off successfully, and what I was to do. I was left waiting with his beautiful, delightful wife; she had herself only lately been married and, being of an ardent, passionate nature, she took the warmest interest in our doings. She tried with feigned gaiety to assure me that everything would go off splendidly, though she was herself so fretted by anxiety that her face was continually changing its expression. We sat together in the window and conversation did not flow easily; we were like children shut up in an empty room as a punishment. Two hours passed in this way.

There is nothing in the world more shattering, more unendurable than inactivity and suspense at such moments. Friends make a great mistake in taking the whole burden off the shoulders of the principal *patient*. They ought to invent duties for him if there are none, to overwhelm him with physical exertions, to distract his mind with bustle and fuss.

At last Astrakov came in, and we rushed to meet him.

'Everything is going marvellously; I saw them gallop off,' he shouted to us from the yard. 'You go out at the Rogozhsky Gate at once; there by the little bridge you will see the horses not far from Perov's tavern. Good luck to you! And change your cab halfway, so that your second cabman may not know where you have come from.'

I flew like an arrow from the bow. . . . And here was the little bridge not far from Perov's; there was no one there, and on the other side of the bridge, too, there was no one. I dismissed the cabman and went forward on foot. I walked backwards and forwards, and eventually saw a carriage on another road. A handsome young coachman was standing by it. 'Hasn't a tall gentleman in a straw hat driven by here,' I asked him, 'and not alone, with a young lady?'

'I have seen no one,' the coachman answered reluctantly.

'With whom did you come here?'

'With gentlefolks.'

'What is their name?'

'What is that to you?'

'What a fellow you are, really! If it was nothing to do with me, I should not be asking you.'

The coachman gave me a searching look and smiled – apparently my appearance disposed him more favourably to me.

'If you have business with them then you ought to know their names yourself: who do you want?'

'You are a regular flint; well, I want a gentleman named Ketscher.'

The coachman smiled again, and pointing towards the graveyard said: 'Over there, do you see something black in the distance? That's himself, and the young lady is with him; she did not bring a hat, so Mr Ketscher gave her his, seeing it was a straw one.'

This time we again met in a graveyard!

With a faint cry she flung herself on my neck.

'And it's for ever!' she cried.

'For ever,' I repeated. Ketscher was touched and tears gleamed in his eyes; he took our hands and said in a trembling voice, 'Friends, be happy!' We embraced him. This was our *real* wedding!

For over an hour we waited in a private room at Perov's tavern, and still the carriage with Matvey did not come! Ketscher frowned. The possibility of trouble had never entered our heads, we were so happy there, the three of us, and as much at home as though we had always been together. There was a wood in front of the windows, and from the storey below came strains of music and a gypsy chorus; the weather was lovely after the storm.

I was not afraid, like Ketscher, of the police being put on our track by the princess; I knew that she had too much pride to involve a policeman in our family affairs. Besides, she never took any step without consulting the Senator, nor the Senator without consulting my father; my father would never consent to the police stopping me in Moscow or near Moscow, which would mean my being sent to Bobruysk or to Siberia for disobedience to the will of His Majesty. The only possible danger was from the secret police, but it had all been done so quickly that it was hard for them to know. Besides, if they had got an inkling of anything, it would never occur to anyone that a man who had secretly returned from exile and was eloping with his bride would be sitting calmly in Perov's tavern where there was a crush of people from morning to night.

At last Matvey appeared with the carriage.

'One more glass,' commanded Ketscher, 'and off you go!'

And then we were alone, that is, the two of us, flying along the Vladimir road.

At Bunkovo while the horses were being changed we went into the inn. The old hostess came to ask us whether we would like anything, and, looking at us good-naturedly, she said: 'How young and pretty your good lady is, and the two of you, God bless you, make a pretty pair.' We blushed up to our ears and did not dare to look at each other, but asked for tea to cover our confusion. After five o'clock the next day we reached Vladimir. There was no time to be lost; leaving Natalie with the family of an old married official, I rushed off to find whether everything was ready. But who was there to get things ready at Vladimir?

There are good-natured people everywhere. A Siberian regiment of Uhlans was stationed at Vladimir at the time; I was only slightly acquainted with the officers but, meeting one of them rather often in the public library, I had taken to bowing to him; he was polite and very nice. A month later he admitted that he knew me and my 1834 story, and told me that he had himself been a student at Moscow University. When I was leaving Vladimir and looking about for someone in whose hands to leave various arrangements, I had thought of this officer, had driven to his place and told him frankly what I was up to. Genuinely touched by my confidence, he pressed my hand, promised to do everything, and kept his word.

He was waiting for me in full uniform, with white facings, his shako with no cover, with a cartridge-belt across his shoulder, and all sorts of cords and lace. He told me that the bishop had given the priest permission to marry us, but had ordered the baptismal certificate to be shown first. I gave the officer the baptismal certificate, while I went off to another young man who had also been a Moscow student. He was serving his two *provincial years* in accordance with the new regulation, in the governor's office, and was almost dying of boredom.

'Would you like to act as groomsman?'

'Whose?'

'Mine.'

'What, yours?'

'Yes, yes, mine.'

'Delighted. When?'

'Now.'

He thought that I was joking, but when I hastily told him how it was he skipped with joy. To be groomsman at a clandestine wedding, to bustle about, possibly to get into trouble, and all that in a little town absolutely without any diversions! He promised at once to get me a carriage and four and ran to his chest of drawers to see whether he had a clean white waistcoat.

As I drove away, I met my Uhlan with a priest sitting on his knee. Imagine an officer, spruced up in his variegated uniform, in a little drozhki with a stout priest, adorned with big, combed-out beard, and arrayed in a silk cassock, which kept catching in all the Uhlan's unnecessary accoutrements. This sight alone might have attracted attention not only in the street that led from the Golden Gate at Vladimir, but on the Paris boulevards, or in Regent Street itself. But the Uhlan had not thought of that, and indeed I only thought of it afterwards. The priest had been going from house to house holding services, since it was St Nicholas's Day, and my cavalry officer had captured him by force somewhere and requisitioned him. We drove off to the bishop's.

To explain the situation I must describe how the bishop came to be involved in it at all. The day before I went away the priest who had agreed to marry us suddenly announced that he would not do it without the bishop's sanction, that he had heard something and was afraid. In spite of all my eloquence, as well as the Uhlan's, the priest was obstinate and stuck to his point. The Uhlan suggested trying the priest of his regiment. This man, a priest with a cropped head and shaven skin, wearing a long, full-skirted coat and trousers tucked into his high boots, and placidly smoking a soldier's pipe, though attracted by certain details in our proposal, yet refused to marry us, declaring, in a mixture of Polish and White Russian, that he was strictly and absolutely forbidden to marry 'civilians'.

'And we are still more strictly forbidden to be witnesses and groomsmen at marriages without permission,' the officer observed. 'And yet I'm going, you see.'

'It's a different matter if one stands before Jesus, as I do; it's a different matter.'

'God helps those who help themselves,' I said to the Uhlan. 'I'll go straight to the bishop. And by the way, why don't you ask permission?'

'That won't do. The Colonel would tell his wife and she'd gossip about it all over the place. Besides, he'd very likely refuse.'

Bishop Parfeny of Vladimir was an intelligent, austere, rough old man; managing and self-willed, he might equally well have been a governor or a general, and, indeed, I think he would have been more in his right place as a general than as a monk; but it had turned out otherwise, and he ran his diocese as he would have a division in the Caucasus. I noticed in him, on the whole, far more of the qualities of an administrator than of one dead to the things of this life. He was, besides, rather severe than ill-natured; like all business-like men, he grasped questions quickly and clearly and was furious when people talked nonsense to him or did not understand him. It is far easier to come to an understanding with a man of this sort than with people who are mild, but weak and irresolute. In accordance with the custom at all provincial towns, on arriving at Vladimir I had gone once after mass to call on the bishop. He received me cordially, gave me his blessing, and regaled me with salmon; then he invited me to come one day and spend the evening talking with him, saying that his eyes were failing and he could not read in the evening. I went two or three times; he talked about literature, knew all the new Russian books and read the magazines, and so we got on splendidly together. Nevertheless, it was with some alarm that I knocked at his episcopal door.

It was a hot day. His Reverence the bishop received me in the garden. He was sitting under a big, shady lime tree, and had taken off his cowl and let his grey locks flow in freedom. A bald, well set up head-priest was standing before him, bareheaded, and right in the sun, reading some document aloud; his face was crimson and big drops of perspiration stood out on his forehead: he screwed up his eyes at the dazzling whiteness of the paper with the sunlight upon it, yet he did not dare to move nor did the bishop tell him to step out of the sun.

'Sit down,' said the bishop after blessing me, 'we are just finishing; these are our little Consistory affairs. Read on,' he added to the head-priest, and he, after mopping his face with a dark blue handkerchief and coughing aside, started reading again.

'What news have you to tell me?' Parfeny asked me, handing the pen to the head-priest, who seized this excellent opportunity to kiss his hand.

I told him of the priest's refusal.

'Have you the necessary papers?'

I showed him the governor's permission.

'Is that all?'

'Yes.'

Parfeny smiled: 'And on the lady's side?'

'There is her baptismal certificate; it will be brought on the day of the wedding.'

'When is the wedding?'

'In two days.'

'You've found a house, then?'

'Not yet.'

'Now see here,' Parfeny said to me, putting his finger on his lip and hooking his mouth towards his cheek, one of his favourite tricks; 'you're an intelligent and well-read man, but you won't catch an old sparrow by putting salt on its tail. There is something shady about it, so, since you have come to me, you had much better tell your business according to your conscience, as if you were at confession. Then I'll tell you straight what can be done and what can't, and in any case my advice will do you no harm.'

My business seemed to me so clear and so just that I told him the whole story, without, of course, going into unnecessary details. The old man listened attentively and often looked into my face. It appeared he was an old acquaintance of the princess's, and therefore could to some extent judge for himself of the truth of my account.

'I understand, I understand,' he said when I had finished. 'Well, let me write a letter to the princess on my own account.'

'I assure you that no peaceful means will lead to anything: her humours and obduracy have gone too far. I have told your Reverence all about it, as you desired; now I must add that if you refuse to help me I shall be forced to do secretly, stealthily, by bribes, what I am now doing quietly, but straightforwardly and openly. I can assure you of one thing: neither prison nor a fresh term of exile shall stop me.'

'You see,' said Parfeny, getting up and stretching, 'what a head-strong fellow you are. Perm has not been enough for you; steep hills have not broken you in yet. Am I saying that I forbid it? Get married if you like, there is nothing unlawful about it; but it would be better to do it peacefully, with the consent of the family.

Send me your priest, I'll prevail on him somehow; only remember one thing: without the proper papers on the bride's side don't even attempt it. So it's a case of "Neither prison nor exile" – upon my word, what are people coming to nowadays! Well, the Lord be with you! Good luck to you, only you'll get me into trouble with the princess.'

And so, in addition to the Uhlan officer, a part in our plot was also taken by his Reverence Parfeny, Bishop of Vladimir and Suzdal.

When as a preliminary measure I had asked the governor's permission, I had in no way represented my marriage as being secret; this was the surest method of avoiding talk about it, and nothing could be more natural than the arrival of my future bride in Vladimir, since I had been deprived of the right to leave it. It was also natural that under the circumstances we should wish our wedding to be as modest as possible.

When we arrived with the priest at the bishop's on the 9th of May, his servitor told us that he had gone to his out-of-town house that morning and would not be back until night. It was already between seven and eight in the evening, one cannot marry after ten, and the next day was Saturday. What was to be done? The priest was scared. We went in to see the head-monk, the bishop's chaplain; he was drinking tea with rum in it and was in the most benign frame of mind. I told him our business, and he poured me out a cup of tea and insisted on my adding rum to it; then he took out some huge silver spectacles, read the baptismal certificate, turned it over, looked at the other side where there was nothing written, folded it up, and giving it back to the priest said: 'In the most perfect possible order.'

The priest still hesitated. I told the chaplain that if I were not married that day it would upset me terribly.

'Why put it off?' he said. 'I will tell his Reverence; marry them, Father Ioann, marry them – in the name of the Father, the Son, and the Holy Ghost, Amen!'

There was nothing for the priest to say; he drove off to write out a certificate that we were not related within the prohibited degrees, while I galloped off for Natalie.

When we were driving alone together out of the Golden Gate, the sun, which had till then been hidden by clouds, shed a dazzling light upon us with its last, bright-red rays, and so triumphantly and joyously that we both said in one breath: 'That's to see us off!' I remember her smile at the words and the pressure of her hand.

The drivers' little church, two miles from the town, was empty; there were neither choristers nor lighted candelabra. Five or six troopers of the Uhlan regiment came in as they were passing, and went out again. The old clerk chanted in a soft, faint voice; Matvey looked at us with tears of joy, the young groomsmen stood behind us with the heavy crowns with which all the Vladimir drivers had been married one after another. The deacon with a shaky hand passed us the silver bowl of union ... it grew dark in the church, with only a few candles glowing here and there; all this was, or seemed to us, extremely fine, from its very simplicity. The bishop drove by, and seeing the church doors open stopped and sent to inquire what was happening. The priest, turning a little pale, went out to him himself, and returned a minute later with a cheerful face, saying to us: 'His Reverence sends you his episcopal blessing and bade me tell you he is praying for you.'

By the time we were driving home the news of our clandestine marriage was all over the town; ladies were waiting on the balconies and the windows were open. I let down the carriage windows and was a little vexed that the darkness prevented me from showing my 'young woman'.

At home we drank two bottles of wine with Matvey and the groomsmen; the latter stayed twenty minutes with us, and then we were left alone, and again, as at Perov's, this seemed so simple and natural that we were not in the least surprised at it, though for months afterwards we could not get over the wonder of it.

We had three rooms. We sat at a little table in the drawing-room, and forgetting the fatigue of the last few days we talked away part of the night....

To have a crowd of outsiders at wedding festivities has always seemed to me something coarse, unseemly, almost cynical; why this premature lifting of the veil from love, this initiation of indifferent casual spectators into the privacy of the family? How all these hackneyed greetings, commonplace vulgarities, stupid allusions, must wound the poor girl who is thrust into the public eye in the part of the bride ... not one delicate feeling is spared: the luxury of the bridal chamber, the charm of the night attire, are displayed not only for the visitors to admire but for every idle vagrant. And after this the first days of the new life that is beginning, in which every minute is precious, which ought to be spent far away in solitude, are, as though in mockery, passed in endless dinners and exhausting balls, among a crowd.

Next morning we found in the *salon* two rose-bushes and an immense bouquet. Nice, kind Yuliya Fëdorovna (the governor's wife), who had taken a warm interest in our romance, had sent them. I embraced and kissed her footman and then we went off to see her. Since the bride's trousseau consisted of two dresses, the one in which she had travelled and the other in which she had been married, she put on the wedding dress.

From Yuliya Fëdorovna's we drove to the bishop's; the old man himself led us into the garden, with his own hands cut a nosegay of flowers, told Natalie how I had tried to frighten him with the prospect of my own ruin, and in conclusion advised her to study housekeeping. 'Do you know how to salt cucumbers?' he asked Natalie.

'I do,' she answered, laughing.

'Oh, that's hard to believe. And you know, it is essential!'

In the evening I wrote a letter to my father. I begged him not to be angry at the accomplished fact, and, 'since God had united us', to forgive me and add his blessing. My father as a rule wrote me a few lines once a week; he did not write one day earlier or later in reply, and even began his letter as usual: 'I received your letter of the 10th of May at half-past five the day before yesterday, and from it learned, not without distress, that God had united you with Natasha. I do not cross the will of God in anything, but submit blindly to the trials which He lays upon me. But since the money is mine and you have not thought it necessary to comply with my wishes, I must inform you that I shall not add one kopeck to your present allowance of one thousand silver roubles a year.'

How whole-heartedly we laughed at this distinction between the spiritual and temporal power.

And yet how we needed something more! The money I had borrowed was all spent. We had nothing, absolutely nothing, no clothes, no linen, no crockery. We sat like prisoners in a little flat because we had nothing to go out in. Matvey with a view to economy made a desperate effort to transform himself into a cook, but except beef-steaks and chops he could cook nothing, and so for the most part confined himself to provisions that needed little cooking: ham, salt fish, milk, eggs, cheese, and extremely hard cakes flavoured with mint and not in their first youth. Dinner was an endless source of amusement to us; sometimes we had milk by way of soup, and sometimes last by way of dessert. Over this Spartan fare we used to smile as we recalled the long procession at the celebration of dinner at the princess's and at my father's, where half a dozen flunkeys ran about the room with bowls and dishes, cloaking under the magnificent *mise en scène* the really very uninventive fare.

So we struggled along in poverty for a year. The Chemist sent us ten thousand paper roubles; more than six thousand of this had to go to pay our debts, but what remained was a great help. At last even my father was tired of trying to take us by hunger, like a fortress, and without adding to my allowance he began sending us presents of money, though I never dropped a hint about money after his famous *distinguo*!

I began looking for other quarters. A big, neglected manor-house with a garden was to let on the other side of the river Lybed. It belonged to the widow of a prince who had ruined himself at cards, and it was being let particularly cheaply because it was far away and inconvenient, and chiefly because the princess bargained to keep part of it, in no way separated from the rest, for her son, a spoilt fellow of thirty, and for his servants. No one would agree to this partial possession; I at once accepted it, for I was fascinated by the loftiness of the rooms, the size of the windows, and the big, shady garden. But this very loftiness and spaciousness made a very amusing contrast with our complete lack of movable belongings and articles of the first necessity. The princess's housekeeper, a good-natured old woman, who was greatly attracted by Matvey, provided us, at her own risk, first with a table-cloth, then with cups, then with sheets, then with knives and forks.

What bright and untroubled days we spent in the little three-roomed flat at the Golden Gate and in the princess's huge house! ... There was a big, scarcely furnished *salon*, in which we were sometimes over-taken by such childishness that we raced about it, jumped over the chairs, lit candles in all the candelabra fastened to the wall, and after illuminating the room *a giorno*, recited poetry. Matvey and our maid, a young Greek girl, took part in everything and played the fool as much as we did. Order 'did not triumph'[8] in our household.

And for all this childishness our life was full of a profound earnestness. Cast away in the quiet, peaceful little town, we were completely devoted to each other. From time to time came news of one of our friends, a few words of warm sympathy, and then again we were alone, absolutely alone. But in this solitude our hearts were not closed by our happiness; on the contrary, they were more open to every interest than ever before; we led a full and many-sided life, we thought and read, gave ourselves up to every pursuit and again concentrated on our love; we compared our thoughts and dreams, and saw with amazement how endless was our sympathy, how in all the subtlest, evanescent intricacies and ramifications of feeling and thought, tastes and antipathies, all was kinship and harmony. The only difference was that Natalie brought into our union a gentle, mild, gracious element, the characteristics of a young girl with all the poetry of a loving woman, while I brought lively activity, my *semper in motu*, infinite love and, moreover, a medley of earnest ideas, laughter, 'dangerous' thoughts, and unfeasible projects.

'... My desires had reached a standstill, I was satisfied, I lived in the present, I expected nothing from the morrow, I trusted carelessly that it would take nothing from me. Personal life could give nothing more, this was the limit; any change could but diminish it, in some way or another.

'In the spring Ogarëv came from his exile for a few days. He was then at the very height of his powers; he too was soon to pass through painful experiences; at moments he seemed to feel that misfortune was at hand, but he could still turn aside and take the lifted hand of destiny for a dream. I myself thought then that these storm-clouds would be dissipated; carelessness is characteristic of everyone young and not devoid of strength; it expresses a confidence in life and in oneself. The feeling of complete mastery over one's fate lulls us asleep ... while dark powers and black-hearted people draw us without a word to the edge of the precipice.

'And well it is that man either does not suspect, or can shut his eyes and forget. Where there is apprehension there can never be complete happiness; complete happiness is serene as the sea in the calm of summer. Apprehension gives its morbid, feverish intoxication which pleases like the thrill of suspense at cards, but this is far from the feeling of harmonious, everlasting peace. And so, whether it be a dream or not, I prize most highly that trust in life, before life itself has called it in question and woken one up. . . . Chinese die of a crude intoxication with opium. . . .'

So I ended this chapter in 1853 and so I end it now.

CHAPTER VI

THE 13TH OF JUNE 1839[1]

ONE long, winter evening towards the end of 1838 we were sitting alone, as always, reading and not reading, talking and being silent, and in silence continuing the talk. There was a hard frost outside, and in the room it was not at all warm. Natasha did not feel well and was lying on the sofa, covered with a cloak, and I was sitting on the floor near her. My reading did not get on: she was absent-minded, thinking of something else, absorbed in something, and her face kept changing.

'Alexander,' she said, 'I have a secret: come nearer and I will tell you in your ear. Or no: guess it yourself.'

I did guess, but insisted on her telling me. I longed to hear this news from her: she told me; we looked at each other in excitement and with tears in our eyes.

How rich is the human heart in the capacity for happiness, for joy, if only people know how to give themselves up to it without being distracted by trifles. The present is usually disturbed by external worries, empty cares, irritable obstinacy, all the rubbish which is brought upon us in the midday of life by the vanity of vanities, and the stupid ordering of our everyday life. We waste our best minutes; we let them slip through our fingers as though we had heaven knows how many of them. We are generally thinking of tomorrow, of next year, when we ought to be clutching with both hands the brimming cup which life itself, unbidden, with her customary lavishness, holds out to us, and to drink and drink of it until the cup passes into other hands. Nature does not care to spend a long time offering us her treat.

One would have thought that nothing could have been added to our happiness, and yet the news of the coming child opened new tracts of the heart, new raptures, hopes and alarms of which we had before known nothing.

When rather frightened and alarmed, love grows more tender, is more anxious in its solicitude: from the selfishness of two it becomes not a mere selfishness of three but the self-denial of two for a third; family life begins with children. A new element is entering into life, a mysterious person is knocking at the door, a guest who as yet is not, and who is indispensable, who is passionately awaited. Who is he? No one knows but, whoever he is, he is a fortunate stranger: with what love he is met on the threshold of life!

*

The fatal day was approaching and everything became more and more frightening. I looked at the doctor and the mysterious face of the midwife with slavish reverence. Neither Natasha nor I nor our young maid knew anything about this sort of thing; luckily, at my father's request, an elderly lady, Praskovya Andreyevna, a sensible, practical and capable woman, came from Moscow to stay with us. Seeing our helplessness she autocratically took the reins of management into her own hands and I obeyed her like a negro.

One night I felt a hand touch me: I opened my eyes, and Praskovya Andreyevna was standing before me in a nightcap and dressing-gown with a candle in her hand; she told me to send for the doctor and the midwife. I was petrified, just as though the news were something I had not in the least expected. If I could have drunk some opium, turned over on the other side and slept through the danger ... but there was no help for it: I dressed with trembling hands and hurried to wake Matvey.

A dozen times I ran out from the bedroom into the front hall to listen for a carriage coming in the distance. Everything was still but for the faint, faint rustle of the breeze of morning in the warm June air of the garden; the birds were beginning to sing, the crimson dawn threw a light flush over the leaves, and again I hurried back to the bedroom, pestered kind Praskovya Andreyevna with stupid questions, squeezed Natasha's hands convulsively, did not know what to do, trembled and was in a fever ... but at last the drozhki rattled on the bridge across the Lybed – thank God, it was in time!

At eleven o'clock in the morning I started as though from a violent electric shock when the loud cry of a new-born baby reached my ear. 'A boy,' Praskovya Andreyevna called to me as she went towards the cradle; I would have taken the baby from the pillow, but I could not, my hands trembled so violently. The thought of danger (which often begins only at this stage) that had constricted my chest vanished at once, a turbulent joy took possession of my heart as though in it all the bells were pealing for a festival of festivals! Natasha smiled at me, smiled at the baby, wept and laughed, and only her broken, spasmodic breathing, her weary eyes, and deathly pallor reminded me of the struggle, the agony that she had just been through.

Then I left the room, for I could bear no more. I went into my study and flung myself on the sofa, quite at the end of my strength, and lay for half an hour without any definite thought, without any definite emotion, in a sort of agony of bliss.

*

MOSCOW, PETERSBURG
AND NOVGOROD
1840–1847

CHAPTER I

RETURN TO MOSCOW AND
INTELLECTUAL DEBATE

AT the beginning of 1840 we left Vladimir and the poor, narrow river Klyazma. With anxiety and a heavy heart I left the little town where we were married. I foresaw that the same simple, profound intimate life would be no more, and that we should have to furl many of our sails.

Our long, solitary walks outside the town where, lost among the meadows, we felt so keenly the spring in nature and the spring in our hearts, would never come again....

The winter evenings when, sitting side by side, we closed the book and listened to the crunch of sledge-runners and the jingle of bells, that reminded us of the 3rd of March, 1838 and our journey of the 9th of May,[1] would never come again....

They would never come again!

In how many keys and for how many ages men have known and repeated that 'The May of life blossoms once and never again',[2] and yet the June of mature age with its hard, harvest-time work, with its stony roads, catches a man unawares. Youth, all unheeding, floats along in a sort of algebra of ideas, emotions, and yearnings, is little interested in the particular, little touched by it; and then comes love, the unknown quantity found; all is concentrated on one person, through whom everything passes, in whom the universal becomes dear, in whom the elegant becomes beautiful; then, too, the young are untouched by the external, they are *given* to each other, and about them let no grass grow!

But it does grow, together with the nettles and the thistles, and sooner or later they begin to sting or hook on to you.

We knew that we could not take Vladimir with us, but still we thought that our May was not yet over. I even fancied that in going back to Moscow I was going back once more to my student days. All the surroundings helped to maintain the illusion. The same house, the same furniture – here was the room where Ogarëv and I, shut in together, used to conspire two paces away from the

Senator and my father, and here was my father himself, grown older and more bent, but just as ready to scold me for coming home late. 'Who is lecturing tomorrow? When is the rehearsal? I am going from the university to Ogarëv's. . . .' It was 1833 over again!

Ogarëv was actually there.

He had received permission to go to Moscow a few months before me. Again his house became a centre where old and new friends met. And although the old unity was no more, he was surrounded by all the nice people.

Ogarëv, as I have had occasion to observe already, was endowed with a peculiar magnetism, a feminine quality of attraction. For no apparent reason others are drawn to such people and cling to them; they warm, unite, and soothe them, they are like an open table at which everyone sits down, renews his powers, rests, grows calmer and more stout-hearted, and goes away a friend.

His acquaintances swallowed up a great deal of his time; he suffered from this at times, but he kept his door open, and met everyone with his gentle smile. Many people thought it a great weakness. Yes, time was lost and wasted, but love was gained, not only of intimate friends, but of outsiders, of the weak: and that is worth as much as reading and other interests.

I have never been able to understand clearly how it is that people like Ogarëv can be accused of idleness. The standards of the factory and the workhouse hardly apply here. I remember that in our student days Vadim[3] and I were once sitting over a glass of Rhine wine when he became more and more gloomy, and suddenly with tears in his eyes, repeated the words of Don Carlos[4] (who quoted them from Julius Caesar): 'Twenty-three and nothing done for immortality!' This so mortified him that he brought his open hand down with all his might on the green wine-glass and cut it badly. All that is so, but neither Caesar nor Don Carlos and Posa, nor Vadim and I explained why we must do something for *immortality*. There is work and it has to be done, and is it to be done for the sake of the work, or for the sake of being remembered by mankind?

All that is somewhat obscure: and what is work?

Work, *business*[5]... Officials recognise as such only civil and criminal affairs; the merchant regards as work nothing but commerce; military men call it their work to strut about like cranes and to be armed from head to foot in time of peace. To my thinking, to serve as the link, as the centre of a whole circle of people, is a very great work, especially in a society both disunited and fettered. No one has reproached me for idleness, and many people have liked some of the things I have done; but do they know how much of all that I have done has been the reflection of our talks, our arguments, the nights we spent idly strolling about the streets and fields, or still more idly sitting over a glass of wine?

But soon in these surroundings a draught blew which reminded us that spring was over. When the joy of meeting had subsided and festivities were over, when what was most important had been said, and said again, and we had to continue on our way, we saw that the carefree luminous life that we thought to find in accordance with our memories, was no longer to be found in our circle, and especially not in Ogarëv's house. Friends were noisy, disputes boiled up, sometimes wine flowed, but there was no gaiety, not such gaiety as in old days. Everyone had an ulterior motive, something unspoken; there was a feeling of strain: Ogarëv looked sad and Ketscher raised his eyebrows ominously. An alien note made a clamant discord in our harmony; all the warmth, all the friendliness of Ogarëv was not enough to hush it.

What I had feared a year before had come to pass, and it was worse than I had thought.

Ogarëv had lost his father in 1838, and had married not long before his father's death. The news of his marriage frightened me: it had all happened so quickly and unexpectedly. The rumours that had reached me about his wife were not altogether favourable to her, yet he wrote with enthusiasm and was happy; I put more faith in him, but still I was uneasy.

At the beginning of 1839 they had come for a few days to Vladimir. It was our first meeting since the auditor Oransky had read us our sentence.[6] We were in no mood to be critical. I only remember that in the first minutes her voice struck me unpleasantly; but that momentary impression passed in the radiance of our joy. Yes, those were the days of fullness and personal happiness, when a man all unsuspecting reaches the highest limit, the furthest boundary of personal happiness. There was not the shadow of a sad memory, not the faintest dark foreboding; it was all youth, friendship, love, exuberant strength, energy, health, and an endless road before us. Even the mood of mysticism, which then had not yet passed away, gave a festive solemnity to our meeting, like chiming bells, choristers, and the lustres in a church.

There was a small iron crucifix on a table in my room. 'On your knees!' said Ogarëv, 'and let us give thanks that we are all four here together.' We knelt down beside him and embraced, wiping away our tears.

But one of the four scarcely needed to wipe them away. Ogarëv's wife looked at the proceedings with some astonishment. I thought at the time that this was *retenue*, but she told me herself afterwards that this scene had struck her as affected and childish. Of course it might strike one so looking on at it as an outsider, but why was she looking on at it as an outsider? Why was she so sober in that moment of intoxication, so grown-up in the midst of our youthfulness?

Ogarëv went back to his estate, and she went to Petersburg to solicit for permission for him to return to Moscow.

A month later she passed through Vladimir again, alone. Petersburg and two or three aristocratic drawing-rooms had turned her head. She wished for outward glitter, she found pleasure in the thought of wealth. Would she manage to come to terms with this? I wondered. Many misfortunes might grow from such a conflict of tastes. But wealth was something new to her and so was Petersburg and its drawing-rooms; perhaps it was a momentary inclination; she was sensible, she loved Ogarëv – and I hoped.

In Moscow people were apprehensive that she would not digest it so easily. An artistic and literary circle rather flattered her vanity, but her chief attention was not turned in that direction. She would have consented to have a shrine for artists and savants in an aristocratic drawing-room, and she tried to force Ogarëv into an empty world in which he was bored to death. His more intimate friends began to notice this, and Ketscher, who had long been scowling over it, menacingly proclaimed his *veto*. Hot-tempered, vain, and unaccustomed to using the curb on herself, she wounded a vanity as irritable as her own. Her stiff, rather frigid manners, and the sarcasms that she uttered in the voice which at our first meeting had so strangely jarred on me, provoked a sharp opposition. After quarrelling for two months with Ketscher who, though at bottom he was right, was constantly wrong in procedure, and arousing the hostility of several persons who, owing to their material situation, were perhaps too ready to take offence, she eventually found herself face to face with me.

She was afraid of me. In me she wanted to take the measure of herself and to discover once for all which was to have the upper hand, friendship or love, as though one or the other must take this upper hand. There was more involved in this than the desire to have her own way in a capricious dispute: there was the consciousness that I opposed her views more strongly than any of them; there was envious jealousy in it too, and a feminine love of power. With Ketscher she argued till she shed tears, and she quarrelled with him, as angry children quarrel every day, but without bitterness; but me she could not look at without turning pale and trembling with hatred. She reproached me for destroying her happiness through a selfish claim to Ogarëv's exclusive friendship, and for my repulsive pride. I felt that this was unjust, and became cruel and merciless in my turn. She confessed to me herself, five years later, that she had had thoughts of poisoning me, her hatred went to such a length. She broke off her acquaintance with Natalie because of her love for me and the affection all our friends had for her.

Ogarëv suffered. No one spared him, neither she nor I nor the others. We chose his heart (as he expressed it himself in a letter) 'for our field of battle', and did not consider that whichever side won he suffered equally. He swore to reconcile us, he tried to soften the awkwardness of the situation, and we were reconciled; but wounded vanity voiced its savage cry and smarting susceptibility flared into warfare at a word. Ogarëv saw with horror that everything he prized was falling to pieces, that to the woman he loved his sacred things were not sacred, that she was a stranger – but he could not stop loving her. We were his own people, but he saw with grief that even we did not spare him one drop of the bitterness in the cup that fate presented to him. He could not rudely tear apart the bonds of *Naturgewalt* that bound him to her, nor the firm bonds of sympathy that bound him to us; in either event his heart could not but bleed, and, conscious of that, he tried to keep both her and us – convulsively gripped and held her hands and ours – while we savagely strained apart, quartering him like executioners!

Man is cruel, and is tamed only by long trials; the child is cruel in its ignorance, the young man is cruel in the pride of his purity, the priest is cruel in the pride of his holiness, and the doctrinaire in the pride of his learning – we are all merciless, and most merciless of all when we are in the right. The heart usually melts and softens in consequence of deep scars, of scorched wings, of acknowledged downfalls; in consequence of the fear that makes a man cold all over when, alone and without witnesses, he begins to suspect what a weak, vile creature he is. His heart grows gentler; as he wipes away the sweat of shame and horror, afraid of an eye-witness, he seeks justification for *himself* and finds it for *another*. The part of judge, of executioner, from that moment inspires aversion in him.

In those days I was far from that!

The feud was carried on intermittently. The exasperated woman, pursued by our intolerance, got more and more deeply entrammelled, could not walk, floundered about, fell down – and did not change. Feeling herself powerless to conquer, she burned with vexation and *dépit*, with jealousy in which there was no love. Her dishevelled ideas, taken disconnectedly from George Sand's novels and from our conversations, and never attaining clarity in anything, carried her from one absurdity to another, to eccentricities, which she took for originality and independence, to that form of feminine emancipation in virtue of which women arbitrarily deny all that they dislike in the existing and accepted order, while they obstinately cling to all the rest.

A rupture was becoming unavoidable, but for a long time yet Ogarëv was sorry for her, for a long time he still tried and hoped to save her. And whenever for a minute some tender feeling was awakened or poetic chord was touched in her, he was ready to forget the past for ever and begin a new life of harmony, peace, and love; but she could not restrain herself, she lost her balance and every time sank lower. Thread by thread their union was painfully torn apart till the last thread wore through – and they parted for ever.

In all this one question presents itself that is not quite easily answered. How was it that the powerful, sympathetic influence that Ogarëv exercised on everything round him, which drew outsiders into higher spheres and common interests, slid over this woman's heart without leaving on it a vestige of wholesomeness? And yet he loved her passionately and put more energy and heart into trying to save her than into all the rest; and she herself loved him at first, there is no doubt of that.

I have thought a great deal about this. At first, of course, I put the blame on one side only, but afterwards I began to understand that this strange, monstrous fact has an explanation and that properly there is no contradiction in it. To have an influence on a sympathetic circle is far easier than to have an influence on one woman. To preach from the pulpit, to sway men's minds from the platform, to teach from the lecturer's desk, is far easier than to bring up one child. In the lecture-room, in the church, in the club, similarity of interests and aspirations takes the foremost place; men meet there for the sake of them, and it is only a matter of developing them further. Ogarëv's circle consisted of his old comrades of the university, young *savants*, artists, and men of letters; they were united by a common religion, a common language, and still more by a common hatred. Those for whom this religion was not really a living question gradually dropped off, and others came to fill their places, and the circle itself, as well as its thinking, grew the stronger from this free play of elective affinity and the community of conviction that bound them together.

Intimacy with a woman is a purely personal matter, based on some secret physiological affinity, unaccountable, passionate. We are first intimate, afterwards we become acquainted. Among people the cards of whose life are not 'fixed' before they are dealt, whose life is not dominated by one idea, equilibrium is easily established; everything with them happens casually, he yields half and she half, and if they do not it does not much matter. On the other hand, a man devoted to his idea discovers with horror that it is alien to the creature he has set so close to him. He goes to work in haste to awaken her, but as a rule only frightens or muddles her. Torn away from the traditions from which she has not freed herself, and flung as it were, across a ravine filled with nothing, she believes that she is emancipated – presumptuously, egotistically rejects the old at random and accepts the new without discrimination. There is disorder and chaos in her head and in her heart ... the reins are flung down, egoism is unbridled ... and we imagine that we have accomplished something, and preach to her as though we were in a lecture-room.

The gift for education, the gift of patient love, of complete, of persevering devotion is more rarely met with than any other. No mother's passionate love alone, nor mere dialectical skill can replace it.

Is not this the reason why people torment children, and some-times grown-up people too : that it is so hard to educate them and so easy to flog them? When we punish, are we not avenging our-selves for our own incapacity?

Ogarëv understood this even then; that was why everyone (and I among them) reproached him for being too gentle.

The circle of young people that formed itself round Ogarëv was not our old circle. Only two of his old friends, besides ourselves, were in it. Tone, interests, pursuits, all had changed. Stankevich's[7] friends took the lead in it; Bakunin[8] and Belinsky[9] stood at their head, each with a volume of Hegel's philosophy in his hand, and each filled with the youthful intolerance inseparable from vital, passionate convictions.

German philosophy had been grafted on Moscow University by M. G. Pavlov.[10] The Chair of Philosophy had been abolished since 1826. Pavlov gave us an introduction to philosophy instead of physics and agriculture. It would have been hard to learn physics at his lectures, impossible to learn agriculture; but they were extremely profitable. Pavlov would stand at the door of the Faculty of Physics and Mathematics and stop a student with the question: 'You want to acquire a knowledge of nature? But what is nature? what is knowledge?'

This was extremely valuable: our young students enter the university entirely without philosophical preparation; only the divinity students have any conception of philosophy, yet that is an utterly distorted one.

By way of answer to these questions, Pavlov expounded the doctrines of Schelling and of Oken[11] with such a formative clarity as no teacher of natural philosophy had shown before. If he did not attain complete lucidity in anything it was not his fault, but was due to the cloudiness of Schelling's teaching. Pavlov may more justly be blamed for stopping short at this Mahabharata of philo-sophy and not going through the severe ordeal of Hegelian logic. But even in his science he went no further than the introduction and general conception, or at any rate he led others no further. Such a halt at the beginning, such a failure to top out one's build-ing, these houses without roofs, foundations without houses, and ostentatious front halls leading to a modest dwelling, are quite in the spirit of the Russian people. Are we not perhaps satisfied with front halls because our history is still knocking at the gate?

What Pavlov did not do was done by one of his pupils – Stanke-vich.

Stankevich, also one of the *idle* people who accomplish *nothing*, was the first disciple of Hegel in the circle of young people in Moscow. He had made a profound study of German philosophy, which appealed to his aesthetic sense: endowed with exceptional abilities, he drew a large circle of friends into his favourite pursuit. This circle was extremely remarkable: from it came a regular legion of *savants*, writers and professors, among whom were Belinsky, Bakunin, and Granovsky.[12]

Before our exile there had been no great sympathy between our circle and Stankevich's. They disliked our almost exclusively political tendency, while we disliked their almost exclusive specu-lative interests. They considered us to be *Frondeurs* and French, we thought them sentimentalists and German. The first man who was acknowledged both by us and by them, who held out the hand of friendship to both and by his warm love for both and his con-ciliating character removed the last traces of mutual misunder-standing, was Granovsky; but when I arrived in Moscow he was still in Berlin, and poor Stankevich at the age of twenty-seven was dying on the shore of the Lago di Como.

Sickly in constitution and gentle in character, a poet and a dreamer, Stankevich was naturally bound to prefer contemplation and abstract thought to living and purely practical questions; his artistic idealism suited him; it was 'the crown of victory' set on the pale, youthful brow that bore the imprint of death. The others had too much physical vigour and too little poetical feeling to remain long absorbed in speculative thought without passing on into life. An exclusively speculative tendency is utterly opposed to the Russian temperament, and we shall soon see how the *Russian spirit* transformed Hegel's teaching and how the vitality of our nature asserted itself in spite of all those who took the tonsure of philosophical monasticism. But at the beginning of 1840 the young people surrounding Ogarëv had as yet no thought of rebel-ling against the letter on behalf of the spirit, against the abstract on behalf of life.

My new acquaintances received me as people do receive exiles and old champions, people who come out of prison or return from captivity or banishment, that is, with respectful indulgence, with a readiness to receive us into their alliance, though at the same time refusing to yield a single point and hinting at the fact that they are 'today' and we are already 'yesterday', and exacting an unconditional acceptance of Hegel's *Phenomenology* and *Logic*, and their interpretation of them, too.

They discussed these subjects incessantly; there was not a paragraph in the three parts of the *Logic*, in the two of the *Aesthetic*, the *Encyclopaedia*, and so on, which had not been the subject of desperate disputes for several nights together. People who loved each other avoided each other for weeks at a time because they disagreed about the definition of 'all-embracing spirit', or had taken as a personal insult an opinion on 'the absolute personality and its existence in itself'. Every insignificant pamphlet published in Berlin or other provincial or district towns of German philosophy was ordered and read to tatters and smudges, and the leaves fell out in a few days, if only there was a mention of Hegel in it. Just as Francœur in Paris wept with emotion when he heard that in Russia he was taken for a great mathematician and that all the younger generation made use of the same letters as he did when they solved equations of various powers, tears might have been shed by all those forgotten Werders, Marheinekes, Michelets, Ottos, Watkes, Schallers, Rosenkranzes, and even Arnold Ruge himself,[13] whom Heine so wonderfully well dubbed 'the gatekeeper of Hegelian philosophy', if they had known what bloodshed, what declarations they were exciting in Moscow between the Maroseyka and the Mokhovaya,[14] how they were being read, and how they were being *bought*.

Pavlov's great value lay in the unusual clarity of his exposition, a clarity in which none of the depth of German thought was lost; the young philosophers, on the contrary, adopted a conventional language; they did not translate philosophical terms into Russian, but transferred them whole, even, to make things easier, leaving all the Latin words *in crudo*, giving them orthodox terminations and the seven Russian cases.

I have the right to say this because, carried away by the current of the time, I wrote myself exactly in the same way, and was actually surprised when Perevoshchikov, the well-known astronomer, described this language as the 'twittering of birds'.[15] No one in those days would have hesitated to write a phrase like this: 'The concretion of abstract ideas in the sphere of plastics presents that phase of the self-seeking spirit in which, defining

itself for itself, it passes from the potentiality of natural immanence into the harmonious sphere of pictorial consciousness in
beauty.' It is remarkable that here Russian words, as in the celebrated dinner of the generals of which Yermolov[16] spoke, sound
even more foreign than Latin ones.

German learning – and it is its chief defect – has become accustomed to an artificial, heavy, scholastic language of its own, just
because it has lived in academies, that is, in the monasteries of
idealism. It is the language of the priests of learning, a language
for the *faithful*, and none of the catechumens understood it. A key
was needed for it, as for a letter in cipher. The key is now no
mystery; when they understood it, people were surprised that very
sensible and very simple things were said in this strange jargon.
Feuerbach was the first to begin using a more human language.

The mechanical copying of the German ecclesiastico-scientific
jargon was the more unpardonable as the leading characteristic of
our language is the extraordinary ease with which everything is
expressed in it – abstract ideas, the lyrical emotions of the heart,
'life's mouse-like flitting',[17] the cry of indignation, sparkling mis-
chief, and shaking passion.

Another mistake, far graver, went hand in hand with this
distortion of language. Our young philosophers distorted not
merely their phrases but their understanding; their attitude to life,
to reality, became schoolboyish and literary; it was that learned
conception of simple things at which Goethe mocks with such
genius in the conversation of Mephistopheles with the student.
Everything that in reality was direct, every simple feeling, was
exalted into abstract categories and came back from them without
a drop of living blood, a pale, algebraic shadow. In all this there
was a *naïveté* of a sort, because it was all perfectly sincere. The
man who went for a walk in Sokolniky went in order to give himself up to the pantheistic feeling of his unity with the cosmos; and
if on the way he happened upon a drunken soldier, or a peasant
woman who got into conversation with him, the philosopher did
not simply talk to them, but defined the essential substance of the
people in its immediate and fortuitous manifestation. The very
tear that started to the eye was strictly referred to its proper
classification, to *Gemüth* or 'the tragic in the heart'.

It was the same thing in art. A knowledge of Goethe, especially
of the second part of *Faust* (either because it is inferior to the first

or because it is more difficult), was as obligatory as the wearing of clothes. The philosophy of music had a place in the foreground. Of course, no one ever spoke of Rossini; to Mozart they were indulgent, though they did think him childish and poor. To make up for this they carried out philosophical investigations into every chord of Beethoven and greatly respected Schubert, not so much, I think, for his superb melodies as for the fact that he chose philosophical themes for them, such as 'The Omnipotence of God' and 'Atlas'. French literature – everything French in fact, and, incidentally, everything political also – shared the interdict laid on Italian music.

From this it is easy to see on what field we were bound to meet and do battle. So long as we were arguing that Goethe was objective but that his objectivity was subjective, while Schiller as a poet was subjective but that his subjectivity was objective, and *vice versa*, everything went peaceably. Questions that aroused more passion were not slow to make their appearance.

While Hegel was Professor in Berlin, partly from old age, but twice as much from satisfaction with his position and the respect he enjoyed, he purposely screwed his philosophy up above the earthly level and kept himself in an ambience where all contemporary interests and passions became somewhat indistinguishable, like buildings and villages seen from an air-balloon; he did not like to be entangled in these accursed practical questions with which it is difficult to deal and which must receive a positive answer. How clamant this violent and insincere dualism was, in a doctrine which set out from the elimination of dualism, can be understood readily. The real Hegel was the modest Professor at Jena, the friend of Hölderlin, who hid his *Phenomenology* under his coat when Napoleon entered the town; then his philosophy did not lead to Indian quietism, nor to the justification of the existing forms of society, nor to Prussian Christianity; then he had not given his lectures on the Philosophy of Religion, but had written things of genius such as the article on the executioner and the death penalty, printed in Rosenkranz's biography.

Hegel confined himself to the sphere of abstractions in order to avoid the necessity of touching upon empirical deductions and practical applications; the one domain which he, very adroitly, selected for the practical application of his theories was the calm, untroubled ocean of aesthetics. He rarely ventured into the light of day, and then only for a minute, wrapped up like an invalid; and even then he left behind in the dialectic maze just those questions that were most interesting to the modern man. The extremely feeble intellects (Gans[18] is the only exception), who surrounded him, accepted the letter for the thing itself and were pleased by the empty play of dialectics. Probably at times the old man felt sad and ashamed at the sight of the limited outlook of his excessively complacent pupils. If the dialectic method is not the development of the reality itself, the educating of it to think, so to speak, it becomes a purely external means of making a farrago of things run the gauntlet of a system of categories, an exercise in logical gymnastics, as it was with the Greek Sophists and the medieval schoolmen after Abelard.

The philosophical phrase which did the greatest harm, and in virtue of which the German conservatives strove to reconcile philosophy with the political régime of Germany – 'all that is real is rational' – was the principle of sufficient reason and of the correspondence of logic and facts expressed in other words. Hegel's phrase, wrongly understood, became in philosophy what the words of the Christian Girondist Paul once were: 'There is no power but from God.' But if all powers are from God, and if the existing social order is justified by reason, the struggle against it, if only it exists, is also justified. These two sentences accepted in their formal meaning are pure tautology; but, tautology or not, Hegel's phrase led straight to the recognition of the sovereign authorities, led to a man's sitting with folded arms, and that was just what the Berlin Buddhists wanted. However contrary such a view may be to the Russian spirit. our Moscow Hegelians were genuinely misled and accepted it.

Belinsky, the most active, impulsive, and dialectically passionate, fighting nature, was at that time preaching an Indian stillness of contemplation and theoretical study instead of conflict. He believed in that view and did not flinch before any of its consequences, nor was he held back by considerations of moral propriety nor the opinion of others, which has such terrors for the weak and those who lack independence. He was free from timidity for he was strong and sincere; his conscience was clear.

'Do you know that from your point of view,' I said to him, thinking to impress him with my revolutionary ultimatum, 'you can prove that the monstrous tyranny under which we live is rational and ought to exist?'

'There is no doubt about it,' answered Belinsky, and proceeded to recite to me Pushkin's 'Anniversary of Borodino'.

That was more than I could stand and a desperate battle raged between us. Our falling out reacted upon the others, and the circle fell apart into two camps. Bakunin wanted to reconcile, to explain, to *exorcise*, but there was no real peace. Belinsky, irritated and dissatisfied, went off to Petersburg, and from there fired off his last furious salvo at us in an article which he likewise called 'The Anniversary of Borodino'.

Then I broke off all relations with him. Bakunin, though he argued hotly, began to reconsider things; his sound revolutionary judgement pushed him in another direction. Belinsky reproached him for weakness, for concessions, and went to such exaggerated extremes that he scared his own friends and admirers. The chorus were on Belinsky's side, and looked down upon us, haughtily shrugging their shoulders and considering us to be behind the times.

In the midst of this intestine strife I saw the necessity *ex ipso fonte bibere* and began studying Hegel in earnest. I even think that a man who has not *lived through* Hegel's *Phenomenology* and Proudhon's *Contradictions of Political Economy*, who has not passed through that furnace and been tempered by it, is not complete, not modern.

When I had grown used to Hegel's language and mastered his method, I began to perceive that he was much nearer to our viewpoint than to that of his followers; he was so in his early works, he was so everywhere where his genius had taken the bit between its teeth and had dashed forward oblivious of the Brandenburg Gate. The philosophy of Hegel is the algebra of revolution; it emancipates a man in an unusual way and leaves not one stone upon another of the Christian world, of the world of tradition that has outlived itself. But, perhaps with intention, it is badly formulated.

Just as in mathematics – only there with more justification – men do not go back to the definition of space, movement, force, but continue the dialectical development of their laws and qualities; so also in the formal understanding of philosophy, after once becoming accustomed to the first principles, men go on merely drawing deductions. Anyone new to the subject, who has not stupefied himself by the method's being turned into a habit, grasps at just these traditions, these dogmas which have been accepted as thoughts. To people who have long been studying the subject, and are consequently not free from predilections, it seems astonishing that others should not understand things that are 'perfectly clear'.

How can anyone fail to understand such a simple idea as, for instance, 'that the soul is immortal and that what perishes is only the personality', a thought so successfully developed in his book by the Berlin Michelet; or the still simpler truth that the absolute spirit is a personality, conscious of itself through the world, and at the same time having its own self-consciousness?

All these things seemed so easy to our friends, they smiled so condescendingly at 'French' objections, that for some time I was stifled by them and worked and worked to reach a precise understanding of their philosophic jargon.

Fortunately scholasticism is as little natural to me as mysticism, and I stretched its bow until the string snapped and the blindfold dropped from my eyes. Strange to say, it was an argument with a lady that brought me to this.

At Novgorod a year later I became acquainted with a general.[19] I made his acquaintance because no one could have been less like a general.

There was a painful feeling in his house, there were tears in the air, and it was obvious that death had passed through it. His hair was prematurely grey and his kindly, mournful smile more than his wrinkles, was expressive of suffering. He was about fifty. The traces of a fate that had cut off living branches was still more clearly visible on the pale, thin face of his wife. It was too quiet in their house. The general occupied himself with mechanics, and in the mornings his wife gave French lessons to some poor girls; when they had gone she took up a book, and the only things that recalled a different, bright, fragrant life were the flowers, of which there were many, and the toys in a cupboard – only no one played with them.

They had had three children: two years before I knew them an unusually gifted boy of nine had died; a few months later another child died of scarlatina; the mother hastened into the country to save the last child by a change of air and came back a few days later with a little coffin in the carriage with her.

Their life had lost its meaning: it was ended, and continued

without object, without need. Their existence was maintained by
the compassion of each for the other; the one comfort available to
them was the deep conviction that each was essential to the other,
in order that the cross might somehow be borne. I have seen few
more harmonious marriages, though indeed it was no marriage,
for it was not love that bound them together but a deep comrade-
ship in misfortune; their fate was tightly bound and held
together, along with the little, cold hands of those three, and the
hopeless emptiness around and before them.

The bereaved mother was completely devoted to mysticism; she
found release from her anguish in the world of mysterious reconciliations; she was deceived by the flattery that religion pays the
human heart. For her mysticism was no light thing, no mere
dream; it was having her children again, and she was defending
them when she defended her religion. But, since she had an extraordinarily active intelligence, she challenged discussion and knew
her strength. I have met, both before and since, many mystics of
various kinds, from Vitberg and the followers of Towjanski,[20]
who took Napoleon for the military incarnation of God and raised
their caps when they passed the Vendôme Column, to the now-
forgotten 'Ma-Pa',[21] who told me himself of his interview with
God which took place on the high-road between Montmorency
and Paris. They were all for the most part hysterical people who
worked on the nerves, impressed the fancy or the heart, mixed up
philosophical conceptions with an arbitrary symbolism, and did
not care to come out into the open field of logic.

But it was upon that field that L— D—[22] took a firm and
fearless stand. Where and how she had succeeded in obtaining
such artistic skill in argument I do not know. Altogether women's
development is a mystery; there is nothing: just dress and dances,
mischievous back-biting and novel-reading, making eyes and shedding tears – and all at once there appears a titanic will, mature
thought, and colossal intelligence. The young girl carried away by
her passions vanishes, and before you stands a Théroigne de Méricourt,[23] the beauty of the tribune, swaying multitudes of the
people, or an eighteen-year-old Princess Dashkov, sword in
hand, on horseback, in the midst of a mutinous crowd of
soldiers.[24]

In L— D— everything was complete: she had no doubts,
no wavering, no theoretical weakness; even the Jesuits or the

Calvinists can hardly have been so harmoniously consistent in their doctrines as she.

Deprived of her little ones, she had come, instead of hating death, to hating life. That is just what is needed for Christianity, for that complete apotheosis of death: contempt for earth, contempt of the body, has no other meaning. Hence the attack upon everything living and realistic, enjoyment, health, gaiety, the abundant sense of existence. And L— D— had reached the point of disliking both Goethe and Pushkin.

Her attacks on my philosophy were peculiar. She used ironically to declare that all our dialectical subtleties and elaborate constructions were just the beating of the drum, the noise with which cowards try to drown the terrors of their conscience.

'You will never,' she used to say, 'get to a personal god, nor to the immortality of the soul, by any philosophy, and none of you has the courage to be an atheist and reject the life beyond the grave. You are too human not to be horrified by these consequences; an inner repulsion rejects them, so you invent your logical miracles to divert your eyes and to arrive at what is given by religion in a simple and childlike way.'

I objected, I argued, but I was inwardly conscious that I had no complete proofs and that she had a firmer footing on her ground than I on mine.

To complete my distress the inspector of the Medical Board must needs turn up; he was a good-natured man, but one of the most ridiculous Germans I have ever met. A desperate admirer of Oken and Carus,[25] he argued by means of quotations, had a ready-made answer for everything, never had doubts about anything, and imagined that he was completely in agreement with me.

The doctor lost his temper, grew furious the more readily as he could not hold his own by other means, looked upon L— D—'s views as feminine caprice, took refuge in Schelling's lectures on academic doctrine, and read extracts from Burdach's *Physiology* to prove that there is an eternal and spiritual element in man, and that some personal *Geist* is hidden in nature.

L— D—, who had long ago passed through these 'back premises' of pantheism, confuted him and glanced with a smile, from me to him. She was, of course, more in the right than he was, and I conscientiously racked my brains, and was vexed when the good doctor laughed triumphantly. These arguments interested me so much that I set to work upon Hegel with new zest. The torture of uncertainty did not last long; the truth flashed before my eyes and began to grow clearer and clearer; I inclined to my opponent's side, but not in the way she wished.

'You are perfectly right,' I said to her, 'and I am ashamed of having argued against you; of course there is no personal spirit, nor immortality of the soul, and that is why it has been so hard to prove that there is. See how simple and natural it all becomes without these previous assumptions.'

She was troubled by my words but quickly recovered herself and said:

'I am sorry for you, but perhaps it is for the best; you will not long remain in that position, for it is too empty and difficult there, while,' she added, smiling, 'our doctor is incurable, he has no fears, and he is in such a fog that he does not see one step before him.'

Her face was paler than usual, however.

Two or three months later, Ogarëv passed through Novgorod. He brought me Feuerbach's *Wesen des Christenthums*; after reading the first pages I leapt up with joy. Down with the trapping of masquerade; away with the stammering allegory! We are free men and not the slaves of Xanthos;[26] there is no need for us to wrap the truth in myth.

In the heat of my philosophic ardour I began my series of articles on 'Dilettantism in Science', in which, among other things, I paid the doctor out.

Now let us go back to Belinsky.

A few months after his departure to Petersburg in 1840 we arrived there too. I did not go to see him. Ogarëv took my quarrel with Belinsky very much to heart; he knew that Belinsky's absurd opinion was a passing malady, and indeed I knew it too, but Ogarëv was kinder. At last by his letters he almost forced a meeting on us. Our interview was at first cold, unpleasant, and strained, but neither Belinsky nor I was very diplomatic and in the course of trivial conversation I mentioned the article on 'The Anniversary of Borodino'. Belinsky jumped up from his seat and, flushing crimson, said with great simplicity,

'Well, thank God, we've come to it at last. Otherwise I am so stupid I should not have known how to begin. ... You've won; three or four months in Petersburg have done more to convince me than all the arguments. Let us forget this nonsense. It is enough to tell you that the other day I was dining at a friend's and there was an officer of the Engineers there; my friend asked him if he would like to make my acquaintance. "Is that the author of the article on 'The Anniversary of Borodino?'" the officer asked him in his ear. "Yes." "No, thank you very much," he answered dryly. I heard it all and could not restrain myself. I pressed the officer's hand warmly and said to him: "You're an honourable man, I respect you. . . ." What more would you have?'

From that moment up to Belinsky's death we went hand in hand.

Belinsky, as was to be expected, fell upon his former opinion with all the stinging vehemence of his language and all his furious energy. The position of many of his friends was not very much to be envied. *Plus royalistes que le roi*, with the courage of misfortune they tried to defend their theories, while not averse to an honourable truce. All those with sense and vitality went over to Belinsky's side; only the obstinate formalists and pedants held aloof. Some of them reached such a point of German suicide through dead, scholastic learning that they lost all living interest and were themselves lost without a trace. Others became orthodox Slavophils. Strange as the combination of Hegel and Stefan Yavorsky[27] may appear, it is more possible than might be supposed; Byzantine theology is just such a superficial casuistry and play with logical formulas as Hegel's dialectics, formally accepted. Some of the articles in the *Muscovite*[28] are a triumphant demonstration of the extremes to which, with talent, the sodomitical union of philosophy and religion can go.

Belinsky by no means abandoned Hegel's philosophy when he renounced his one-sided interpretation of it. Quite the contrary, it is from this point that there begins his living, apt, original combination of philosophical with revolutionary ideas. I regard Belinsky as one of the most remarkable figures of the period of Nicholas. After the liberalism which had somehow survived 1825 in Polevoy,[29] after the gloomy article of Chaadayev, Belinsky appears on the scene with his caustic scepticism, won by suffering, and his passionate interest in every question. In a series of critical articles he touches in season and out of season upon everything, true everywhere to his hatred of authority and often rising to poetic inspiration. The book he was reviewing usually served him as a starting-point, but he abandoned it half-way and plunged into some other question. The line 'That's what kindred are' in *Onegin* is enough for him to summon family life before the judgment seat and to pick blood relationships to pieces down to the last thread. Who does not remember his articles on the 'Tarantas', on Turgenev's 'Parasha', on Derzhavin, on Mochalov,[30] and *Hamlet*? What fidelity there is to his principles, what dauntless consistency, what adroitness in navigating between the shoals of the censorship, what boldness in his attacks on the literary aristocracy, on the writers of the first three grades,[31] on the secretaries of state of literature who were always ready to defeat an opponent by foul means if not by fair, if not by criticism then by delation? Belinsky scourged them mercilessly, tearing to pieces the petty vanity of the conceited, limited writers of eclogues, lovers of culture, benevolence and tenderness; he turned into derision their dear, their heartfelt notions, the poetical dreams flowering under their grey locks, their *naïveté*, hidden under an Anna ribbon.[32]

How they hated him for it!

The Slavophils on their side began their official existence with the war upon Belinsky; he drove them by his taunts to the *murmolka* and the *zipun*.[33] It is worth remembering that Belinsky had formerly written in *Notes of the Fatherland*, while Kireyevsky[34] began publishing his excellent journal under the title of *The European*; no better proof than these titles could be found to show that at first the difference was only between shades of opinion and not between parties.

Belinsky's articles were awaited with feverish expectation by the young people in Moscow and Petersburg from the 25th of every month. Half a dozen times the students would call in at the coffee-houses to ask whether *Notes of the Fatherland* had been received; the heavy volume was snatched from hand to hand. 'Is there an article by Belinsky?' 'Yes,' and it was devoured with feverish interest, with argument ... and three or four cherished convictions and reputations were no more.

Skobelev, the governor of the Peter-Paul fortress, might well say in jest to Belinsky when he met him on the Nevsky Prospect: 'When are you coming to us? I have a nice warm little cell all ready that I am keeping for you.'

I have spoken in another book of Belinsky's development and of his literary activity; here I will only say a few words about the man himself.

Belinsky was very shy and quite lost his head in an unfamiliar or very numerous company; he knew this and did the most absurd things in his desire to conceal it. Ketscher tried to persuade him to go to visit a lady; the nearer they came to her house the gloomier Belinsky became; he kept asking whether they could not go another day, and talked of having a headache. Ketscher, who knew him, would accept no evasions. When they arrived Belinsky set off running as soon as he got out of the sledge, but Ketscher caught him by the overcoat and led him to be introduced to the lady.

He sometimes put in an appearance at Prince Odoyevsky's[35] literary-diplomatic evenings. At these there were crowds of people who had nothing in common except a certain fear of and aversion from each other: clerks from the embassies and Sakharov[36] the archaeologist, painters and A. Meyendorf,[37] several councillors of state of the cultured sort, Ioakinth Bichurin[38] from Pekin, people who were half gendarmes and half literary men, others who were wholly gendarmes and not at all literary men. A— K—[39] hammered away to a point where generals took him for an authority. The hostess concealed her affliction at her husband's vulgar tastes, and gave way to them much as Louis-Philippe at the beginning of his reign indulged his electors by inviting to the balls at the Tuileries whole *rez-de-chaussée* of suspender-craftsmen, chandlers, shoe-makers, and other worthy citizens.

Belinsky was utterly lost at these evenings, between a Saxon ambassador who did not understand a word of Russian and an official of the Third Division who understood even words that were not uttered. He was usually ailing for two or three days afterwards and cursed the man who had persuaded him to go.

One Saturday, since it was New Year's Eve, Odoyevsky took it into his head to mix a punch *en petit comité* when the principal guests had dispersed. Belinsky would certainly have gone away, but he was prevented by a barricade of furniture; he was somehow stuck in a corner and a little table was set before him with wine and glasses on it; Zhukovsky[40] in the white trousers of his uniform, with gold lace on them, sat down obliquely opposite him. Belinsky stood it for a long time but, seeing no chance of his lot improving, he began moving the table a little; the table yielded at first, but then lurched over and crashed to the floor, while the bottle of Bordeaux very deliberately began to empty itself over Zhukovsky. He jumped up, and the red wine trickled down his trousers; there was an uproar: one servant rushed up with a napkin to daub the wine on to the other parts of the trousers, and another picked up the broken wine-glasses ... while this hubbub was going on Belinsky disappeared and, near to death as he was, ran home on foot.

Dear Belinsky! for what a long time he was angry and upset at such incidents, with what horror he used to recall them, walking up and down the room and shaking his head without the trace of a smile!

But in that shy man, that frail body, there dwelt a mighty spirit, the spirit of a gladiator! Yes, he was a powerful fighter! he could not preach or lecture; what he needed was a quarrel. If he met with no objection, if he was not stirred to irritation, he did not speak well, but when he felt stung, when his cherished convictions were called in question, when the muscles of his cheeks began to quiver and his voice to burst out, then he was worth seeing; he pounced upon his opponent like a panther, he tore him to pieces, made him a ridiculous, a piteous object, and incidentally developed his own thought, with unusual power and poetry. The dispute would often end in blood, which flowed from the sick man's throat; pale, gasping, with his eyes fixed on the man with whom he was speaking, he would lift his handkerchief to his mouth with shaking hand and stop, deeply mortified, crushed by his physical weakness. How I loved and how I pitied him at those moments!

Persecuted financially by the sharks of literature, morally persecuted by the censorship, surrounded in Petersburg by people for whom he had little sympathy, and consumed by a disease to which the Baltic climate was fatal, he became more and more irritable. He shunned outsiders, was *farouche*, and sometimes spent weeks together in melancholy inactivity. Then the publishers sent note after note demanding copy, and the enslaved writer, grinding his teeth, took up his pen and wrote the venomous articles quivering with indignation, the indictments which so impressed their readers.

Often, utterly exhausted, he would come to us to rest, and lie on the floor with our two-year-old child; he would play with him for hours together. While we were only the three of us things went swimmingly, but if there came a ring at the bell, a spasmodic grimace passed over his face and he would look about him un-easily, trying to find his hat; then, with the weakness of a Slav, he would often remain. Here one word, a remark that was not to his liking, would lead to the most extraordinary scenes and arguments...

Once he went in Holy Week to dine with a literary man, and Lenten dishes were served.

'Is it long,' he asked, 'since you became so devout?'

'We eat Lenten fare,' answered the literary gentleman, 'simply and solely for the sake of the servants.'

'*For the sake of the servants*,' said Belinsky, and he turned pale. 'For the sake of the servants,' he repeated, and flung down his dinner napkin. 'Where are your servants? I'll tell them that they are deceived. Any open vice is better and more humane than this contempt for the weak and uneducated, this hypocrisy in support of ignorance. And do you imagine that you are free people? You are on the same level as all the tsars and priests and slave-owners. Goodbye. I don't eat Lenten fare for the edification of others; I have no servants!'

Among the Russians who might be classified as inveterate Germans, there was one, a *magister*[41] of our university, who had lately arrived from Berlin; he was a good-natured man in dark-blue spectacles, stiff and decorous; he had come to a standstill for ever after upsetting and enfeebling his faculties with philosophy and philology. A doctrinaire and something of a pedant, he was fond of holding forth in edifying style. On one occasion, at a literary evening in the house of the novelist who kept the fasts for the sake of his servants, the *magister* was preaching some sort of *honnête et modéré* twaddle. Belinsky was lying on a sofa in the corner and as I passed him he took me by the tail of my coat and said:

'Do you hear the rubbish that monster is talking? My tongue has long been itching, but my chest hurts a bit and there are a lot of people. Be a father to me, make a fool of him somehow, squash him, crush him with ridicule, you can do it better – come, cheer me up.'

I laughed and told Belinsky that he was setting me on like a bull-dog at a rat. I scarcely knew the gentleman and had hardly heard what he said.

Towards the end of the evening, the *magister* in the blue spectacles, after abusing Koltsov[42] for having abandoned the national costume, suddenly began talking of Chaadayev's famous 'Letter', and concluded his commonplace remarks, uttered in that didactic tone which of itself provokes derision, with the following words: 'Be that as it may, I consider his action contemptible and revolting: I have no respect for such a man.'

There was in the room only one man closely associated with Chaadayev, and that was I. I shall have a great deal to say about Chaadayev later on; I always liked and respected him and was liked by him; I thought it was unseemly to let pass this savage remark. I asked him dryly whether he supposed that Chaadayev had had ulterior aims in writing his letter, or had been insincere.

'Certainly not,' answered the *magister*.

An unpleasant conversation followed; I demonstrated to him that the epithets 'revolting and contemptible' were themselves revolting and contemptible when applied to a man who had boldly expressed his opinion and had suffered for it. He expatiated to me on the oneness of the people, the unity of the fatherland, the crime of destroying that unity, and of sacred things that must not be touched.

Suddenly Belinsky mowed down the speech I was making: he leapt up from his sofa, came up to me as white as a sheet, slapped me on the shoulder and said:

'Here you have them, they have spoken out – the inquisitors, the censors – keeping thought in leading-strings ...' and so he went on and on.

He spoke with formidable inspiration, seasoning serious words with deadly sarcasms:

'We are strangely sensitive: men are flogged and we don't resent it, sent to Siberia and we don't resent it; but here Chaadayev, you see, has rubbed the people's honour the wrong way: he mustn't dare to talk; to speak is insolence – a flunkey must never speak! Why is it that in more civilised countries, where one would expect susceptibilities, too, to be more developed than in Kostroma and Kaluga, words are not resented?'

'In civilised countries,' replied the *magister*, with inimitable self-complacency, 'there are prisons in which they confine the senseless creatures who insult what the whole people respect ... and a good thing too.'

Belinsky seemed to tower: he was terrifying, great at that moment. Folding his arms over his sick chest and looking straight at the *magister*, he answered in a hollow voice:

'And in still more civilised countries there is a guillotine to deal with those who think that a good thing.'

Having said this, he sank exhausted in an easy-chair and spoke no more. At the word 'guillotine' our host turned pale, the guests were disquieted and a pause followed. The *magister* had been annihilated, but it is just at such moments that human vanity takes the bit between its teeth. I. Turgenev advises a man, when he has gone such lengths in argument that he begins to feel frightened himself, to move his tongue ten times round the inside of his mouth before uttering a word.

The *magister*, unaware of this homely advice, went on babbling feeble trivialities, addressing himself rather to the rest of the company than to Belinsky.

'In spite of your intolerance,' he said at last, 'I am certain that you will agree with one . . .'

'No,' answered Belinsky; 'whatever you said I shouldn't agree with anything!'

Everyone laughed and went in to supper. The *magister* picked up his hat and went away.

Suffering and privation soon completely undermined Belinsky's sickly constitution. His face, particularly the muscles about his lips, and the mournfully fixed look in his eyes, testified equally to the intense workings of his spirit and the rapid dissolution of his body.

I saw him for the last time in Paris in the autumn of 1847; he was in a very bad way and afraid of speaking aloud; it was only at moments that his former energy revived and its ebbing fires glowed brightly. It was at such a moment that he wrote his letter[43] to Gogol.

The news of the revolution of February found him still alive; he died taking its glow for the flush of the rising dawn!

So this chapter ended in 1854; since that time much has changed. I have been brought much *closer* to that time, closer because of my increasing remoteness from people here, and through the arrival of Ogarëv[44] and by two books: Annenkov's *Biography of Stankevich* and the first parts of Belinsky's complete works. From the windows suddenly thrown open the fresh air of the fields, the young breath of spring was wafted into the hospital wards. . . .

Stankevich's correspondence was unnoticed when it came out. It appeared at the wrong moment. At the end of 1857 Russia had not yet come to herself after the funeral of Nicholas; she was expectant and hopeful; that is the worst mood for reminiscences ... but the book is not lost. It will remain in the pauper's burial-ground one of the rare memorials of its times from which any man who can read may learn what in those days was buried without a word. The pestilential streak, running from 1825 to 1855, will soon be completely cordoned off; men's traces, swept away by the police, will have vanished, and future generations will often come to a standstill in bewilderment before a waste land rammed smooth, seeking the lost channels of thought which actually were never interrupted. The current was apparently checked: Nicholas tied up the main artery – but the blood flowed along side-channels. It is just these capillaries which have left their trace in the works of Belinsky and the correspondence of Stankevich.

Thirty years ago the Russia of the future existed exclusively among a few boys, hardly more than children, so insignificant and unnoticed that there was room for them between the soles of the great boots of the autocracy and the ground – and in them was the heritage of the 14th of December, the heritage of a purely national Russia, as well as of the learning of all humanity. This new life sprouted like the grass that tries to grow on the lip of a still smouldering crater.

In the very jaws of the monster these children stand out unlike other children; they grow, develop, and begin to live an utterly different life. Weak, insignificant, unsupported – nay, on the contrary, persecuted by all, they may easily perish, leaving not the smallest trace, but they survive, or, if they die half-way, not everything dies with them. They are the rudimentary germs, the embryos of history, barely perceptible, barely existing, like all embryos in general.

Little by little groups of them are formed. What is more nearly akin to them gathers round their centre-points; then the groups repel one another. This dismemberment gives them width and many-sidedness for their development; after developing to the end, that is to the extreme, the branches unite again by whatever names they may be called – Stankevich's circle, the Slavophils, or our little *coterie*.

The leading characteristic of them all is a profound feeling of alienation from official Russia, for their environment, and at the same time an impulse to get out of it – and in some a vehement desire to get rid of it.

The objection that these circles, unnoticed both from above and from below, form an exceptional, an extraneous, an unconnected phenomenon, that the education of the majority of these young people was exotic, strange, and that they sooner express a translation into Russian of French and German ideas than anything of their own, seems to us quite groundless.

Possibly at the end of the last century and the beginning of this there was in the aristocracy a fringe of Russian foreigners who had sundered all ties with the national life; but they had neither living interests, nor *coteries* based on convictions, nor a literature of their own. They were sterile and became extinct. Victims of Peter's break with the people, they remained eccentric and whimsical, they were not merely superfluous but undeserving of pity. The war of 1812 set a term to them – the older generation were living out their time, and none of the younger developed in that direction. To include among them men of the stamp of P. Ya. Chaadayev would be a most fearful mistake.

Protest, rejection, hatred of one's country if you will, has a completely different significance from indifferent aloofness. Byron, lashing at English life, fleeing from England as if from the plague, remained a typical Englishman. Heine, trying, from anger at the abominable political condition of Germany, to turn Frenchman, remained a genuine German. The highest protest against Judaism – Christianity – is filled with the spirit of Judaism. The rupture of the states of North America with England could lead to war and hatred, but it could not make the North Americans un-English.

As a rule it is with great difficulty that men abandon their physiological memories and the mould in which they are cast by heredity; to do so a man must be either peculiarly unpassioned and featureless or absorbed in abstract pursuits. The impersonality of mathematics and the unhuman objectivity of nature do not call forth those sides of the soul and do not awaken them; but as soon as we touch upon questions of life, of art, of morals, in which a man is not only an observer and investigator but at the same time himself a participant, then we find a physiological limit – which it is very hard to cross with one's old blood and brains unless one can erase from them all traces of the songs of the cradle, of the fields and the hills of home, of the customs and whole setting of the past.

The poet or the artist in his truest work always belongs to the people. Whatever he does, whatever aim and thought he may have in his work, he expresses, whether he will or not, some elements of the popular character and expresses them more profoundly and more clearly than the very history of the people. Even when renouncing everything national, the artist does not lose the chief features from which it can be recognised to what people he belongs. Both in the Greek *Iphigenia* and in the Oriental *Divan* Goethe was a German. Poets really are, as the Romans called them, prophets; only they utter not what is not and what will be by chance, but what is unrecognised, what exists in the dim consciousness of the masses, what is already slumbering in it.

Everything that has existed from time immemorial in the soul of the Anglo-Saxon people is held together, as if by a ring, by personality alone; and every fibre, every hint, every attempt, which has slowly come down from generation to generation, unconscious of itself, has taken on form and language.

Probably no one supposes that the England of the time of Elizabeth – particularly the majority of the people – had a precise understanding of Shakespeare; they have no precise understanding of him even now – but then they have no precise understanding of themselves either. But when an Englishman goes to the theatre he understands Shakespeare instinctively, through sympathy, of that I have no doubt. At the moment when he is listening to the play, something becomes clearer and more familiar to him. One would have thought that a people so capable of rapid comprehension as the French might have understood Shakespeare too. The character of Hamlet, for instance, is so universally human, especially in the stage of doubts and irresolution, in the consciousness of some black deeds being perpetrated round about them, some betrayal of the great in favour of the mean and trivial, that it is hard to imagine that he should not be understood; but in spite of every trial and effort, Hamlet remains alien to the Frenchman.

If the aristocrats of the last century, who systematically despised everything Russian, remained in reality incredibly more Russian than the house-serfs remained peasants, it is even more impossible that the younger generation could have lost their Russian character because they studied science and philosophy from French and German books. A section of the Slavs at Moscow, with Hegel in their hands, attained the heights of ultra-Slavism.

The very appearance of the circles of which I am speaking was a natural response to a profound, inward need in the Russian life of that time.

We have spoken many times of the stagnation that followed the crisis of 1825. The moral level of society sank, development was interrupted, everything progressive and energetic was struck out of life. Those who remained – frightened, weak and bewildered – were petty and insignificant; the trash of the generation of Alexander occupied the foremost place; little by little they changed into cringing officials, lost the savage poetry of junketing and lordliness together with any shadow of independent dignity; they served tenaciously, they served until they reached high positions, but they never became great personages. Their day was over.

Below this great world of society, the great world of the people maintained an indifferent silence; nothing was changed for them: their plight was bad, but no worse than before, the new blows fell not on their bruised backs. Their time had not yet come. Between this roof and this foundation the first to raise their heads were children, perhaps because they did not suspect how dangerous it was; but, let that be as it might, with these children Russia, stunned and stupefied, began to come to herself.

What halted them was the complete contradiction of the *words* they were taught with the *facts* of life around them. Their teachers, their books, their university spoke one language and that language was intelligible to heart and mind. Their father and mother, their relations, and their whole environment spoke another with which neither mind nor heart was in agreement – but with which the dominant authorities and financial interests were in accord. This contradiction between education and custom nowhere reached such dimensions as among the nobility and gentry of Russia. The shaggy German student with his round cap covering a seventh part of his head, with his world-shaking pranks, is far nearer to the German *Spiessbürger* than is supposed, and the French *collégien*, lank from vanity and emulation, is already *en herbe l'homme raisonnable qui exploite sa position*.

The number of educated people amongst us has always been extremely small; but those who were educated have always received an education, not perhaps very comprehensive, but fairly general and humane: it made men of all with whom it succeeded. But a man was just what was not wanted either for the hierarchical pyramid or for the successful maintenance of the landowning régime. The young man had either to dehumanise himself again – and the greater number did so – or to stop short and ask himself: 'But is it absolutely essential to go into the service? Is it really a good thing to be a landowner?' After that there followed for some, the weaker and more impatient, the idle existence of a cornet on the retired list, the sloth of the country, the dressing-gown, eccentricities, cards, wine; for others a time of ordeal and inner travail. They could not live in complete moral disharmony, nor could they be satisfied with a negative attitude of withdrawal; the stimulated mind required an outlet. The various solutions of these questions, all equally harassing for the younger generation, determined their distribution into various circles.

Thus our *coterie*, for instance, was formed, and at the university it met Sungurov's,[45] already in existence. His, like ours, was concerned rather with politics than with learning. Stankevich's circle, which came into being at the same time, was equally near both and equally remote from both. He went by another path: his interests were purely theoretical.

Between 1830 and 1840 our convictions were too youthful, too ardent and passionate, not to be exclusive. We could feel a cold respect for Stankevich's circle, but we could not be intimate with its members. They traced philosophical systems, were absorbed in self-analysis, and found peace in a luxurious pantheism from which Christianity was not excluded. We were dreaming how to get up a new league in Russia on the pattern of the Decembrists and looked upon learning itself as a means to our end. The government did its best to strengthen us in our revolutionary tendencies.

In 1834 all Sungurov's circle was sent into exile and – vanished.

In 1835 we were exiled. Five years later we came back, tempered by our experience. The dreams of youth had become the irreversible determination of maturity. This was the most brilliant period of Stankevich's circle. Stankevich himself I did not find in Moscow – he was in Germany; but it was just at that moment that Belinsky's articles were beginning to attract the attention of everyone.

On our return we measured our strength with them. The battle was an unequal one; basis, weapons, and language – all were different. After fruitless skirmishes we saw that it was our turn now to undertake serious study and we too set to work upon Hegel and the German philosophy. When we had sufficiently assimilated that, it became evident that there was no ground for dispute between us and Stankevich's circle.

The latter was inevitably bound to break up. It had done its work, and had done it most brilliantly; its influence on the whole of literature and academic teaching was immense – it is enough to mention the names of Belinsky and Granovsky; Koltsov was formed in it, Botkin, Katkov, and others belonged to it.[46] But it could not remain a closed circle without passing into German doctrinairism – men who are alive and are Russian are not capable of that.

Close to Stankevich's circle, as well as ours, there was another, formed during our exile and in the same relationship to them as we were; its members were afterwards called Slavophils. The Slavophils approached from the opposite direction the vital questions which occupied us, and were far more deeply immersed in living work and real conflict than Stankevich's circle.

It was natural that Stankevich's society should split up between them and us. The Aksakovs and Samarin joined the Slavophils, that is, Khomyakov and the Kireyevskys.[47] Belinsky and Bakunin joined us. The closest friend of Stankevich, the most nearly akin to him in his whole nature, Granovsky, was one of us from the day he came back from Germany.

If Stankevich had lived, his circle would still have broken up. He would himself have gone over to Khomyakov or to us.

By 1842 the sifting in accordance with natural affinity had long been complete, and our camp stood in battle array face to face with the Slavophils. Of that conflict we shall speak in another place.

In conclusion I shall add a few words about the elements of which Stankevich's circle was composed; this will throw a light of its own on the strange underground currents which were silently undermining the compact crust of the Russo-German régime.

Stankevich was the son of a wealthy landowner of the province of Voronezh, and was at first brought up in all the ease and freedom of a landowner's life in the country; then he was sent to the school at Ostrogozhsk (and that was something quite out of the way). For fine natures a wealthy and even aristocratic education is very good. A sufficiency gives unfettered freedom and space for growth and development of every sort; it does not constrict the young mind with premature anxiety and apprehension of the future, and it provides complete freedom to pursue the subjects to which it is drawn.

Stankevich's development was broad and harmonious; his artistic, musical, and at the same time reflective and contemplative nature showed itself from the very beginning of his university career. His special faculty, not only for deeply and warmly understanding, but also for reconciling, or as the Germans say 'removing' contradictions, was based on his artistic temperament. The need for harmony, proportion, and enjoyment makes such people indulgent as to the means; to avoid seeing the well, they cover it over with canvas. The canvas will not stand a push, but the eye is not bothered by a yawning gulf. In this way the Germans attained to pantheistic quietism and rested upon it; but such a gifted Russian as Stankevich could not remain 'at peace' for long.

This is evident from the first question which involuntarily troubled him immediately after he left the university.

His pressing business was finished, he was left to himself, he was no longer led by others, *but he did not know what he should do.* There was nothing to go on with, there was no one and nothing around him that appealed to a lively man. A youth, when his mind had cleared and he had had time to look about him after school, found himself in the Russia of those days in the position of a traveller waking up in the steppe; one might go where one would – there were traces, there were bones of those who had perished, there were wild beasts and the empty desert on all sides with its dumb threat of danger, in which it is easy to perish and impossible to struggle. The one thing which could be pursued honourably and heartily was study.

And so Stankevich persevered in the pursuit of learning. He imagined that it was his vocation to be an historian, and began studying Herodotus; it could be foreseen that nothing would come of that pursuit.

He would have liked to be in Petersburg, where there was such ebullition of activity *of a sort* and to which he was attracted by the theatre and by nearness to Europe; he would have liked to be an honorary superintendent of the school at Ostrogozhsk. He determined to be of use in that 'modest career' – which was to be even less successful than Herodotus. He was in reality drawn to Moscow, to Germany, to his own university circle, to his own interests. He could not exist without intimate friends (another proof that there were at hand no interests very near to his heart). The need for sympathy was so strong in Stankevich that he sometimes invented intellectual sympathy and talents, and saw and admired in people qualities in which they were completely lacking.*

But – and in this lay his personal power – he did not often need to have recourse to such fictions; at every step he met wonderful people – he had the faculty of meeting them – and everyone to whom he opened his heart remained his passionate friend for life; and to every such friend Stankevich's influence was either an immense benefit or an alleviation of his burden.

In Voronezh Stankevich used sometimes to go to the one local library for books. There he used to meet a poor young man of humble station, modest and melancholy. It turned out that he was the son of a cattle-dealer who had business with Stankevich's father over supplies. Stankevich befriended the young man; the cattle-dealer's son was a great reader and fond of talking of books. Stankevich got to know him well. Shyly and timidly the youth confessed that he had himself tried his hand at writing verses and, blushing, ventured to show them. Stankevich was amazed at the immense talent not conscious nor confident of itself. From that minute he did not let him go until all Russia was reading Koltsov's songs with enthusiasm. It is quite likely that the poor cattle-dealer, oppressed by his relations, unwarmed by sympathy or recognition, might have wasted his songs on the empty steppes beyond the Volga over which he drove his herds, and Russia would never have heard those wonderful, truly native songs, if Stankevich had not crossed his path.

When Bakunin finished his studies at the school of artillery, he received a commission as an officer in the Guards. It is said that his father was angry with him and himself asked that he should be transferred into the army of the line. Cast away in some God-forsaken village in White Russia with his guns, he grew *farouche*

---

* Klyushnikov vividly expressed this in the following image: 'Stankevich is a silver rouble that envies the size of a copper piece.' – Annenkov, *Biography of Stankevich*, p. 133.

and unsociable, left off performing his duties, and would lie for whole days together on his bed wrapped in a sheepskin coat. His commanding officer was sorry for him; he had, however, no alternative but to remind him that he must either carry out his duties or go on the retired list. Bakunin had not suspected that he had a right to take the latter course and at once asked to be relieved of his commission. On receiving his discharge he came to Moscow, and from that date (about 1836) for him life began in earnest. He had studied nothing before, had read nothing, and hardly knew any German. With great dialectical abilities, with a gift for obstinate, persistent thinking, he had strayed without map or compass in a world of fantastic projects and efforts at self-education. Stankevich perceived his talents and set him down to philosophy. Bakunin learnt German from Kant and Fichte and then set to work upon Hegel, whose method and logic he mastered to perfection – and to whom did he not preach it afterwards? To us and to Belinsky, to ladies and to Proudhon.

But Belinsky drew as much from the same source; Stankevich's views on art, on poetry and its relation to life, grew in Belinsky's articles into that powerful modern critical method, that new outlook upon the world and upon life which impressed all thinking Russia and made all the pedants and doctrinaires recoil from Belinsky with horror. It was Stankevich's lot to initiate Belinsky into the mysteries; but the passionate, merciless, fiercely intolerant talent that carried Belinsky beyond all bounds wounded the aesthetically harmonious temperament of Stankevich.

And at the same time it was Stankevich who encouraged the gentle, loving, pensive, and in those days hypochondriac, Granovsky. Stankevich was a support and an elder brother to him. His letters to Granovsky are exquisitely delightful – and how Granovsky loved him.

'I have not yet recovered from the first shock,' wrote Granovsky soon after Stankevich's death; 'real grief has not touched me yet; I am afraid of it in the future. Now I am still unable to believe that my loss is possible – only at times there is a stab at my heart. He has taken with him something essential to my life. To no one in the world was I so much indebted. His influence over us was always unbounded and beneficent.'

And how many could say that – perhaps have said it!

In Stankevich's circle only he and Botkin were well-to-do and

completely free from financial anxieties. The others made up a
very mixed proletariat. Bakunin's relations gave him nothing;
Belinsky, the son of a petty official at Chembary, expelled from
Moscow University for 'lack of ability', lived on the scanty pay he
got for his articles. Krasov,[48] on taking his degree, went to a
situation at a landowner's in some province, but life with this
patriarchal slave-owner so frightened him that he came back on
foot to Moscow with a wallet on his back, in the winter, together
with some peasants in charge of a train of wagons. Probably the
father and mother of each one of them when giving him their
blessing had said – and who will presume to reproach them for it –
'Come, mind you work hard at your books; and when you've
done with your studying you must make your own way, there's
nobody you can expect to leave you anything, and we've nothing
to give you either; you must make a career for yourself and think
about us too.' On the other hand Stankevich had probably been
told that in all likelihood he could occupy an honourable position
in society, that he was called by wealth and birth to play an
important part – while in Botkin's household everyone, from his
old father down to the clerks, urged upon him by word and
example the necessity of making money, of piling up more and
more.

What was it touched these men? Whose inspiration re-created
them? They had no thought, no care for their social position, for
their personal advantage or for security; their whole life, all their
efforts were bent on the public good, regardless of all personal
profit; some forgot their wealth, others their poverty, and went
forward, without looking back, to the solution of theoretical
questions. The interests of truth, the interests of learning, the
interests of art, *humanitas*, swallowed up everything.

And observe that the renunciation of this world was by no
means confined to their time at the university and two or three
years of youth. The best men of Stankevich's circle are dead; the
others have remained what they were to this day. Belinsky, worn
out by work and suffering, fell a fighter and beggar. Granovsky,
delivering his message of learning and humanity, died as he
mounted his platform. Botkin did not, in fact, become a merchant
. . . not one of them 'distinguished themselves' in the government
service.

It was just the same in the two contiguous circles, the Slavophils and ours. Where, in what corner of the Western world of today, will you find such groups of anchorites of thought, of ascetics of learning, of fanatics of conviction, whose hair is turning grey but whose enthusiasms are for ever young?

Where? Point to them. I boldly throw down the glove – and I only except for the moment one country, Italy – and I shall pace out the field of combat: that is, I shall not let my opponent escape from statistics into history.

We know how great was the interest in theory and the passion for truth and religion in the days of such martyrs for science and reason as Bruno, Galileo, and the rest; we know, too, what the France of the Encyclopaedists was in the second half of the eighteenth century; but later? Later – *sta viator!*

In the Europe of today there is no youth and there are no young men. The most brilliant representative of the France of the last years of the Restoration and of the July dynasty, Victor Hugo, has protested against my saying this. He is speaking, properly, of the young France of the 'twenties, and I am ready to admit that I have been too sweeping;* but beyond that I will not yield one step even to him. I have their own admissions. Take *Les mémoires d'un enfant du siècle* and the poems of Alfred de Musset, reconstruct the France which is visible between the lines of George Sand's journal, of the contemporary drama and novels, and of the cases in the law courts.

But what does all this prove? A great deal; and in the first place that the Chinese shoes of German manufacture in which Russia has hobbled for a hundred and fifty years, though they have caused many painful corns, have evidently not crippled her bones, since whenever she has had a chance of stretching her limbs, such fresh, young powers have appeared. This does not in any way guarantee the future, but it does make it extremely *possible*.

---

* Victor Hugo, after reading *My Past and Thoughts* in the French translation of Delaveau, wrote me a letter in defence of the youth of France at the time of the Restoration.

CHAPTER II

PETERSBURG AND THE SECOND BANISHMENT

THOUGH we were so comfortable in Moscow, we had to move to Petersburg. My father insisted upon it. Count Strogonov, the Minister for Home Affairs, commanded me to enter his secretariat, and we set off there at the end of the summer of 1840.

I had, however, been in Petersburg for two or three weeks in December 1839.

It had happened in this way. When I was relieved from police supervision and received the right to visit the 'residence and the capital', as K. Aksakov expressed himself, my father definitely preferred the 'residence' on the Neva to the ancient capital. Count Strogonov, the director of the university, wrote to his brother and I had to present myself to him. But that was not all. I had been recommended by the governor of Vladimir for the grade of collegiate assessor; my father wanted me to receive this grade as soon as possible. In the Heralds' College the provinces take their turn; this turn comes with the pace of a tortoise, if there is no special solicitation. There almost always is; the cost of it is high, because a whole province may be taken outside its regular turn, but a single name must not. Therefore all have to be paid for, 'or else some would be getting to the head of the queue for nothing'. Usually the officials to be promoted get up a subscription and send a delegate to represent them; but on this occasion my father took all the expense upon himself, and in that way several of the titular councillors of Vladimir were indebted to him for becoming assessors eight months before the proper time.

When he sent me off to Petersburg to attend to this business, my father repeated once more, as he said goodbye to me, 'For God's sake, be careful; be on the alert with everyone, from the guard of the *diligence* to the acquaintances to whom I am giving you letters. Do not trust anyone. Petersburg nowadays is not what it was in our time. There is sure to be a spy or two in every company. *Tiens-toi pour averti.*'

With this commentary on Petersburg life I got into a *diligence* of the original pattern, that is, one which had all the defects subsequently eliminated by different ones, and drove off.

When I reached Petersburg at nine o'clock in the evening, I took an *izvozchik* and drove to St Isaac's Square,[1] I wanted that to be the place with which I was to begin my acquaintance with Petersburg. Everything was covered with deep snow, only Peter I on his horse, gloomy and menacing, stood out sharply against the grey background in the darkness of the night.

> 'And looming black through mists of night
> With stately poise and haughty mien,
> Pointing afar with outstretched hand,
> A warrior on a horse is seen,
> A mighty figure, bold and free.
> The steed is reined. It rears aloft
> And paws the air imperiously,
> So that its lord might further see. . . .'
>
> OGARËV: *Humorous Verse* (tr. Juliet M. Soskice)

Why was it that the conflict of the 14th of December took place on that square? Why was it from that pedestal that the first cry of Russian freedom rang out? Why did the square of soldiers press close round Peter I? Was it his reward . . . or his punishment? The 14th of December, 1825, was the sequel of the work interrupted on the 21st of January, 1725.[2] Nicholas's guns were turned upon the insurrection and upon the statue alike; it is a pity that the grapeshot did not shoot down the bronze Peter. . . .

Returning to my hotel I found one of my cousins awaiting me and, after talking to him of one thing and another, I touched, without thinking, upon St Isaac's Square and the 14th of December.

'How is uncle?' asked my cousin. 'How did you leave him?'

'Thank God, just as usual; he sends you his greetings.'

My cousin, without changing his expression in the least, telegraphed reproach, advice, warning with his eyes alone; the direction of his eyes made me look round. A man was putting wood into the stove; when he had made it up, himself performing the duty of bellows as he did so, and making a pool on the floor from the snow that melted off his boots, he took an oven fork, the length of a Cossack's lance, and went out.

My cousin at once fell to scolding me for having touched upon such a 'scabrous' subject, and in Russian too, before the man. As he went away he said to me in an undertone: 'By the way, before I forget it, there is a barber comes here to the hotel, who sells all sorts of rubbish, combs and pomatums that have gone bad; please be on your guard with him. I am certain that he is connected with the police: he talks all sorts of nonsense. While I was staying here I bought some trifles from him to get rid of him quickly.'

'To encourage him. Well, and is the laundress in the ranks of the gendarmes too?'

'You may laugh, you may laugh; you'll come to grief before anyone; you're only just back from exile, and they will put a dozen nurses to keep watch on you.'

'And yet they say that seven are enough for the child to grow up with one eye.'

On the next day I went to see the official who used in old days to look after my father's affairs: he was a Little Russian, who spoke Russian with a howling accent, never listened to what was said to him, and showed his surprise at everything by shutting his eyes and holding up his plump little paws in a way that reminded one of a mouse. ... He could not restrain himself either, and seeing that I had taken up my hat, led me aside to the window, looked about him, and said to me: 'You mustn't be angry. Just for the sake of my old acquaintance with the family of your father and his late brothers, don't you say much about what has happened to you. Upon my word, just think yourself, what's the use of it? Now it has all passed like smoke. You said something before my cook; she is a Finnish woman. Who can tell what she is, and I was a little ... more than a little in fact ... frightened.'

A pleasant town, I thought as I left the frightened clerk. ... The soft snow was falling in big flakes, the damp, cold wind penetrated to the very bones, and tore at one's hat and coat. My driver, who could scarcely see a step before him, screwed up his eyes and bent his head before the snow, shouting, "Ware, 'ware!" I remembered my father's advice. I thought of my cousin, of the clerk, and of the travelling sparrow in George Sand's fable who asked the half-frozen wolf in Lithuania why he lived in such a horrid climate. 'Freedom,' answered the wolf, 'makes one forget the climate.'

The driver was right – beware, beware! and how I longed to make haste and get away.

My stay was, in fact, brief on my first visit. In three weeks I had finished all my business, and had galloped back to Vladimir by the New Year.

The experience I had gained in Vyatka was extremely useful to me at the Heralds' College. I knew already that it was something after the style of old St Giles's in London, the den of a gang of officially recognised thieves, which no inspection, no reform could change. To clear St Giles's they took it by assault, bought up the houses, and razed them to the ground. That is what should be done with the Heralds' College. Moreover, it is utterly useless – a sort of parasitic service, the office of official promotion, a Ministry of the Table of Ranks, an archaeological society for the investigation of patents of nobility, a secretariat of secretariats. It need hardly be said that the abuses there were bound to be *second-rate.*

My father's agent brought me a tall old man in a uniform tail-coat, every button of which was hanging by a thread; he was anything but clean, and had already had a drop, though it was early in the day. This was the proof-corrector of the Senate Printing Press; after correcting grammatical errors he used to help various senior secretaries with other mistakes behind the scenes. Within half an hour I had come to terms with him, after bargaining exactly as though we were discussing the purchase of a horse or a piece of furniture. He could not, however, give me a positive answer himself, but ran round to the Senate for instructions and, after getting them at last, asked for 'a trifle of earnest-money'.

'But they will keep their promise?'

'No, permit me: they are not that sort of people. It never happens that after taking a gratuity they do not discharge a debt of honour,' answered the proof-corrector in a tone so greatly offended that I thought it necessary to soften him with a slight additional gratuity.

'There used,' he observed, when I had thus disarmed him, 'to be a secretary in the Heralds' College who was a wonderful man. You've maybe heard of him; he was a reckless bribe-taker, and got away with everything. Once a provincial official came to the office to talk about his business, and as he said goodbye he gave him a grey note on the sly, under cover of his hat.

' "But why do you make a secret of it?" the secretary said to him – "upon my word, as though you were giving me a love-letter. If it's a grey one – all the better. Let the other petitioners see it: it will stimulate them when they know that I have accepted two hundred roubles and settled your business for it."

'And he smoothed out the note, folded it up and put it in his waistcoat pocket.'

The proof-corrector was right. The secretary discharged his debt of honour.

I left Petersburg with a feeling very close to hatred, and yet there was no help for it. I had to move to that unfriendly town.

I was not long in the service. I got out of my duties in every possible way, and so I have not a great deal to tell about the service. The secretariat of the Ministry of Home Affairs had the same relationship to the secretariat of the Governor of Vyatka as boots that have been cleaned have to those that have not; the leather is the same, the sole is the same, but the one sort show mud, and the others polish. I did not see clerks drunk in Petersburg. I did not see twenty kopecks taken for looking up a reference, but yet I somehow fancied that under those close-fitting dress-coats and carefully combed heads there dwelt such vile, black, petty envious, cowardly little souls that the head-clerk of my table at Vyatka seemed to me more of a man than any of them. As I looked at my new colleagues I recalled how, on one occasion, after having a drop too much at supper at the district surveyor's, he played a dance tune on the guitar, and at last could not resist leaping up with his instrument and beginning to join in the dance; but these Petersburg men are never carried away by anything: their blood never boils, and wine does not turn their heads. In some dancing class, in company with young German ladies, they can walk through a French quadrille, pose as disillusioned, repeat lines from Timofeyev or Kukolnik[3] ... they were diplomats, aristocrats, and Manfreds. It is only a pity that Dashkov, the Minister, could not train these Childe Harolds not to stand at attention and bow even at the theatre, at church, and everywhere.

The Petersburghers laugh at the costumes seen in Moscow; they are outraged by the caps and Hungarian jackets, the long hair and civilian moustaches. Moscow certainly is an unmilitary city, rather dishevelled and unaccustomed to discipline, but whether that is a good quality or a defect is a matter of opinion. The harmony of uniformity; the absence of variety, of what is personal, whimsical, and wayward; the obligatory wearing of uniform, and outward good form – all develop to the highest degree in the most inhuman condition in which men live – in barracks. Uniform and uniformity are passionately loved by despotism. Nowhere are fashions so respectfully observed as in Petersburg, and that shows the immaturity of our civilisation; our clothes are alien. In Europe people dress, but we dress up, and so are frightened if a sleeve is too full, or a collar too narrow. In Paris all that people are afraid of is being dressed without taste; in London all that they are afraid of is catching cold; in Italy everyone dresses as he likes. If one were to show him the battalions of exactly similar, tightly buttoned frock-coats of the fops on the Nevsky Prospect, an Englishman would take them for a squad of 'policemen'.

I had to do violence to my feelings every time I went to the Ministry. The chief of the secretariat, K. K. von Paul, a *Herrn-huter*,[4] and a virtuous and lymphatic native of the island of Dagö, induced a kind of pious boredom into all his surroundings. The heads of the sections ran anxiously about with portfolios and were dissatisfied with the head-clerks of the tables; the latter wrote and wrote and certainly were overwhelmed with work, and had the prospect before them of dying at those tables, or, at any rate, if not particularly fortunate, of sitting there for twenty years. In the Registry there was a clerk who for thirty-three years had been keeping a record of the papers that went out, and sealing the parcels.

My 'literary exercises' gained me some exemption here too; after experience of my incapacity for anything else the head of the section entrusted me with the composition of a general report on the Ministry from the various provincial secretariats. The foresight of the authorities had found it necessary to propound certain findings in advance, not leaving them to the mercy of facts and figures. Thus, for instance, in the draft of the proposed report appeared the statement: 'From the examination of the number and nature of crimes' (neither their number nor their nature was yet known) 'your Majesty may be graciously pleased to perceive the progress of national morality, and the increased zeal of the officials for its improvement.'

Fate and Count Benckendorf[5] saved me from taking part in this spurious report. It happened in this way.

At nine o'clock one morning, early in December, Matvey told me that the superintendent of the local police-station wished to see me. I could not guess what had brought him to me, and bade

Matvey show him in. The superintendent showed me a scrap of paper on which was written that he *invited* me to be at the Third Division of His Majesty's Own Chancellery at ten o'clock that morning.

'Very well,' I answered. 'That is by Tsepnoy Bridge, isn't it?'

'Don't trouble yourself,' he answered. 'I have a sledge downstairs. I will go with you.'

It is a bad business, I thought, with a pang at my heart.

I went into the bedroom. My wife was sitting with the baby, who had only just begun to recover after a long illness.

'What does he want?' she asked.

'I don't know, some nonsense. I shall have to go with him. . . . Don't worry.'

My wife looked at me and said nothing; she only turned pale as though a cloud had passed over her face, and handed me the child to say goodbye to it.

I felt at that moment how much heavier every blow is for a man with a wife and children; the blow does not strike him alone, he suffers for all, and involuntarily blames himself for their sufferings.

The feeling can be restrained, stifled, concealed, but one must recognise what it costs. I went out of the house in black misery. Very different was my mood when I had set off six years before with Miller, the *politsmeyster*, to the Prechistensky policestation.

We drove over the Tsepnoy Bridge and through the Summer Garden and turned towards what had been Kochubey's house; in the lodge there the secular inquisition founded by Nicholas was installed: people who went in at its back gates, before which we stopped, did not always come out of them again, or if they did, it was perhaps to disappear in Siberia or perish in the Alexeyevsky ravelin. We crossed all sorts of courtyards and little squares, and came at last to the office. In spite of the presence of the commissar, the gendarme did not admit us, but summoned an official who, after reading the summons, left the policeman in the corridor and asked me to follow him. He took me to the Director's room. At a big table near which stood several armchairs a thin, grey-headed old man, with a sinister face, was sitting quite alone. To maintain his importance he went on reading a paper to the end, and then got up and came towards me. He had a star on his breast, from which I concluded that he was some sort of commanding officer in the army of spies.

'Have you seen General Dubelt?'

'No.'

He paused. Then, frowning and knitting his brows, without looking me in the face, he asked me in a sort of threadbare voice (the voice reminded me horribly of the nervous, sibilant notes of Golitsyn junior at the Moscow commission of inquiry):

'I think you have not very long had permission to visit Petersburg or Moscow?'

'I received it last year.'

The old man shook his head. 'And you have made a bad use of the Tsar's graciousness. I believe you'll have to go back again to Vyatka.'

I gazed at him in amazement.

'Yes,' he went on, 'you've chosen a fine way to show your gratitude to the government that permitted you to return.'

'I don't understand in the least,' I said, lost in surmises.

'You don't understand? That's just what is bad, too! What connections! What pursuits! Instead of showing your zeal from the first, effacing the stains left from your youthful errors, using your abilities to good effect – no! not at all: it's nothing but politics and tattling, and all to the detriment of the government. This is what your talk has brought you to! How is it that experience has taught you nothing? How do you know that among those who talk to you there isn't each time some scoundrel[6] who asks nothing better than to come *here* a minute later to give information?'

'If you can explain to me what all this means, you will greatly oblige me. I am racking my brains and cannot understand what your words are leading up to, or what they are hinting at.'

'What are they leading to? Hm. . . . Come, did you hear that a sentry at the Blue Bridge killed and robbed a man at night?'

'Yes, I did,' I answered with great simplicity.

'And perhaps you repeated it?'

'I believe I did repeat it.'

'With comments, I dare say?'

'Very likely.'

'With what sort of comments? There it is: a propensity to censure the government. I tell you frankly, the one thing that does you credit is your sincere avowal: it will certainly be taken into consideration by the Count.'

'Upon my word,' I said, 'what is there to avow? All the town was talking of the story; it was talked of in the secretariat of the Ministry of Home Affairs and in the shops. What's surprising in my having spoken about the incident?'

'The diffusion of false and mischievous rumours is a crime that the laws do not tolerate.'

'You seem to be charging me with having invented the affair.'

'In the note of information to the Tsar it is merely stated that you assisted in the propagation of this mischievous rumour, upon which followed the decision of His Majesty concerning your return to Vyatka.'

'You are simply trying to frighten me,' I answered. 'How is it possible, for such a trivial business, to send a man with a family a thousand miles away, and, what's more, to condemn and sentence him without even inquiring whether it is true or not?'

'You have admitted it yourself.'

'But how was it the report was submitted and the matter settled before you spoke to me?'

'Read for yourself.'

The old man went over to a table, fumbled among a small heap of papers, composedly pulled one out and handed it to me. I read it and could not believe my eyes: such complete absence of justice, such insolent, shameless disregard of the law was amazing, even in Russia.

I did not speak. I fancied that the old gentleman himself felt that it was a very absurd and extremely silly business, so that he did not think it necessary to defend it further, but after a brief silence asked:

'I believe you said you were married?'

'I am married.'

'It is a pity that we did not know that before. However, if anything can be done the Count will do it. I shall tell him of our conversation. *In any case* you will be banished from Petersburg.'

He looked at me. I did not speak, but felt that my face was burning. Everything I could not utter, everything held back within me, could be seen in my face.

The old gentleman dropped his eyes, considered for a moment, and suddenly, in an apathetic voice, with an affectation of urbane delicacy, said to me:

'I shall not venture to detain you further. I sincerely wish you – however, you will hear later.'

I rushed home. My heart boiled with a consuming fury – that feeling of impotence; of having no rights, the condition of a caged beast, jeered at by a sneering street-boy, who knows that all the tiger's strength is not enough to break the bars.

I found my wife in a fever; she had been taken ill that day and, having another fright in the evening, was prematurely confined a few days later.[7] The baby only lived a day, and after three or four years she had hardly recovered her strength.

They say that that tender paterfamilias, Nicholas Pavlovich, wept when his daughter died. . . .

And passionately fond they are of raising a turmoil, galloping hell for leather, kicking up a dust, and doing everything at head-long speed, as though the town were on fire, the throne were tottering, or the dynasty in danger – and all this without the slightest necessity! It is the romanticism of the gendarmes, the dramatic exercises of the detectives, the lavish setting for the dis-play of loyal zeal . . . the *oprichniki*,[8] the whippers-in, the hounds!

On the evening of the day on which I had been to the Third Division we were sitting sorrowfully at a small table – the baby was playing with his toys on it, and we were saying little; sud-denly someone pulled the bell so violently that we could not help starting. Matvey rushed to open the door, and a second later an officer of gendarmes darted into the room, clashing his sabre and jingling his spurs, and began in choice language apologising to my wife. He could not have imagined, he had had no suspicion, no idea that there was a lady and children in the case. It was extremely unpleasant. . . .

Gendarmes are the very flower of courtesy; if it were not for their duty, for the sacred obligations of the service, they would never make secret reports, or even fight with post-boys and drivers at departures. I know this from the Krutitsky Barracks where the *désolé* officer was so deeply distressed at the necessity of searching my pockets.[9]

Paul Louis Courier[10] observed in his day that executioners and prosecutors are the most courteous of men. 'My dear executioner,' writes the prosecutor, 'if it is not disturbing you too much, you will do me the greatest service if you will kindly take the trouble to chop off So-and-so's head tomorrow morning.' And the executioner hastens to answer that 'he esteems himself fortunate indeed that he can by so trifling a service do something agreeable for the prosecutor and remains, always his devoted and obedient servant, the executioner'; and the other man, the third, remains devoted without his head.

'General Dubelt asks you to see him.'

'When?'

'Upon my word! now, at once, this minute.'

'Matvey, give me my overcoat.'

I pressed my wife's hand – her face was flushed, her hand was burning. Why this hurry at ten o'clock in the evening? Had a plot been discovered? Had someone run away? Was the precious life of Nicholas Pavlovich in danger? I really had been unfair to that sentry, I thought. It was not surprising that with a government like this one of its agents should murder two or three passers-by; were the sentries of the Second and Third grades any better than their comrade on the Blue Bridge? And what about the head sentry of all?

Dubelt had sent for me in order to *tell me* that Count Bencken-dorf required my presence at eight o'clock the next morning to inform me of the decision of His Majesty!

Dubelt was an unusual person; he was probably more intelligent than the whole of the Third Division – indeed, than all three divisions of His Majesty's Own Chancellery. His sunken face, shaded by long, fair moustaches, his fatigued expression, particularly the furrows in his cheeks and forehead, clearly witnessed that his breast had been the battlefield of many passions before the pale-blue uniform had conquered, or rather concealed, everything that was in it. His features had something wolfish and even foxy about them, that is, they expressed the subtle intelligence of beasts of prey; there was at once evasiveness and arrogance in them. He was always courteous.

When I went into his study he was sitting in a uniform coat without epaulettes, and smoking a pipe as he wrote. He rose at once, asked me to sit down facing him and began with the following surprising sentence:

'Count Alexander Khristoforovich has given me the oppor-

tunity of making your acquaintance. I believe you saw Sakhtynsky this morning?'

'Yes, I did.'

'I am very sorry that the reason I have had to ask you to see me
is not an entirely pleasant one for you. Your imprudence has once
more brought His Majesty's anger upon you.'

'I will say to you, General, what I said to Count Sakhtynsky: I
cannot imagine that I shall be exiled simply for having repeated
a street rumour, which you, of course, heard before me, and
possibly spoke of just as I did.'

'Yes, I heard the rumour, and I spoke of it, and so far we are
evens; but this is where the difference begins: in repeating the
absurd story I swore that there was nothing in it, while you made
the rumour a ground for accusing the whole police force. It is all
this unfortunate passion *de dénigrer le gouvernement* – a passion
that has developed in all of you gentlemen from the pernicious
example of the West. It is not with us as in France, where the
government is at daggers drawn with the parties, where it is
dragged in the mud. Our government is paternal: everything is
done as privately as possible. ... We do our very utmost that
everything shall go as quietly and smoothly as possible, and here
men, who in spite of painful experience persist in a fruitless
opposition, alarm public opinion by stating verbally and in
writing that the soldiers of the police murder men in the streets.
Isn't that true? You have written about it, haven't you?'

'I attach so little importance to the matter that I don't think it
at all necessary to conceal that I have written about it, and I will
add to whom – to my father.'

'Of course it is not an important matter, but see what it has
brought you to. His Majesty at once remembered your name, and
that you had been at Vyatka, and commanded that you should be
sent back there, and so the Count has commissioned me to inform
you that you are to go to him tomorrow at eight o'clock and he
will announce to you the will of His Majesty.'

'And so it is left that I am to go to Vyatka with a sick wife and
a sick child on account of something that you say is not
important?'

'Why, are you in the service?' Dubelt asked me, looking
intently at the buttons of my half-dress uniform coat.

'In the office of the Minister of Home Affairs.'

'Have you been there long?'

'Six months.'

'And all the time in Petersburg?'

'All the time.'

'I had no idea of it.'

'You see,' I said, smiling, 'how discreetly I have behaved.'

Sakhtynsky did not know that I was married, Dubelt did not know that I was in the service, but both knew what I said in my own room, what I thought and what I wrote to my father. . . . The trouble was that I was just beginning to be friendly with Petersburg literary men, and to publish articles and, worse still, had been transferred by Count Strogonov from Vladimir to Petersburg, the secret police having no hand in it, and when I arrived in Petersburg I had not reported either to Dubelt or to the Third Division, which kindly persons had hinted that I should do.

'To be sure,' Dubelt interrupted me, 'all the information that has been collected about you is entirely to your credit. Only yesterday I was speaking to Zhukovsky and should be thankful to hear my sons spoken of as he spoke of you.'

'And yet I am to go to Vyatka?'

'You see it is your *misfortune* that the report had been handed in already, and that many circumstances had not been taken into consideration. Go you must: there's no altering that, but I imagine that another town might be substituted for Vyatka. I will talk it over with the Count: he is going to the Palace again today. We will try and do all that can be done to make things easier; the Count is a man of angelic kindness.'

I got up and Dubelt escorted me to the door of his study. At that point I could not restrain myself: I stopped and said to him:

'I have one small favour to ask of you, General. If you want me, please do not send constables or gendarmes. They are noisy and alarming, especially in the evening. Why should my sick wife be more severely punished than anyone on account of the sentry business?'

'Oh! good heavens, how unpleasant that is,' replied Dubelt, 'how clumsy they all are! You may rest assured that I will not send a policeman again. And so till tomorrow; don't forget, eight o'clock at the Count's; we shall meet there.'

It was exactly as though we were agreeing to go to Smurov's to eat oysters together.

At eight o'clock next morning I was in Benckendorf's reception room. I found five or six petitioners waiting there; they stood gloomy and anxious by the wall, started at every sound, squeezed themselves together even more closely, and bowed to every adjutant that passed. Among their number was a woman in deep mourning, with tear-stained eyes. She sat with a paper rolled up in her hand, and the roll trembled like an aspen leaf. Three paces from her stood a tall, rather bent old man of seventy or so, bald and sallow, in a dark-green army great-coat, with a row of medals and crosses on his breast. From time to time he sighed, shook his head and whispered something under his breath.

Some sort of 'friend of the family', a flunkey, or a clerk on duty, sat in the window, lolling at his ease. He got up when I went in, and looking intently at his face I recognised him; that loathsome figure had been pointed out to me at the theatre as one of the chief street spies, and his name, I remember, was Fabre. He asked me:

'Have you come with a petition to the Count?'

'I have come at his request.'

'Your surname?'

I mentioned it.

'Ah,' he said, changing his tone as though he had met an old acquaintance, 'won't you be pleased to sit down? The Count will be here in a quarter of an hour.'

It was horribly still and *unheimlich* in the room; the daylight hardly penetrated through the fog and frozen window-panes, and no one said a word. The adjutants ran quickly to and fro, and the gendarme standing at the door sometimes jingled his accoutrements as he shifted from foot to foot. Two more petitioners came in. A clerk on duty ran to ask each what he had come about. One of the adjutants went up to him and began telling him something in a half-whisper, assuming a desperately roguish air as he did so. No doubt it was something nasty, for they frequently interrupted their talk with noiseless, flunkeyish laughter, during which the worthy clerk, affecting to be quite helpless and ready to burst, repeated: 'Do stop, for God's sake stop, I can't bear it.'

Five minutes later Dubelt appeared, with his uniform unbuttoned as though he were off duty, cast a glance at the petitioners, at which they all bowed, and seeing me in the distance said: '*Bonjour, Monsieur Herzen. Votre affaire va parfaitement bien* ... very well indeed.'

They would let me stay, perhaps! I was on the point of asking, but before I had time to utter a word Dubelt had disappeared. Next there walked into the room a general, scrubbed and decorated, tightly laced and stiffly erect, in white breeches and a scarf: I have never seen a finer general. If ever there is an exhibition of generals in London, like the Baby Exhibition at Cincinnati at this moment, I advise sending this very one from Petersburg. The general went up to the door from which Benckendorf was to enter and froze in stiff immobility; with great curiosity I scrutinised this sergeant's ideal. He must have flogged soldiers in his day for the way they paraded. Where do these people come from? He was born for military rules and regulations and files on parade. He was attended by the most elegant cornet in the world, probably his adjutant, with incredibly long legs, fair-haired, with a tiny face like a squirrel's, and that good-natured expression which often persists in mamma's darlings who have never studied anything, or at any rate have never succeeded in learning anything. This honeysuckle in uniform stood at a respectful distance from the model general.

Dubelt darted in again, this time assuming an air of dignity, and with his buttons done up. He at once addressed the general, and asked him what he could do for him. The general, with the correctness with which orderlies speak when reporting to their superior officers, announced:

'Yesterday I received through Prince Alexander Ivanovich His Majesty's command to join the active Army in the Caucasus, and esteemed it my duty to report to his Excellency before leaving.'

Dubelt listened with religious attention to this speech, and with a slight bow as a sign of respect went out and returned a minute later.

'The Count,' he said to the general, 'sincerely regrets that he has not time to receive your Excellency. He thanks you and has commissioned me to wish you a good journey.' Upon this Dubelt flung wide his arms, embraced the general, and twice touched his cheeks with his moustaches.

The general retreated at a solemn march, the youth with a squirrel's face and the legs of a crane set off after him. This scene compensated me for much of the bitterness of that day. The general's standing at attention, the farewell by proxy, and finally the sly face of *Reineke Fuchs* as he kissed the brainless countenance of his Excellency – all this was so ludicrous that I could only just contain myself. I fancied that Dubelt noticed this and began to respect me from that time.

At last the doors were flung open *à deux battants* and Benckendorf came in. There was nothing unpleasant in the exterior of the chief of gendarmes; his appearance was rather typical of the Baltic barons and of the German aristocracy generally. His face looked creased and tired, he had the deceptively good-natured expression which is often found in evasive and apathetic people.

Possibly Benckendorf did not do all the harm he might have done, being the head of that terrible police, being outside the law and above the law, and having a right to meddle in everything. I am ready to believe it, especially when I recall the vapid expression of his face. But he did no good either; he had not enough will-power, energy, or heart for that. To shrink from saying a word in defence of the oppressed is as bad as any crime in the service of a man as cold and merciless as Nicholas.

How many innocent victims passed through Benckendorf's hands, how many perished through his lack of attention, through his absent-mindedness, or because he was engaged in gallantry – and how many dark images and painful memories may have haunted his mind and tormented him on the steamer on which, having prematurely collapsed and grown decrepit, he sailed off to seek, in betrayal of his own religion, the intercession of the Catholic Church with its all-forgiving indulgences. . . .[11]

'It had come to the knowledge of his Imperial Majesty,' he said to me, 'that you take part in the diffusion of rumours injurious to the government. His Majesty, seeing how little you have reformed, deigned to order that you should be sent back to Vyatka; but I, at the request of General Dubelt, and relying upon information collected about you, have reported to His Majesty about the illness of your wife, and His Majesty has been pleased to alter his decision. His Majesty forbids you to visit Petersburg and Moscow, and you will be under police supervision again, but it is left to the Ministry of Home Affairs to appoint the place of your residence.'

'Allow me to tell you frankly that even at this moment I cannot believe that there has been no other reason for exiling me. In 1835 I was exiled on account of a supper-party at which I was not present! Now I am being punished for a rumour about which the whole town was talking. It is a strange fate!'

Benckendorf shrugged his shoulders and, turning out the palms of his hands like a man who has exhausted all the resources of argument, interrupted me.

'I make known to you the Imperial will, and you answer me with criticisms. What good will come of all that you say to me, or that I say to you? It is a waste of words. Nothing can be changed now. What will happen later partly depends on you, and, since you have referred to your first trouble, I particularly recommend you not to let there be a third. You will certainly not get off *so easily* a third time.'

Benckendorf gave me a benevolent smile and turned to the petitioners. He said very little to them; he took their petition, glanced at it, and then handed it to Dubelt, interrupting the petitioners' observations with the same graciously condescending smile. For months together these people had been pondering and preparing themselves for this interview, upon which their honour, their fortune, their family depended; what labour, what effort had been employed before they were received; how many times they had knocked at the closed door and been turned away by a gendarme or porter. And how great, how poignant must the necessities have been that brought them to the head of the secret police; no doubt all legal channels had been exhausted first. And this man gets rid of them with commonplaces, and in all probability some Head of a Table proposed *some* decision, in order to pass the case on to *some* other secretariat. And what was he so absorbed in? Where was he in a hurry to go to?

When Benckendorf went up to the old man with the medals, the latter fell on his knees and said:

'Your Excellency, put yourself in my place.'

'How abominable!' cried the Count; 'you are disgracing your medals,' and full of noble indignation he passed by without taking his petition. The old man slowly got up, his glassy eyes were full of horror and craziness, his lower lip quivered and he babbled something.

How inhuman these people are when the whim takes them to be human!

Dubelt went up to the old man and said: 'Whatever did you do that for? Come, give me your petition. I'll look through it.'

Benckendorf had gone to see the Tsar.

'What am I to do?' I asked Dubelt.

'Settle on any town you choose with the Minister of Home Affairs; we shall not interfere. We will send the whole case on there tomorrow. I congratulate you on its having been so satisfactorily settled.'

'I am very much obliged to you!'

From Benckendorf I went to the Ministry. Our Director, as I have mentioned, belonged to that class of Germans who have something of the lemur about them, lanky, sluggish, and dilatory. Their brains work slowly, they do not catch the point at once and they labour a long time if they are to reach any sort of conclusion. My account unfortunately arrived before the communication from the Third Division; he had not expected it at all, and so was completely bewildered, uttered incoherent phrases, noticed this himself, and in order to recover himself said to me: '*Erlauben Sie mir deutsch zu sprechen.*' Possibly his remarks came out more correct grammatically in German, but they did not become any clearer or more definite in meaning. I distinctly perceived two feelings struggling in him: he grasped all the injustice of the affair, but considered himself bound as Director to justify the action of the government; at the same time, he did not want to show himself a barbarian before me, nor could he forget the hostility which invariably reigned between the Ministry and the secret police. So the task of expressing all this jumble was in itself not easy. He ended by admitting that he could say nothing until he had seen the Minister, and by going off to see him.

Count Strogonov sent for me, inquired into the matter, listened attentively to the whole thing, and said to me in conclusion:

'It's a police trick, pure and simple – well, all right: I'll pay them out for it.'

I imagined, I confess, that he was going straight off to the Tsar to explain the business to him; but ministers do not go so far as that.

'I have received His Majesty's command concerning you,' he went on: 'here it is. You see that it is left to me to select the place of your exile and to employ you in the service. Where would you like to go?'

'To Tver or Novgorod,' I answered.

'To be sure. . . . Well, since the choice of a place is left to me, and it probably does not matter to you to which of those towns I appoint you, I shall give you the first councillor's vacancy in the provincial government. That is the highest position that you can receive with your seniority, so get yourself a uniform made with an embroidered collar,' he added jocosely.

So that was how I recouped myself, though not in my own suit.

A week later Strogonov recommended me to the Senate for an appointment as councillor at Novgorod.

It really is very funny to think how many secretaries, assessors, and district and provincial officials had been long soliciting, passionately and persistently soliciting, to get that post; bribes had been given, the most sacred promises had been received, and here, all at once, a Minister, to carry out His Majesty's will and at the same time to have his revenge on the secret police, *punished* me with this promotion and, by way of gilding the pill, flung this post, the object of ardent desires and ambitious dreams, at the feet of a man who accepted it with the firm intention of throwing it up at the first opportunity.

From Strogonov I drove to see a lady; I must say a few words about this acquaintance.

Among the letters of introduction given me by my father when I first went to Petersburg was one which I had picked up a dozen times, turned over and hid again in the table drawer, putting off my visit until another day. The letter was addressed to a lady of seventy, of high rank and great wealth, whose friendship with my father dated from time immemorial; he had first made her acquaintance when she was at the Court of Catherine II; then they had met in Paris, had travelled here and there together, and at last both had come home to rest, some thirty years before.

I disliked persons of consequence as a rule, particularly when they were women, and even more so when they were seventy; but my father had inquired for the second time whether I had called upon Olga Alexandrovna Zherebtsov, so eventually I resolved to swallow this pill. A footman led me into a rather gloomy drawing-room, poorly decorated – blackened, as it were, and faded; the furniture, the hangings, all had lost their colour, and all had evidently been in the same places for a long time. To me it had the atmosphere of Princess Meshchersky's house; old age, no less than youth, stamps its footprints on all about it. I waited with resignation for my hostess to make her appearance, preparing myself for tedious questions, for deafness, for a cough, for attacks on the younger generation, and perhaps for moral exhortations.

Five minutes later a tall old lady, with a stern face that bore traces of great beauty, walked in with a firm step; an unswerving will, a strong character, and a powerful intellect were apparent in her carriage, her behaviour, and her gestures. She scanned me from head to foot with a penetrating gaze, went up to the sofa, pushed back a table with one movement of her arm, and said to me: 'Sit in this arm-chair here, nearer to me. I am a great friend of your father's, you know, and I'm fond of him.' She opened the letter, and handed it to me, saying: 'Please read it to me; my eyes are bad.'

The letter was written in French, with various compliments, reminiscences, and allusions. She listened with a smile, and when I had finished she said: 'His mind shows no signs of age, he is just the same as ever; he was very nice and very caustic. And now, I suppose, he keeps his room, in a dressing-gown, and plays the invalid? Two years ago I was passing through Moscow and then I went to see your father."I can hardly see anyone," he said. "I am breaking up," and then he began talking and forgot his ailments. It's all softness: he is not much older than I am, two or three years at the most, though I doubt if he is that, and I am a woman, but I'm still on my feet. Yes, yes, much water has flowed by since those days your father talks of. Why, only fancy, he and I were among the leading dancers. The English dances were the fashion in those days; Ivan Alexeyevich and I used to dance at the late Empress's. Can you imagine your father in a full-skirted, light-blue French *caftan*, wearing powder, and me in a hoop and *décolletée*? It was very pleasant to dance with him, *il était bel homme*; he was better-looking than you – let me have a good look at you – yes, he certainly was rather better. ... Don't be angry: at my age one may tell the truth. Besides, I believe you don't care about that – of course, you are literary and learned. Ah, my goodness, by the way, do tell me, please, what was all that business of yours? Your father wrote to me when you were sent to Vyatka. I did try to speak to Bludov,[12] but he didn't do anything. They don't say what it was you were exiled for. Everything for them is a *secret d'état*.'

There was so much simplicity and genuineness in her manner

that, contrary to my expectation, I was at ease and unconstrained with her. I answered between jest and earnest and told her about our affair.

'He makes war on students,' she observed; 'he has nothing in his head but conspiracies, and, to be sure, they are pleased to oblige him, they think of nothing but nonsense. They are such wretched little creatures about him! Where did he get hold of them? – no standing or family. Well, *mon cher conspirateur*, how old were you then? – about sixteen?'

'Just one and twenty,' I answered, laughing genuinely at her utter contempt for our political activities, that is mine and Nicholas's, 'but then I was the eldest.'

'Four or five students scared *tout le gouvernement*: think of it – what a disgrace!'

After talking in this way for half an hour, I got up to go.

'Stay a little,' said Olga Alexandrovna in a still more friendly tone. 'I have not finished my catechism; how was it you carried off your bride?'

'How do you know?'

'Oh, my dear, the world is full of rumour – youth, *des passions*. I talked to your father at the time. He was still angry with you, but there, he is a sensible man, he understood ... provided you live happily, what more does he want? "Well," he said to me, "the boy came to Moscow against orders. If he had been caught he would have been sent to the fortress." And to that I said: "But you see he wasn't caught, so you ought to be glad of it; and what is the use of talking nonsense and imagining what might have happened?" "Oh, you were always reckless," he told me, "and lived at break-neck speed." "Well, my dear sir, I am ending my days no worse than other people," I answered him. "And what's all this – leaving the young people without money? That's beyond anything." "Well," he said, "I'll send them some. I'll send them some. Don't be angry." You'll bring your wife to see me, won't you?'

I thanked her, and said that for the time being I had come to Petersburg alone.

'Where are you staying?'

'At Demouthe's.'

'And do you dine there?'

'Sometimes there; sometimes at Dumais'.'

'Why at restaurants? It's expensive, and besides it's not nice for a married man. If it won't bore you to dine with an old woman, come here. I am really very glad to have made your acquaintance. I'm grateful to your father for sending you to me; you are a very interesting young man, and have a good understanding of things although you are young – so you and I will have a talk about one thing and another, for you know I am bored with these courtiers – always the same thing: the court, and who has received a decoration; it is all futile.'

In one volume of Thiers' *History of the Consulate* he gives a fairly detailed and correct account of the murder of Paul.[13] There are two references in his story to a woman, the sister of Count Zubov, who was the last of Catherine's favourites. The beautiful young widow of a general (killed in time of war, I believe), a passionate and vigorous nature, spoilt by her high position, endowed with exceptional intellect and masculine strength of will, she became the centre of the discontented during the savage and senseless reign of Paul. The conspirators met at her house; she goaded them on, their relations with the English Embassy passed through her. Paul's police suspected her at last and she, warned in time, perhaps by Pahlen[14] himself, went abroad before it was too late. The plot was by then matured, and while dancing at a ball at the court of the king of Prussia she received the news that Paul had been killed. She made no attempt to conceal her joy and enthusiastically announced the news to everyone in the ballroom. This so scandalised the king of Prussia that he ordered her to be expelled from Berlin within twenty-four hours.

She went to England. Brilliant, spoilt by her life at court, and devoured by a craving for a great career, she made her appearance in London as a lioness of the first magnitude, and played a notable part in the closed, inaccessible society of English aristocracy. The Prince of Wales, the future King George IV, was at her feet, and soon more than that. .... The years of her life abroad were spent amidst noisy pomp, but they passed, the blooms fell one after another.

With old age came emptiness, the blows of fate, loneliness, and the sad life of memories. Her son was killed at Borodino; her daughter died leaving her a grandchild, Countess Orlov. Every August the old lady went from Petersburg to Mozhaysk to visit her son's grave. Loneliness and misfortune had not broken her strong character, but had only made it more surly and angular. Like a tree in winter she retained the outline of her branches, the leaves had fallen, and the bare twigs were cold and stiff as dry bones; but the gigantic stature and bold dimensions were but the more distinctly visible, and the trunk, grey with hoar-frost, stood proud and sombre, and no wind, no weather could bend it.

Her long life, full of travels, its immense wealth of meetings and collisions, had formed her disdainful view of the world, by no means without its melancholy truth. She had her own philosophy, based on a profound contempt for her fellow-creatures, though, owing to her active disposition, she could not abandon them altogether.

'You don't know them yet,' she would say to me, nodding her head towards various stout and thin senators and generals. 'I've had my fill of looking at them. It is not so easy to take me in as they imagine; before I was twenty my brother was in the highest favour, and the Empress was very kind to me and very fond of me. So, would you believe it, old men, beribboned and decorated, who could hardly drag one leg after the other, struggled with one another to get into the hall and hand me my pelisse or my warm shoes. The Empress died, and the next day my house was deserted. They avoided me like the plague in that madman's days, you know, and the very same personages, too. I went my own way; I had no need of anyone, and I crossed the sea. After my return God visited me with great misfortunes, but I met with no sympathy from anyone. There were two or three old friends, and they did stay by me. Well, a new reign came on. Orlov,[15] you see, is in power, though indeed I don't know how far that is true ... they imagine it is, at least. They know that he is my heir, and my granddaughter loves me; so now they are so friendly – once more they are ready to hand me my fur coat and my goloshes! Ugh! I know them, but one is sometimes bored with being alone; my eyes are bad, it is hard to read, and besides one does not always care to, so I let them come: they babble all sorts of nonsense; it's a distraction, and serves to pass an hour or two. . . .'

She was a strange, eccentric ruin of another age, surrounded by degenerate successors that had sprung up on the mean and barren soil of Petersburg court life. She felt superior to it, and she was right. If she had shared the Saturnalia of Catherine and the orgies of George IV, she had also shared the dangers of the conspirators in the reign of Paul.

Her mistake lay not in her contempt for worthless people, but in her taking this produce of the court kitchen-garden for the whole of our generation. In the reign of Catherine the court and the Guards really did include all that was cultured in Russia; and this continued more or less until 1812. Since then Russian society has taken terrific strides; the war evoked consciousness, and consciousness evoked the 14th of December. Society was divided in two from within: it was not the best part that remained on the side of the court; some were estranged by the executions and savage punishments, others by the new tone prevailing. Alexander had carried on the cultural traditions of the reign of Catherine; under Nicholas the worldly aristocratic tone was replaced by one of dry formality and insolent despotism on the one hand and absolute servility on the other – a blend of the abrupt and boorish manner of Napoleon with the soullessness of bureaucracy. A new society rapidly developed, the centre of which was in Moscow.

There is a wonderful book which one cannot help recalling when one speaks of Olga Alexandrovna – I mean the *Memoirs* of Princess Dashkov,[16] published twenty years ago in London. To the book are appended the *Memoirs* of the two Wilmot sisters who lived with Princess Dashkov between 1805 and 1810. They were highly cultured Irishwomen, with a great gift of observation. I should very much like their letters and *Memoirs* to be known in Russia.

When I compare Moscow society before 1812 with that which I left in 1847 my heart throbs with joy. We have taken a tremendous step forward. In those days there was a society of the discontented – that is, those who had been retired, dismissed, or laid on the shelf; now there is a society of independent people. The lions of those days were capricious oligarchs, such as Count A. G. Orlov and Ostermann, 'a society of shadows' as Miss Wilmot says, a society of statesmen who had died fifteen years before in Petersburg, but went on powdering their heads, covering themselves with ribbons, and appearing at dinners and festivities in Moscow, pouting and giving themselves airs, and having neither power nor significance. After 1825 the lions of Moscow were Pushkin, M. Orlov,[17] Chaadayev, Yermolov. Then society with cringing servility thronged the house of Count Orlov, ladies 'in other people's diamonds',* and their cavaliers who dared not sit down without permission; the Count's serfs danced before them in masquerade attire. Forty years later I saw the same society crowding about the platform in one of the lecture-rooms at Moscow University; the daughters of those ladies in other people's jewels, the sons of the men who had not dared to sit down, were following with passionate sympathy the profound, vigorous words of Granovsky, greeting with outbursts of applause every word which with its audacity and nobility went straight to the heart.

It was just this society, that gathered from all parts of Moscow and crowded round the platform on which the young champion of learning delivered his earnest message and foretold the future from the past – it was just this society of whose existence Mme Zherebtsov had no suspicion. She was particularly kind and attentive to me because I was the first specimen of a world unknown to her; she was surprised at my language and at my ideas. She appreciated in me the growing shoots of another Russia, not that Russia whose only light fell on to it from the frosted windows of the Winter Palace. Thanks to her for that!

. . .

It was natural that from Count Strogonov I should go straight to Olga Alexandrovna and tell her all that had happened.

'Lord! What stupidity; they go from bad to worse,' she observed when she had heard my story. 'How can a man with a family be dragged off to exile for such trifles? Let me talk to Orlov. I hardly ever ask him for anything; they all dislike it; but there: for once he may do something. Come and see me in a couple of days, and I'll tell you his answer.'

Two mornings later she sent for me. I found several visitors with her. She had a white batiste kerchief round her head instead of a cap; this was usually a sign that she was out of humour; she screwed up her eyes and hardly took notice of the privy councillors and public generals who had come to pay their respects to her.

. . .

[. . .] When we were left alone she said to me:

'I asked you to come here to tell you that I have made a fool of

* Miss Wilmot's words.

myself in my old age. I gave you a promise, and I've accomplished nothing; you know the peasants' proverb: "Don't step into the water till you know how deep it is." I spoke to Orlov about your case yesterday and you've nothing to expect. . . .

. . .

[. . .]He regularly dug his toes in. "That's Benckendorf's affair," he told me. "I'll talk to him perhaps, but as for reporting on it to the Tsar, I can't; he doesn't like it – besides, it isn't done!" "What's so marvellous," I said, "about talking to Benckendorf? I can do that myself. Besides, he is in his dotage; he doesn't know what he is doing. His head is full of actresses, it seems, though I should have thought his flirting days were over; some wretched secretary of his lays all sorts of information, and he gives it in. What would he do? No!"I said, "you had better not demean yourself by asking favours of Benckendorf; the whole nasty business is his doing." "It is the rule with us," he said to me, and began telling me all about it. . . . Well, I saw that he was simply afraid to go to the Tsar . . . "Whatever is he – a wild beast, or what, that you are afraid to approach him, though you see him half a dozen times a day?" I said, and turned away in disgust; it is no use talking to them. Look,' she added, pointing to Orlov's portrait. 'What a conquering hero he is there; yet he is afraid to say a word!'

. . .

'Why, I suppose down there where you've been, in that Vologda, the clerks imagine Count Orlov is a man in favour, that he has power. . . . That's all nonsense. I'll be bound it is his subordinates who spread the rumour. None of them have any influence, they don't behave so as to have influence, and they are not on that footing. . . . You must forgive me for meddling in what isn't my business. Do you know what I advise you? What do you want to go to Novgorod for? You had better go to Odessa; it is further away from them and almost like a foreign town; besides, if Vorontsov[18] hasn't got corrupted, he is a man of a different stamp.'

Olga Alexandrovna's confidence in Vorontsov, who was at that time in Petersburg and came to see her every day, was not fully justified. He was willing to take me with him to Odessa *if* Benckendorf would give his consent.

Meanwhile the months passed, the winter was over, and no one reminded me about going away. I was forgotten and I gave up being *sur le qui-vive*, particularly after the following meeting. Bolgovsky, the military governor of Vologda, was at that time in Petersburg; being a very intimate friend of my father, he was rather fond of me and I was sometimes at his house. He had taken part in the killing of Paul, as a young officer in the Semënovsky Regiment, and was afterwards mixed up in the obscure and un-explained Speransky[19] affair in 1812. He was at that time a colonel in the army at the front. He was suddenly arrested, brought to Petersburg, and then sent to Siberia. Before he had time to reach his place of exile Alexander pardoned him, and he returned to his regiment.[20]

One day in the spring I went to see him; a general was sitting in a big easy-chair with his back towards the door so that I could not see his face, but only one silver epaulette.

'Let me introduce you,' said Bolgovsky, and then I recognised Dubelt.

'I have long enjoyed the pleasure of Leonty Vasilyevich's atten-tion,' I said, smiling.

'Are you going to Novgorod soon?' he asked me.

'I supposed I ought to ask you about that.'

'Oh! not at all! I had no idea of reminding you. I simply asked the question. We have handed you over to Count Strogonov, and we are not trying to hurry you, as you see. Besides, with such a legitimate reason as your wife's illness. . . .'

He really was the politest of men!

At last, at the beginning of June, I received the Senate's *ukaz*, confirming my appointment as councillor in the Novgorod Pro-vincial Government. Count Strogonov thought it was time for me to set off, and about the 1st of July I arrived in Novgorod, the 'City in the keeping of God and of Saint Sophia', and settled on the bank of the Volkhov, opposite the very barrow from which the Voltaireans of the twelfth century threw the wonder-working statue of Perun[21] into the river.

CHAPTER III

COUNCILLOR AT NOVGOROD

BEFORE I went away Count Strogonov told me that the military governor of Novgorod, Elpidifor Antiokhovich Zurov was in Petersburg; he said that he had spoken to him about my appointment, and advised me to call upon him. I found him a rather simple and good-natured general, short, middle-aged and with a very military exterior. We talked for half an hour, he graciously escorted me to the door and there we parted.

When I arrived in Novgorod I went to see him, and the change of *décor* was amazing. In Petersburg the governor had been a visitor, here he was at home; he actually seemed to me to be taller in Novgorod. Without any provocation on my part, he thought it necessary to inform me that he did not permit councillors to voice their opinions, or put them in writing; that it delayed business, and that, if anything were not right, they could talk it over, but that if it came to giving opinions, one or another would have to take his discharge. I observed with a smile that it was hard to frighten me with a threat of discharge, since the sole object of my service was to get my discharge from it; and I added that while bitter necessity forced me to serve in Novgorod I should probably have no occasion for giving my opinion.

This conversation was quite enough for both of us. As I went away I made up my mind to avoid coming into close contact with him. So far as I could observe, the impression I made on the governor was much the same as that which he made upon me, that is, we could not bear each other, so far as this was possible on so brief and superficial an acquaintance.

When I looked a little into the work of the provincial government I saw that my position was not only very disagreeable but also extraordinarily dangerous. Every councillor was responsible for his own department and shared the responsibility for all the rest. To read the papers concerning all the departments was absolutely impossible, so one had to sign them on trust. The governor, in accordance with his theory that a councillor should never give counsel, put his signature, contrary to the law and good sense, next after that of the councillor whose department the file concerned. For me personally this was excellent; in his signature I found something of a safeguard, since he shared the responsibility, and also because he often, with a peculiar expression, talked of his lófty honesty and Robespierre-like incorruptibility. As for the signatures of the other councillors, they were very little comfort to me. They were case-hardened old clerks who by dozens of years of service had worked their way up to being councillors, and lived only by the service, that is, only by bribes. There was nothing to blame them for in this; a councillor, I think, received 1,200 paper roubles a year: a man with a family could not possibly live on that. When they understood that I was not going to share with them in dividing the common spoil, nor to plunder on my own account, they began to look upon me as an uninvited guest and a dangerous witness. They did not become very intimate with me, especially when they had discovered that there was very slight friendship between the governor and myself. They stood by one another and watched over one another's interests, but they did not care about me.

Moreover, my worthy colleagues were not afraid of big monetary penalties or of deficiencies in their accounts, because they had nothing. They could risk it, and the more readily the more important the affair was; whether the deficit was of 500 roubles or of 500,000, it was all the same to them. In case of a deficit a fraction of their salary went to the reimbursement of the Treasury, and this might last for two or three hundred years, if the official lasted so long. Usually either the official died or the Tsar did, and then in his rejoicing the heir forgave the debts. Such manifestoes are also published during the lifetime of the same Tsar, by reason of a royal birth or coming of age, and odds and ends like that; the officials counted on them. In my case, on the contrary, the part of the family estate and the capital which my father had assigned to me would have been seized.

If I could have relied on my own head-clerks, things would have been easier. I did a great deal to gain their attachment, treated them politely and helped them with money, but my efforts only resulted in their ceasing to obey me. They feared only those councillors who treated them as though they were schoolboys; and they took to coming to the office half-drunk. They were very poor men with no education and no expectations. All the imaginative side of their lives was confined to little pot-houses and strong drink, so I had to be on my guard in my own department, too.

At first the governor gave me Department Four, in which all business dealing with contracts and money matters was dealt with. I asked him to exchange me; he would not, saying that he had no right to make an exchange without the consent of the other councillor. In the governor's presence I asked the councillor in charge of Department Two: he consented and we exchanged. My new department was less attractive; its work was concerned with passports, circulars of all sorts, cases of the abuse of power by landowners, schismatics, counterfeiters and people under the supervision of the police.

Anything sillier and more absurd cannot be imagined; I am certain that three-quarters of the people who read this will not believe it,* and yet it is the downright truth that I, as a councillor in the provincial government, head of the Second Department, counter-signed every three months the *politsmeyster's* report on *myself*, as a man under police supervision. The *politsmeyster* from politeness made no entry in the column for 'behaviour', and in the column for 'occupation' wrote: 'Engaged in the government service.' Such are the Hercules' pillars of insanity that can be reached when there are two or three police forces antagonistic to one another, official forms instead of laws, and a sergeant-major's conception of discipline in place of a governing intelligence.

. . .

For six months I pulled in harness in the provincial government. It was disagreeable and extremely tedious. Every morning at eleven o'clock I put on my uniform, buckled on my civilian sword, and went to the office. At twelve o'clock the military governor arrived; taking no notice of the councillors, he walked straight to a corner and stood his sword there. Then, after looking out of the window and straightening his hair, he went towards his arm-chair and bowed to those present. As soon as the sergeant, with fierce, grey moustaches that stood up at right angles to his lips, had solemnly opened the door and the clank of the sword had

* This is so true that a German who has abused me a dozen times in the *Morning Advertiser* [29 November and 6 December 1855] adduced as proof that I had never been exiled the fact that I had the post of councillor in a provincial government.

become audible in the office, the councillors got up and remained standing with backs bent until the governor bowed to them. One of my first acts of protest was to take no part in this collective rising and reverential expectation, but to sit quietly and to bow only when he bowed to us.

There were no great discussions or heated arguments; it rarely happened that a councillor asked the governor's opinion in advance, still more rarely that the governor put some business question to the councillors. Before everyone lay a heap of papers and everyone signed his name: it was a signature factory.

Remembering Talleyrand's celebrated injunction, I did not try to make any particular show of zeal and attended to business only so far as was necessary to escape reprimand or avoid getting into trouble. But there were two kinds of work in my department towards which I considered I had no right to take so superficial an attitude: these were matters relating to schismatics and to the abuse of power by the landowners.

Schismatics are not consistently persecuted in Russia, but something suddenly comes over the Synod or the Ministry of Home Affairs, and they make a raid on some hermitage, or some community, plunder it, and then subside again. The schismatics usually have intelligent agents in Petersburg who warn them from there of coming danger; the others at once collect money, hide their books and their icons, stand the Orthodox priest a drink, stand a drink to the Orthodox police-captain and buy themselves off; and with that the matter ends for ten years or so.

. . .

The business about the schismatics was of such a kind that it was much best not to stir them up again. I looked through the documents referring to them and left them in peace. On the contrary the cases of the abuse of landowners' power needed a thorough overhauling. I did all I could, and scored several victories in those sticky lists; I delivered one young girl from persecution and put her under the guardianship of a naval officer. This I believe was the only service I did in my official career.

A certain lady was keeping a servant-girl in her house without any documentary evidence of ownership; the girl petitioned that her rights to freedom should be inquired into. My predecessor had very sagaciously thought fit to leave her, until her case should be decided, in complete bondage with the lady who claimed her. I had to sign the documents; I approached the governor and observed that the girl would not be in a very enviable situation in her lady's house after lodging this petition against her.

'What's to be done with her?'

'Keep her in the police-station.'

'At whose expense?'

'At the expense of the lady, if the case is decided against her.'

'And if it is not?'

Luckily at that moment the provincial prosecutor came in. A prosecutor from his social position, from his official relationships, from the very buttons on his uniform, is bound to be an enemy of the governor, or at least to thwart him in everything. I purposely continued the conversation in his presence. The governor began to get angry and said that the whole question was not worth wasting a couple of words on. The prosecutor was quite indifferent to what would happen, and what became of the girl, but he immediately took my side and advanced a dozen different points from the code of laws in support of it. The governor, who in reality cared even less, said to me, smiling ironically:

'It's much the same whether she goes to her mistress or to prison.'

'Of course it's better for her to go to prison,' I observed.

'It will be more consistent with the intention expressed in the code,' observed the prosecutor.

'Let it be as you like,' the governor said, laughing more than ever. 'You've done your *protégée* a service: when she has been in prison for a few months she will thank you for it.'

I did not continue the argument; my object was to rescue the girl from domestic persecution; I remember that a couple of months later she was released and received her complete freedom.

. . .

In halls and maids' rooms, in villages and the torture-chambers of the police, are buried whole martyrologies of frightful villainies; the memory of them works in the soul and in course of generations matures into bloody, merciless vengeance *which it is easy to prevent*, but will hardly be possible to stop once it has begun.

. . .

At the beginning of 1842 I was hopelessly weary of provincial government and was trying to invent an excuse to get out of it. While I was hesitating between one means and another, a quite extraneous incident decided in my favour.

One cold winter's morning as I reached the office I found a peasant woman of about thirty standing in the front hall; seeing me in uniform she fell on her knees before me and bursting into tears besought my protection. Her master, Musin-Pushkin, was sending her with her husband to a settlement, while their son, a boy of ten, was to remain behind; she begged to be allowed to take the child with her. While she was telling me this the military governor came in; I motioned her towards him and passed on her petition. The governor explained to her that children of ten or over are kept by the landowners. The mother, not understanding the stupid law, went on entreating him. He was bored; the woman, sobbing, clutched at his legs, and he pushed her away roughly, saying: 'What a fool you are; don't I tell you in plain Russian that I can do nothing? Why do you keep on so?' After this he went with a firm, resolute step to the corner, where he put his sword.

And I went too ... I had had enough. ... Did not that woman take me for one of *them*? It was high time to put an end to the farce.

'Are you unwell?' asked a councillor called Khlopin, who had been transferred from Siberia for some shortcomings or other.

'I am ill,' I answered, and I got up, took my leave and went away. The same day I sent in a declaration that I was ill, and from that day never set foot in the office of the provincial government. Then I asked for my discharge on the ground 'of illness'. The Senate gave me my discharge accompanying it with promotion to the grade of Aulic Councillor; but Benckendorf at the same time informed the governor that I was forbidden to visit Petersburg or Moscow and was commanded to live at Novgorod.

When Ogarëv returned from his first tour abroad, he did his utmost in Petersburg to procure permission for us to move to Moscow. I had little faith in the success of such a patron and was fearfully bored in the wretched little town with the great historical name. Meanwhile Ogarëv managed our business for us. On the 1st of July 1842, the Empress, taking advantage of some family festivity, asked the Tsar to allow me to live in Moscow in consideration of my wife's illness and her desire to move there. The Tsar agreed and three days later my wife received from Benckendorf a letter in which he informed her that I was permitted to accompany her to Moscow in consequence of the Tsaritsa's intercession. He concluded the letter with the agreeable notification that I should remain under police supervision there also.

I felt no regret at leaving Novgorod and made haste to get away as soon as possible. Before I left it, however, there occurred almost the only pleasant event in my sojourn there.

I had no money! I did not want to wait for a remittance from Moscow and so I commissioned Matvey to try to borrow 1,500 paper roubles for me. An hour later Matvey appeared with an innkeeper called Gibin, whom I knew, and at whose hotel I had stayed for a week. Gibin, a stout merchant with a good-natured expression, bowed and handed me a packet of notes.

'How much interest do you want?' I asked him.

'Well, you see,' answered Gibin, 'I don't do this sort of business and I don't lend money at interest, but since I heard from Matvey Savelyevich that you need money for a month or two, and we very much approve of you, and thank God have the money to spare, I've brought it along.'

I thanked him and asked him which he would like, a simple receipt for the money or a promissory note; but to this, too, Gibin answered: 'Extra work; I trust your word more than a piece of stamped paper.'

'Upon my word, but I may die you know.'

'Well then, in my sorrow at your decease I shouldn't worry much about the loss of the money.'

I was touched and pressed his hand warmly instead of giving him a receipt. Gibin embraced me in the Russian fashion and said: 'We know it all, of course; we know you were not serving of your own will and didn't behave yourself like the other officials, the Lord forgive them, but stood up for the likes of us and the ignorant people, so I am glad a chance has come to do you a good turn too.'

As we were driving out of the town late in the evening our driver pulled up the horses at the inn and Gibin gave me a pie the size of a cart-wheel as provision for the journey. . . .

That was my 'medal for good service'.

CHAPTER IV

MOSCOW AND POKROVSKOYE

OUR life at Novgorod had not been a happy one. I had gone there not in a spirit of self-sacrifice and firmness, but with my heart full of annoyance and exasperation. This second exile, a commonplace affair as it was, irritated more than it distressed me; it was not enough of a calamity to rouse the spirit, but was merely a worry, without the interest of novelty or the stimulus of danger. The provincial government alone, with its Elpidifor Antiokhovich Zurov, councillor Khlopin, and vice-governor Pimen Arapov, was more than enough to poison my existence.

I was ill-humoured; Natalie sank into melancholy. Her sensitive nature, accustomed from childhood to tears and sadness, gave way again to poignant anguish. She dwelt long on tormenting thoughts, and readily let pass everything bright and joyful. Life was becoming more complex; there were more chords in it and with them more anxiety. After Sasha's illness had come the fear of the Third Division, her unfortunate confinement and the death of the baby. The death of a baby is scarcely felt by the father: anxiety over the mother makes him almost forget the little creature that has flitted away almost before it had time to cry and take the breast. But to the mother the new-born child is an old acquaintance; she has long been *feeling* him; there has been a physical, chemical, nervous connection between them; moreover the baby repays the mother for the burden of pregnancy, for the sufferings of childbirth; without him her agonies are motiveless and resented, without him the unneeded milk affects the brain.

After Natalie's death I found among her papers a note which I had quite forgotten. It consisted of a few lines I had written an hour or two before Sasha's birth. It was a prayer, a blessing, a dedication of the new-born creature to 'the service of humanity', his 'consecration to the path of hardship'.

On the other side was written in Natalie's hand: 1 January 1841. – Yesterday Alexander gave me this; he could not have made me a better present; these lines at once called up the whole picture of our three years of unbroken, boundless happiness, resting on love alone.

'So we have passed into a new year; whatever awaits us in it, I bow my head and say for both of us, Thy Will be done!

'We welcomed the New Year at home, in solitude; only A. L. Vitberg was with us. Little Alexander was missing from our party; he was sound asleep and neither past nor future exists for him yet. Sleep, my angel, free from care, I pray for you – and for you too, my child unborn, whom I love with all a mother's love. Your movements, your tremors mean so much to my heart, and may your coming into the world be glad and blessed!'

But the mother's blessing was not fulfilled: the babe was executed by Nicholas. The deadly hand of the Russian autocrat intervened here also – and here also destroyed a life!

The baby's death left its mark upon her soul.

With sadness and mounting resentment we moved to Novgorod.

The *truth* of that period, as it was seen at the time, without the artificial perspective given by distance, without the cooling effect of years, and the illumination rectified by passing through a series of other events, is preserved in a notebook of the period. I had meant to keep a diary, had begun it many times, but had never persevered. On my birthday at Novgorod Natalie gave me a white book in which I sometimes wrote down what was in my heart or my head.

This book has survived. On the first page Natalie wrote: 'May all the pages of this book and of all your life be bright and joyful!'

But three years later she added on the last page: 'In 1842 I hoped that all the pages of your diary might be bright and untroubled; three years have passed since then, and looking back I do not regret that my wish has not been fulfilled; both enjoyment and suffering are essential for a full life, and you will find repose in my love, in the love with which my whole being, my whole life is filled. Peace to the past and a blessing on the future! 25 March 1845, Moscow.'

Here is what was written in the book on the 4th of April, 1842:

'Oh Lord, what unbearable misery! Is it weakness or have I a right to feel it? Must I reckon my life finished? Is all my readiness for work, all my need for expression to be kept under a bushel, till my wants are stifled, and am I then to begin a life of emptiness? It might be possible to exist with no object but one's own inner development, but the same awful depression comes over me in the midst of study. I must express myself – perhaps from the same necessity as the grasshopper churrs ... and for years to come I have to drag this weight.'

And as though frightened at my own words I followed this with Goethe's lines:

> 'Gut verloren – etwas verloren,
> Ehre verloren – viel verloren,
> Musst Ruhm gewinnen,
> Da werden die Leute sich anders besinnen.
> Mut verloren – alles verloren,
> Da wär' es besser nicht geboren';[1]

and later:

'My shoulders are breaking but they will still bear!'

'Will those who come after us understand, will they appreciate all the horror, all the tragic side of our existence? And meanwhile our sufferings are the soil from which their happiness will develop; will they understand what makes us slothful, makes us seek all sorts of pleasure, drink and so on. Why do we not lift our hands to great tasks, why at the moment of rapture do we not forget our despondency? Let them pause with musing and sadness before the stones under which we slumber: we have deserved their sad thoughts!

'I cannot go on for long in my situation: I shall be stifled – and I don't care how I get out of it, if only I do get out. I have written to Dubelt (I asked him to try and get leave for me to move to Moscow). Writing a letter like that makes me ill, on se sent flétri. I expect it is what prostitutes feel when they first begin selling themselves.'[2]

And it was just this vexation, this refractory cry of impatience, this fretting for free activity, this feeling of fetters on the limbs that Natalie misunderstood.

Often I found her with tear-stained eyes by Sasha's cot; she assured me that it was nothing but nerves, that I had better take no notice of it, not question her ... I believed her.

One evening I returned home late; she was in bed when I went in. I was feeling sick at heart. Filippovich had asked me to go and see him in order to tell me that he suspected that one of our common acquaintances had dealings with the police. That sort of thing usually sends a pang to the heart, not so much from the possible danger as from the feeling of moral repulsion.

I walked up and down the room in silence, turning over what I had just heard, when all at once I fancied that Natalie was weeping; I took her handkerchief: it was soaked with tears.

'What is it,' I asked, alarmed and agitated.

She took my hand and in a voice full of tears said:

'My dear, I will tell you the truth; perhaps it is self-love, egoism, madness, but I feel, I see, that I cannot distract your mind; you are bored – I understand it, I don't blame you, but it hurts me, it hurts me, and I cry. I know that you love me, that you are sorry for me, but you don't know what makes you depressed, what gives you that feeling of emptiness; you feel the poverty of your life – and, indeed, what can I do for you?'

I was like a man suddenly roused in the middle of the night and told something frightful before he is quite awake: he is already frightened and trembling, but he does not yet understand what is wrong. I was so completely at peace, so sure of our deep, perfect love, that I never spoke about it; it was the great *assumption* upon which all our life rested; a serene consciousness, a boundless conviction of it excluding doubt or even distrust of myself, constituted the basic element of my personal happiness. Peace, repose, the aesthetic side of life, all that – as before our meeting in the graveyard on the 9th of May, 1838, as at the beginning of our life at Vladimir – rested on her, on her, on her!

My deep distress and my astonishment at first dissipated these clouds, but in a month or two they began to return. I soothed and comforted her; she smiled herself at the dark phantoms, and again the sunshine brightened our little corner; but as soon as I had forgotten them they raised their heads again, though evoked by nothing whatever, and when they had passed I began to be afraid of their return.

Such was the state of mind in which in July 1842 we moved to Moscow.

Life in Moscow, at first too full of distractions, could have no beneficial nor soothing effect. Far from helping her at that time I gave, on the contrary, cause for her *Grübelei* to grow deeper and more intense.

When we moved from our exile at Novgorod to Moscow, this is what occurred just before we left.

One morning I had happened to go into my mother's room: a young maid was in it setting things to rights; she was one of the new ones, that is one of those who had come to my father after the Senator's death. I hardly knew her at all. I sat down and took up a book. It seemed to me that the girl was crying: I glanced at her, and she was indeed crying. Suddenly, in fearful agitation, she came over to me and threw herself at my feet.

'What is it? What's up with you? Just tell me plainly,' I said to her, being myself astonished and embarrassed.

'Take me with you. . . . I'll serve you faithfully and truly. You need a maid: take me. Here I shall die of shame for sure . . .' She sobbed like a child.

Only then did I perceive the reason.

With her face flaming from tears and shame, with a look of fear and apprehension, of entreaty, the poor girl stood before me – with that expression that is given to a woman by pregnancy.

I smiled and told her to get her possessions ready. I knew that my father would not care whom I took with me.

She spent a year with us. The time towards the end of our stay at Novgorod was an anxious one – I was vexed at my exile and from day to day awaited with some irritation permission to go to Moscow. Then I suddenly realised that the maid was very pretty to look at. She guessed that I had! . . . and it might all have passed off without any further progress. Opportunity helped. An opportunity is always found, especially when no attempt is made by either side to evade it.

We moved to Moscow, and there was celebration after celebration. . . . One night, when I had come home late, I had to go in through the back of the house. Katerina opened the door to me. It was obvious that she had only just left her bed: her cheeks were burning from sleep. She had a shawl thrown over her, and her thick plait of hair, hardly fastened, was ready to fall in a heavy wave. . . . This was at dawn. She glanced at me, smiled and said:

'How late you are.'

I looked at her, devouring her beauty, and instinctively, half consciously, I put my hand on her shoulder: the shawl fell off . . . she gasped . . . her breast was bared.

'What are you doing?' she whispered, looking excitedly into my eyes; and turned away as though to leave me without a witness of the state I was in. . . . My hand touched her body, still warm from sleep. . . . How good nature is, when a man forgets himself and surrenders to her, loses himself in her. . . .

At that moment I loved this woman, and there was as it were something immoral in this intoxication . . . someone was being wronged, hurt . . . and who was it? The being who was closest, dearest to me on earth. My passionate inclination was too transitory to possess me utterly – there were no roots (neither on one side or on the other: on her side I doubt whether there was even any inclination); and it might all have passed off without a trace, leaving only a smile, a hot memory, and a couple of times, perhaps, a flushed cheek. . . . That is not what happened, for other forces intervened: *thoughtlessly* I set the stone rolling; to stop it or guide it was beyond my power. . . .

I came to believe that Natalie had heard something, suspected something, and I resolved to tell her what had happened. Such confessions are difficult, but this seemed to me a necessary purification, expiation, reintegration of the frank purity of relationship which silence on my part might unsettle or scare away. I thought my very frankness itself would soften the blow, but it struck powerfully and deep. She was greatly mortified: it seemed to her that I had fallen and had drawn her down with me in my fall. Why had I not thought of the consequences? Why had I not stopped, not before my action itself but before the *repulsion* which it must arouse in a being so closely, so indissolubly linked with me? Did I not know the ascetic point of view from which a woman, even the most highly developed, who has long ago put away Christianity, looks upon *betrayal*, making no distinctions, accepting no mitigating circumstances?

To reproach a woman for her exclusive view of things is hardly just. Has anyone ever tried seriously, honestly to shatter their prejudices? Experience shatters them, but from that comes the breaking not of prejudice but of life. People skirt questions which exercise our minds, as old women and children walk round graveyards and places where. . . .

She got over it, but only after she came close to the very grave. She understood everything, but the blow had been unexpected and heavy. Her faith in me was shaken, her idol destroyed: chimerical sufferings yielded to the fact. Did not what had happened confirm the sloth of my heart? Otherwise would it not have withstood the first temptation – and what a temptation! And where had it happened? A few paces away from her. And who was her *rival*? To whom had she been sacrificed? To a woman who would hang round anybody's neck.

I felt that all this was not so; that she had never been sacrificed, that the word 'rival' did not fit the case and that, if *that* woman had not been a light woman, nothing would have happened; but on the other hand I understood also that it might only seem so.

A struggle to the death went on within her, and then, as before and as afterwards, I was amazed. She never once uttered a word which might have hurt Katerina, or by which she could have guessed that Natalie knew of what had happened – her reproaches were for me. Peacefully and quietly the girl left our house. Natalie dismissed her with such gentleness that the simple woman, still an ingenuous child of the people, told her herself, sobbing on her knees before her, what had happened, and begged for forgiveness.

Natalie fell ill. I stood by, a witness of the woes I had inflicted, and more than a witness: I was my own accuser, and ready to become the executioner. My imagination, too, became distorted: my *fall* was taking on even bigger and bigger dimensions. I was humiliated in my own eyes, and near to despair. [...]

* * *

Natalie became more and more engrossed in melancholy: her faith in me had been shaken, her idol was shattered.

It was a crisis, the painful transition from youth to maturity. She could not overcome the thoughts that gnawed at her, she was ill and grew thin. Frightened and reproaching myself I stood by and saw that I had no longer the autocratic power with which I had once been able to exorcise the spirits of gloom. It wounded me to see it, and I was immensely sorry for her.

They say that children grow when they are ill; in this spiritual illness which brought her to the verge of consumption she made colossal strides in growth. From the slanting rays and glow of dawn she passed by this sorrowful path into the clear, bright light of noon. Her health was equal to the strain and that was all that was needed. Without losing one iota of her womanliness she developed intellectually with extraordinary boldness and depth.

Gently and with a smile of self-sacrifice she bowed to the inevitable without romantic complaint, and on the other hand without obduracy or haughty gratification.

It was not in a book, nor through a book, that she found her freedom, but by life and clearness of vision. Unimportant trials, painful shocks, which for many would have passed without a trace, printed deep furrows on her soul and were enough to arouse her mind to profound activity. A slight hint was sufficient for her to pass from conclusion to conclusion, till she reached that fearless grasp of the truth which is a heavy burden even for a man to bear. Mournfully she parted from her shrine in which had stood so many holy mysteries, bathed in tears of grief and joy; she left them without blushing as big girls blush at the sight of their doll of yesterday. She did not turn away from them: she yielded them up with anguish, knowing that she would be the poorer, the more defenceless for the loss, that the soft light of the glimmering icon lamps would be exchanged for the grey dawn, that she must make friends with harsh, indifferent powers, deaf to the murmur of prayer, deaf to the hopes of immortality. She gently took them from her bosom like a dead child, and gently laid them in the grave, valuing them for her earlier life, valuing the poetry they gave and the comfort they furnished at times. Even later she disliked touching them coldly, just as we avoid wantonly stepping on the heaped-up earth of a grave.

With this intense mental activity, with this shattering and rebuilding of all her convictions, she naturally needed rest and solitude.

We went to my father's estate near Moscow.

And as soon as we found ourselves alone, surrounded by trees and fields, we breathed deeply and again looked clearly at life. We stayed in the country until the late autumn. From time to time we had visitors from Moscow. Ketscher stayed a month with us; all our friends arrived for the 26th of August, Natalie's name-day; then stillness once more, stillness and the woods and the fields – and no one but ourselves.

Pokrovskoye, standing solitary, surrounded by huge forested estates, was of quite a different and much more serious character than Vasilevskoye, flung down so cheerfully with its villages on the bank of the Moskva. This difference was noticeable even in the peasants. The Pokrovskoye peasants, hemmed in by woods, were less like people living within reach of Moscow than those of Vasilevskoye, although as a fact they were fifteen miles nearer the city. They were quieter, more unsophisticated, and hung together extremely closely. My father moved a wealthy family of peasants from Vasilevskoye to Pokrovskoye, but the peasants of the latter place never considered the family as belonging to their village, but always called them 'the settlers'.

With Pokrovskoye, too, I had been closely connected throughout my childhood; I used to stay there when I was too young to remember, and from the year 1821 we used to spend a few days there almost every summer on our way to or from Vasilevskoye. There lived old Kashentsov,[3] paralysed, in disgrace since 1813, who dreamed of seeing his master, the Senator, in all his decorations and insignia; there lived – and later in the cholera of 1831 died – the venerable, grey-headed, corpulent village head-man, Vasily Yakovlev, whom I remembered at all his ages, with his beard of all colours, first dark flaxen and afterwards quite grey; there lived my foster-brother Nikifor, who prided himself on the fact that for my benefit he had been deprived of his mother's milk – she died later on in a madhouse. . . .

The little village of some twenty or twenty-five homesteads stood at some distance from our rather large house. On one side lay a semicircular meadow that had been cleared and fenced in, on the other there was a view of the river, dammed up for the sake of a mill which they had intended to build fifteen years before, and of the dilapidated wooden church all on the slant, which my uncle the Senator and my father, who owned the estate in common, had also been intending to repair every year for the last fifteen.

The house, which had been built by the Senator, was a very good one; there were lofty rooms, big windows, and on both sides covered passages that were like verandahs. It was built of choice, thick beams, not faced with anything either outside or in, but with the crevices packed with tow and moss. The walls smelt of resin, which oozed out here and there like drops of amber. Before the house there was a small field and beyond that began a dark wood of building timber, through which ran a ride to Zvenigorod; in the other direction a by-road ran like a narrow, dusty ribbon by the village and was lost in the rye, coming out through the Maykovsky factory and going on to the Mozhayka. There was the forest stillness, there were the forest sounds, the incessant buzzing of flies, honey-bees and bumble-bees ... and the fragrance ... that fragrance of grass and forest, infused with the scents of plants, of leaves, but not of flowers ... which I have so eagerly sought in Italy and in England both in spring and in hot summer, but have hardly ever found. Sometimes one gets a whiff of it in the hay-field, or when the sirocco is blowing, or before a storm ... and it brings back the little place before the house, on which, to the great distress of the village head-man and the house-serfs, I would not have the grass mown close: on the grass a boy of three, rolling in the clover and the dandelions among the grasshoppers, beetles of every kind and ladybirds, and we ourselves and youth and our friends!

The sun has set but it is still very warm; we don't want to go home, so we stay sitting on the grass. Ketscher sorts out the mush-rooms and scolds me for no other reason. Can that be the tinkle of a bell? Is it something for us? Perhaps – it is Saturday.

'It's the police-captain going somewhere,' says Ketscher, suspecting that it is not.

The troika rolls through the village, rumbles over the bridge and disappears behind a knoll; from there the only road is towards us. While we are running to meet it, it drives up to the house; Shchepkin[4] has already rolled off it like an avalanche, laughing, kissing our hands and making us die with laughter, while Belinsky, cursing the distance from Pokrovskoye and the way that Russian carts and Russian roads are made, is still climbing down and stretching himself, and already Ketscher is rebuking them:

'What devil has brought you here at eight o'clock in the even-ing? Couldn't you have come sooner? It is all due to that finical Belinsky; he can't get up early; what were you thinking about?'

'Why, he has become more of a savage than ever at your place,' says Belinsky, 'and what a head of hair he has grown! You would do for the moving forest in *Macbeth*, Ketscher. Wait a bit – don't exhaust your whole stock of swear-words: there are some male-factors who are coming later still.'

Another troika is already turning into the yard: Granovsky and Yevgeny Korsh.[5]

'Have you come for long?'

'Two days.'

'Splendid!' and Ketscher himself is so pleased that he greets them almost as Taras Bulba[6] greeted his sons.

Yes, that was one of the luminous periods of our life. Of past storms nothing remained but a trace of vanishing cloud; at home among our friends there was perfect harmony.

* * *

CHAPTER V

OUR FRIENDS

1

WITH our visit to Pokrovskoye and the quiet summer we spent there begins the gracious, grown-up, active part of our Moscow life, which lasted till my father's death and perhaps until we went abroad.

Our nerves, overstrained in Petersburg and Novgorod, had relaxed, our inner storms had subsided. The agonising analysis of ourselves and of each other, the useless reopening with our words of recent wounds, the incessant return to the same painful subjects were over; and our shaken faith in our own infallibility gave a truer and more earnest quality to our lives. My article 'On a Drama' was the last word of the sickness we had passed through.

Externally the only restriction we suffered from was police supervision; I cannot say it was very tiresome, but the unpleasant feeling of a cane of Damocles, wielded by the local police-constable, was very disagreeable.

Our new friends received us warmly, much better than two years before. Foremost among them stood Granovsky: to him belongs the chief place in those five years. Ogarëv was abroad almost all the time. Granovsky filled his place for us, and we are indebted to him for the happiest moments of that time. There was a wonderful power of love in his nature. With many I was more in agreement in opinion, but to him I was nearer – somewhere deep down in the soul.

Granovsky and all of us were very busy, all hard at work, one lecturing at the university, another contributing to reviews and magazines, another studying Russian history; the first beginnings of all that was done afterwards date from this time.

By now we were far from being children; in 1842 I was thirty; we knew only too well where our work was leading us, but we went on. We went along our chosen path, not rashly but delibera-tely, with the calm, even step to which experience and family life had trained us. This did not mean that we had grown old: no, we were still young, and that is how it was that some speaking in the university lecture-room, others publishing articles or editing a newspaper were every day in danger of being arrested, dismissed, exiled.

Such a circle of talented, cultured, versatile and pure-hearted people I have met nowhere since, neither in the highest ranks of the political nor on the summits of the literary and artistic worlds. Yet I have travelled a great deal, I have lived everywhere and with all sorts of people. I have been thrust by revolution into the extremes of progress, beyond which there is nothing, and conscientiously I am bound to say the same thing.

The finished, self-contained personality of the Western European, which surprises us at first by his specialisation, surprises us later by his one-sidedness. He is always satisfied with himself, and his *suffisance* offends us. He never forgets his personal views, his position is generally cramped and his morals only appropriate to paltry surroundings.

I do not think that men were always like this here; the Western European is not in a normal condition, *he is moulting*. Unsuccessful revolutions have been absorbed and none of them has transformed him, but each has left its trace and confused his ideas, while the natural surge of historical process has splashed up into the foreground the slimy stratum of the *petit bourgeois*, under which the fossilised aristocratic class is buried and the rising masses submerged. *Petite bourgeoisie* is incompatible with the Russian character – and thank God for it!

Whether it is due to our carelessness, or our lack of moral stability and of defined activity, or our youth in the matter of education, or the aristocratic way in which we are brought up, yet we are in our living on the one hand more artists, and on the other far simpler than Western Europeans; we have not their specialised knowledge, but to make up for that we are more versatile than they. Well developed personalities are not common amongst us, but their development is richer, wider in its scope, free from hedges and barriers. It is quite different in Western Europe.

When you are talking to the most likeable people here[1] you immediately reach contradictions where you and they have nothing in common, and it is impossible to convince. In this stubborn obstinacy and unintentional incomprehension you seem to be knocking your head against the frontier of a world that is completed.

Our theoretical differences, on the contrary, brought more living interest into our lives, and a need for active exchange of opinions kept our minds more vigorous and helped us to progress; we grew in this friction against each other, and in reality were the stronger thanks to that 'composite' workmen's association which Proudhon has so superbly described in the field of mechanical labour.

I love to dwell on that time of work in unison, of a full exalted pulse, of harmonious order and virile struggle, on those years in which we were young for the last time! . . .

Our little circle assembled frequently, at the house sometimes of one, sometimes of another, and oftenest of all at mine. Together with chatter, jest, supper and wine, there was the most active, the most rapid exchange of ideas, news and knowledge; everyone handed on what he had read and learned. Opinion was disseminated through arguments and what had been worked out by each became the property of all. There was nothing of significance in any sphere of knowledge, in any literature or in any art, which did not come under the notice of some one of us, and was not at once communicated to all.

It was just this quality of our gatherings that dull pedants and tedious scholars failed to understand. They saw the meat and the bottles, but they saw nothing else. Feasting goes with fullness of life; ascetic people are usually dry and egoistical. We were not monks: we lived on all sides and, sitting round the table, learnt rather more and did no less than those fasting toilers who grub in the backyards of science.

I will not have anything said against you, my friends, nor against that bright, splendid time; I think of it with more than love: almost with envy. We were not like the emaciated monks of Zurbaran;[2] we did not weep over the sins of this world – we only sympathised with its sufferings, and were ready with a smile for anything, and not depressed by a foretaste of our sacrifices to come. Ascetics who are for ever morose have always excited my suspicion; if they are not pretending, either their mind or their stomach is out of order.

'You're right, my friend, you're right. . . .'[3]

Yes, you were right, Botkin – and far more so than Plato – when you sometimes taught us, not in gardens and porticos (it is too cold in Russia without a roof) but round the friendly dinner-table, that a man may equally find 'pantheistic' enjoyment in contemplating the dance of the sea-waves and of Spanish maidens, in listening to the songs of Schubert and 'listening' to the fragrance of a turkey stuffed with truffles. Listening to your sage words, I appreciated for the first time the democratic profoundness of our language which talks of 'hearing an odour', putting smell on a level with sound.

It was not for nothing that you left your Maroseyka and learned in Paris to respect the culinary art, and from the banks of the Guadalquivir the religion not only of feet, but of calves supreme and sovereign, *soberana pantorrilla!*[4]

Yet Redkin[5] was in Spain – but what good did he get from it? He travelled in that land of historical lawlessness for the sake of making juridical commentaries on Puchta and Savigny.[6] Instead of looking at the fandango and the bolero, he looked at the rising in Barcelona (which ended exactly in the same way as every *cachucha* – that is in nothing) and talked so much about it that the curator Strogonov[7] shook his head and began looking at Redkin's lame leg and muttering something about barricades, as though he doubted that the 'radical jurist' had hurt his leg falling out of the *diligence* on to the pavement in loyal Dresden.

'What disrespect for learning! You know I don't like such jokes,' says Redkin severely, not in the least vexed.

'That m-m-m-ay be so,' Korsh stammers, 'but why is it you so identify yourself with learning that one can't make fun of you without insulting it?'

'Come: now it's started and there will be no end to it,' adds Redkin, and with the determination of a man who has read the whole of Rotteck[8] attacks the soup, lightly bestrewn with Kryukov's[9] jests – elegantly modelled on an antique pattern.

But the attention of all has already abandoned them; it is bent upon the sturgeon, which is *expounded* by Schchepkin himself, who has studied the flesh of contemporary fish more thoroughly than Agassiz[10] did the bones of antediluvian ones. Botkin glances at the sturgeon, screws up his eyes and gently shakes his head, not from side to side but backwards and forward; only Ketscher, indifferent on principle to the grandeurs of this world, lights his pipe and speaks of something else.

Do not be angry with these lines of nonsense; I shall not go on with them: they dropped almost involuntarily from my pen when I thought of our Moscow dinners; for a minute I forgot both the impossibility of writing down jokes and the fact that these sketches are alive only for me, and for a few, a very few, survivors. I am frightened when I consider how short a time ago the path seemed so long, so very long before us all! ...

* * *

2

ON THE GRAVE OF A FRIEND

*'Generous and pure in spirit with a heart
Tender as a caress ... And friendship with him
Lives in my memory like a fairy tale.'*[11]

... In 1840 when I was passing through Moscow I met Granovsky[12] for the first time. He had only just come back from foreign parts and was preparing to occupy his Chair of History at the University. He attracted me by his noble, thoughtful appearance, his melancholy eyes with their wrinkled brows, and his mournful good-natured smile; in those days his hair was long, and he wore a dark blue Berlin overcoat of a peculiar cut, with velvet *revers* and cloth fastenings. His features, dress, dark hair – all gave so much grace and elegance to his figure as he stood at the dividing lines between passing youth and a richly developing manhood, that even a man not easily captivated could not have remained indifferent to him. I have always respected beauty, and looked upon it as a talent and a power.

I had but a passing glimpse of him then, and carried away with me to Vladimir a noble image, and a belief, founded on it, that he would one day be my friend. My presentiment did not deceive me. Two years later, after I had been in Petersburg and, at the end of my second exile, returned to live in Moscow, a close, profound friendship was formed between us.

Granovsky was gifted with an amazing *tact* of the heart. His whole nature was so remote from the irritability of diffidence and from pretentiousness, it was so pure, so open, that he was extraordinarily easy to get on with. He did not oppress me with his friendship, and his love was strong and equally free from jealous exigence and unconcerned indifference. I do not remember that Granovsky ever touched roughly or awkwardly upon those delicate 'capillary tissues' that shrink from light and noise and exist in every man who has really lived. That was why one was not afraid to speak to him of the things of which it is hard to speak even with those most near and dear, whom one trusts completely though some scarcely audible chords in them are not tuned to the same pitch.

In contact with his affectionate, serene, indulgent spirit all awkward discord vanished, the voice of over-sensitive vanity was almost mute. He was a link of union among us for many things and many people, and often brought together in their sympathy with him whole circles that were at enmity among themselves, and friends on the brink of separation. Granovsky and Belinsky, completely unlike each other, were among the most luminous and remarkable personalities of our circle.

Towards the end of the oppressive period from which Russia is now emerging, when everything was crushed to the earth, when only the voice of official baseness dared make itself heard, when literature had been brought to a standstill, and instead of humane learning the theory of slavery was taught, when the censorship shook its head over the parables of Christ and blotted out Krylov's *Fables* – in those days, if one saw Granovsky on the lecture platform, one became lighter of heart. 'All is not yet lost, if he still goes on speaking,' everyone thought, and breathed more freely.

And yet Granovsky was not a fighter like Belinsky, nor a dialectician like Bakunin. His strength lay not in keen polemic nor in bold denunciation, but just in positive moral influence, in the absolute confidence which he inspired, in the artistic completeness of his nature, the calm serenity of his spirit, the purity of his character, and in his constant profound protest against the existing order in Russia. Not only his words were effective but also his silence; his thought, denied free utterance, came out so plainly in his features that it was hard not to read it, especially in a land in which a narrow despotism has trained us all to guess and to divine the hidden word. In the dark years of persecution from 1848 down to the death of Nicholas, Granovsky succeeded in keeping not only his chair in the University, but also his independent way of thinking – and that because a feminine delicacy, a softness of expression, and the reconciling power of which we have spoken were harmoniously combined with chivalrous courage and the complete devotion of passionate conviction.

Granovsky reminds me of a number of the calm, reflective preachers and revolutionaries of the Reformation – not those fierce, turbulent spirits who 'feel their life fully in their wrath' like Luther, but the clear, mild reformers who put the crown of glory on their heads as simply as the crown of thorns. Their gentleness nothing can ruffle: they go forward with firm step but with no loud tramping of feet; judges fear these men, they are ill at ease with them; their forgiving smile leaves a sting in their executioner's conscience.

Such was Coligny himself, such were the best of the Girondins; and certainly Granovsky in all the harmonious moulding of his soul, in his romantic bent, in his dislike of extremes, might more readily have been a Huguenot or a Girondin than an Anabaptist or a Montagnard.[13]

Granovsky's influence on the University, and on the whole of the younger generation, was enormous, and it outlived him; he left a long ray of light behind him. I look with peculiar emotion at the books dedicated to his memory by his former students, at the warm, enthusiastic lines about him in their prefaces and in maga-zine articles, at their splendid, youthful desire to relate their new work to the shade of that friend, to touch gently on his grave as they begin, to reckon their intellectual pedigree from him.

Granovsky's development had been different from ours. Educated in Orël, he went to Petersburg University. Since he received little money from his father he was obliged from a very early age to write 'to order' articles for magazines. He and his friend Yevgeny Korsh, whom he met in his university days and with whom he maintained the closest friendship up to his death, used to work for Senkovsky,[14] who needed fresh energies and inexperienced lads in order to transform their conscientious work into the effervescing Russian champagne of 'The Library of Good Reading'.

Strictly speaking, there was no tempestuous period of passion and dissipation in his life. When he had taken his degree the Institute of Pedagogy sent him to Germany. In Berlin he met Stankevich, and that was the most important event of his whole youth.

Anyone who knew them both would understand how immedi-ately Granovsky and Stankevich must have rushed at each other. There was in them so much that was similar, in character, in tendency, in age ... and each bore within him the fatal seed of premature death. But mere resemblance is not enough to give men this close tie, this enduring sense of kinship. Only that love is deep and lasting in which each completes the other: for active love difference is as necessary as resemblance; without it the feeling is languid and passive and passes into a mere habit.

There was a vast difference in the aspirations and the power of the two young men. Stankevich, tempered from early years by the Hegelian dialectic had a keen talent for speculative thought and, if he brought the aesthetic element into this thinking, he certainly brought philosophy as much into his aesthetics, Granovsky, who sympathised deeply with the intellectual tendencies of the day, had neither love nor talent for abstract thought. His choice of history as his chief pursuit showed a clear understanding of his own vocation. He would never have made either a metaphysician or a remarkable natural philosopher. He could never have endured the passionless impartiality of logic, nor the passionless objectivity of nature; he could not have renounced everything for the sake of thought, nor have renounced himself for the sake of observation; the doings of men, on the contrary, interested him passionately. And is not history the same thought and the same nature expressed in a different manifestation? Granovsky thought historically, learned from history, and later on made propaganda through history, while Stankevich, in a poetic way and for love, grafted on to him not only the theory of contemporary learning but also its method.

Pedants who estimate the labour of thought by the drops of sweat and the panting it has cost will doubt this. ... But, we would ask them, what about Proudhon and Belinsky? Had not they a better grasp, if only of the method of Hegel, than all the scholastics who studied it until they went bald and wrinkled? And yet neither of them knew German, neither of them had read one of Hegel's works, nor one of the dissertations of his followers of the left or right wing, but had only talked sometimes about his method with his disciples. ...

Granovsky's life in Berlin with Stankevich was, to judge from the stories of the one and the letters of the other, one of the most radiant periods of his existence, in which the exuberance of youth, of energy, of the first passionate impulses, of fun and irony without malice, went hand in hand with earnest intellectual work, all warmed and fostered by a deep, ardent friendship such as is only found in youth.

Two years later they parted. Granovsky went to Moscow to take the Chair of History at the University; Stankevich went to Italy for his health, and died of consumption. The death of Stankevich was a great shock to Granovsky. Long afterwards in my presence he received a medallion of his dead friend; I have rarely seen such quiet, speechless, overwhelming sorrow.

This happened soon after his marriage. The harmony that smoothly and calmly surrounded his new life was overcast with mourning crape. It was long before the traces of this blow passed away – indeed, I do not know whether they ever passed entirely.

His wife was very young and hardly yet formed; she retained that peculiar element of youthful awkwardness, even of apathy, which is not infrequently met with in young girls with flaxen hair, especially if they are of German descent. These natures, often gifted and strong, are slow to wake and come late to full consciousness when they awaken. The shock that had awakened the young girl had been so tender and so free from pain and conflict, had come so early, that she had scarcely noticed it. Her blood still flowed slowly and serenely through her heart.

Granovsky's love for her was a quiet, gentle affection, rather deep and tender than passionate. There was something peaceful and touchingly quiet in the atmosphere of their youthful household. It did the heart good to see at times, next to Granovsky who was engrossed in his work, the tall, willowy figure of his silent companion, deeply in love and happy. As I looked at them I used to think of the serene, chaste families of the early Protestants who fearlessly sang forbidden psalms, ready to go hand in hand, calmly and firmly, to face the inquisitor.

They seemed to me like brother and sister, the more so since they had no children.

We quickly became friends and saw each other almost every day; we sat through the nights until dawn talking of everything under the sun. . . . It is during these wasted hours, and through them, that people grow together inseparably and irrevocably.

It is frightening and painful to me to think that later on Granovsky and I were for a long time at variance over theoretical convictions; but to us they were not something extraneous but the real foundation of our lives. But I hasten to declare in advance that, if time proved that we could have separate conceptions, could fail to understand and could wound each other, subsequent time has doubly proved that we could neither part nor cease to be friends: that to accomplish this even death was powerless.

It is true that much later a streak of unkindness,[15] over and above theoretical disagreement, thrust itself between Granovsky and Ogarëv, who loved each other ardently and deeply; but we shall see that it too was effaced – late perhaps, but completely.

As for our disputes, Granovsky himself put an end to them; he concluded a letter from Moscow to me at Geneva on 25 August 1849, with the following words. With pride and reverence I repeat them:

'What was best and strongest in my soul has gone into my affection for you two' (that is for Ogarëv and me). 'There is in it something of the passion which in 1846 set me weeping and blaming myself for being unable to break a tie which apparently could not last. Almost with despair I discovered that you were bound fast to my soul with threads which I could not cut without tearing away the living flesh. This interval has not been profitless to me. I have come out of it victorious over the *worse side* of myself. *Of the romanticism for which you blamed me not a trace is left.* To make up for it, all that was romantic in my very nature has gone into my personal attachments. Do you remember my letter about your *Krupov*?[16] It was written on a night that I well remember. A black shroud dropped off my soul, your image rose up before me in all its clarity, and I stretched out my hand to you in Paris as easily and lovingly as I held it out in the happy, holy minutes of our life in Moscow. It was not your talent only that had so great an effect on me. That piece of writing brought the whole of you back to me with a rush. Once you wounded me by saying: "Don't build anything on the personal: believe only in the general," and I always set so much store by the personal. But for me personal and general are blended in you: that is why I love you so warmly and completely.'

Let these lines be remembered when my account of our differences is read. . . .

At the end of 1843 I published my articles on 'Dilettantism in Science'. Their success was a source of childlike pleasure to Granovsky. He used to drive from one house to another with *Notes of the Fatherland*, used to read them aloud himself with comments, and was seriously vexed if anybody did not like them. After that it was my lot to see Granovsky's success, and a success of a very different order. I am speaking of his first public lectures on the 'Medieval History of France and England'.

'Granovsky's lectures,' Chaadayev said to me as we came away from the third or fourth, out of a lecture-hall packed to overflowing with ladies and all the aristocratic society of Moscow, 'are of historical significance.' I entirely agreed with him. Granovsky turned the lecture-hall into a drawing-room, a place for meeting, for social intercourse of the *beau monde*. To do this he did not deck our history in laces and silks; quite the contrary: his language was severe, extremely grave, full of force, daring, and poetry, which vigorously jolted his hearers and woke them up. He escaped the consequences of his boldness, not from any compromise he made but from the mildness of expression which was natural to him, from the absence of sentences *à la française*, which put huge dots on tiny i's, like the moral after a fable. As he laid the events of history before his audience, grouping them artistically, he spoke *in them* so that the thought, unuttered but perfectly clear, was the more readily assimilated by his hearers that it seemed to be their own thought.

The conclusion of his first course of lectures was a regular ovation, a thing unheard of at Moscow University. When at the end, deeply moved, he thanked the audience, everyone leapt up in a kind of intoxication, ladies waved their handkerchiefs, others rushed to the platform, pressed his hands and asked for his portrait. I myself saw young people with flushed cheeks shouting through their tears: 'Bravo! Bravo!' There was no possibility of getting out. Granovsky stood as white as a sheet, with his arms folded and his head a little bent; he wanted to say a few words more but could not. The applause, the shouting, the fury of approbation doubled, the students ranged themselves on each side of the stairs and left the visitors to make a noise in the lecture-room. Granovsky made his way, exhausted, to the council-room; a few minutes later he was seen leaving it, and again there was endless clapping; he turned, begging for mercy with a gesture and, ready to drop with emotion, went into the office. There I flung myself on his neck and we wept in silence. . . .

Tears as happy flowed down my cheeks when the hero Ciceru-acchio,[17] in the Coliseum illuminated by the last rays of the setting sun, dedicated his youthful son to the Roman people, who had risen in armed insurrection, a few months before they both fell shot without trial by the military executioners of the urchin[18] who wore a crown!

Yes, those were precious tears; the first, born of my faith in Russia, the second, of my faith in the Revolution!

Where is that Revolution? Where is Granovsky? Gone together with the boy with the black curls, the broad-shouldered *popolano*, and the others who were so near and dear to us. My faith in Russia is still left. Surely it will not be my lot to lose that too?

And why did blind chance carry off Granovsky, so noble and so active, that deeply suffering spirit, on the very threshold of a new age for Russia, as yet obscure, but different, at all events? Why did not chance let him breathe that fresh air of which we have had a breath and which does not smell so strongly of the torture-chamber and the barracks?

The news of his death was a hard blow for me. I was on my way to the railway station at Richmond when the letter was given to me. I read it as I walked along, and truly at first I did not understand it. I got into the railway carriage. I did not want to read the letter again: I was afraid of it. Strangers with stupid, ugly faces kept coming in and going out, the engine whistled and I looked at it all and thought: 'But it is absurd! What? That man in the flower of his age, whose smile, whose look is before my eyes now – can he be no more? ...' I was overcome by a heavy torpor and felt fearfully cold. In London I met A. Talandier;[19] after greeting him I said I had received a letter with bad news and, as though I had only just heard it myself, I could not restrain my tears.

We had had little intercourse recently, but I needed to know that there, far away, in our native land, that man was living!

Without him Moscow was empty, another tie was snapped! ... Shall I alone, far away from all, ever be able to visit his grave – it has covered up as much strength, as much of the future, as many thoughts, as much love and life, as another, not quite unknown to him, which I have visited!

. . .

Granovsky was not persecuted; Nicholas's *oprichnina*[20] halted before his glance of mournful reproach. He died surrounded by the love of the younger generation, the sympathy of all cultivated Russia, recognised even by his enemies. Nevertheless I adhere to my expression, yes, he knew much suffering. It is not only chains of iron that wear life away; in the one letter Chaadayev wrote to me abroad (20 July 1851), he tells of how he is perishing, growing feeble and with rapid steps approaching the end – 'not from the oppression against which men revolt, but from that which they endure with a touching resignation, and which for that very reason is even more pernicious.'

Before me lie three or four letters which I received from Granovsky in later years; what a consuming, deadly sadness there is in every line!

'Our situation,' he writes in 1850, 'grows more insufferable every day. Every progressive movement in Western Europe has an after-effect here in some repressive measure. People are being denounced by thousands. They have twice started to get up a case against me during the last three months. But what does personal danger matter in comparison with the universal oppression and suffering? It was proposed to shut the universities, but for the present they have confined themselves to the following measures: they have raised the students' fees, and diminished their number by a law according to which there may not be more than three hundred students at a university. In Moscow University there are fourteen hundred students, so we must expel twelve hundred to have the right to admit a hundred new ones. The Institute of Nobility is closed; many institutions are threatened with the same fate, the Lycée for instance. Despotism is crying aloud that it cannot live in harmony with enlightenment. New programmes have been drawn up for the Cadet Schools. The Jesuits might envy the military pedagogue who drew up this programme. The priest is instructed to instil into the cadets that the greatness of Christ was comprised pre-eminently in submission to authority. He is depicted as a model of submission and discipline. The teacher of history is to unmask the trumpery virtues of the ancient republics and to bring out the grandeur of the Roman Empire, which has not yet been understood by historians, and which lacked only one thing, hereditary succession! ...

'It is enough to drive one mad. It is a blessing for Belinsky that he died in time. Many decent people have sunk into despair and look with blank apathy at what is being done – when will this world fall to pieces?

'I have made up my mind not to resign, but to wait at my post for the achievement of the fates. I can do a little; let them turn me out themselves.'

* * *

One of his last letters he ends like this: 'On all sides a general, obscure murmur can be heard, but where are the forces? where is the resistance? It is painful, brother – and there is no escape *in this life.*'

In our North the savage autocracy wears men out quickly. With a pang of dread I look back – it is like a battlefield: there lie the dead and the maimed. . . .

Granovsky was not alone: he was one of a group of young professors who came back from Germany while we were in exile. They did a great deal for the advancement of Moscow University, and history will not forget them. Men of conscientious erudition, they were pupils of Hegel, Gans, Ritter,[21] and others, just at the period when the dry bones of dialectic began to be clothed with flesh, when learning ceased to consider itself antagonistic to life, when Gans used to come to his lectures not with an ancient folio in his hand, but with the latest number of a review from Paris or London. They were trying at that time to solve historical questions of the day by the dialectic method; it was an impossible task, but it put the facts in a clearer light.

Our professors brought with them their cherished dreams, their ardent faith in learning, and in men: they preserved all the fire of youth, and the lecturer's chair was for them a sacred lectern from which they were called to preach the truth. They took their stand in the lecture-room not as mere professional savants, but as missionaries of the religion of humanity.

And what has become of that *Pléiade* of young professors, beginning with the best of them, Granovsky? Dear Kryukov, brilliant, intelligent, learned, died at thirty-five. Pechërin, the Hellenic scholar, struggled and struggled in the terrible conditions of Russian life till, unable to endure it, he went away without aim, without means, ill and shattered, to foreign lands, wandered homeless and forlorn, became a Jesuit priest and is burning Protestant Bibles in Ireland.[22] Redkin became a secular monk, goes on with his work at the Ministry of Home Affairs, and writes divinely inspired articles, interspersed with texts. Krylov[23] – but enough. *La toile! La toile!*

CHAPTER VI

OUR 'OPPONENTS'

> 'Yes, we were their opponents, but very strange ones. We
> had the same love. but not the same way of loving – and
> like Janus or the two-headed eagle we looked in different
> directions, though the heart that beat within us was but
> one.' The Bell, sheet 90, 15 January 1861. (On the death
> of K. S. Aksakov.)

1

SIDE by side with our circle were our opponents, *nos amis les
ennemis*, or more correctly, *nos ennemis les amis*[1] – the Moscow
Slavophils.

The conflict between us ended long ago and we have held out
our hands to each other; but in the early 'forties we could not but
be antagonistic – without being so we could not have been true to
our principles. We might have been able not to quarrel with them
over their childish homage to the childhood of our history; but
accepting their Orthodoxy as meant in earnest, seeing their ecclesiastical intolerance on both sides – in relation to learning and in
relation to sectarianism – we were bound to take up a hostile
attitude to them. We saw in their doctrines fresh oil for anointing
the Tsar, new chains laid upon thought, new subordination of
conscience to the servile Byzantine Church.

The Slavophils are to blame for our having so long failed to
understand either the Russian people or its history; their icon-
painter's ideals and incense smoke hindered us from seeing the
realities of the people's existence and the foundations of village life.

The Orthodoxy of the Slavophils, their historical patriotism and
over-sensitive, exaggerated feeling of nationality were called forth
by the extremes on the other side. The importance of their outlook, what was true and essential in it, lay not in Orthodoxy, and
not in exclusive nationalism, but in those elements of Russian
life which they unearthed from under the manure of an artificial
civilisation.

· · ·

Their passionate and generally polemical character developed specially in consequence of the appearance of Belinsky's critical articles; and even before that they had had to close their ranks and take a definite stand on the appearance of Chaadayev's *Letter* and the commotion it caused.

That *Letter* was in a sense the last word, the limit. It was a shot that rang out in the dark night; whether it was something foundering that proclaimed its own wreck, whether it was a signal, a cry for help, whether it was news of the dawn or news that there would not be one – it was all the same: one had to wake up.

What, one may wonder, is the significance of two or three pages published in a monthly review? And yet such is the might of speech, such is the power of the spoken word in a land of silence, unaccustomed to free speech, that Chaadayev's *Letter* shook all thinking Russia. And well it might. There had not been one literary work since Woe from Wit[2] which made so powerful an impression. Between that play and the *Letter* there had been ten years of silence, the 14th of December, the gallows, penal servitude, Nicholas. The Petrine period was broken off at both ends. The empty place left by the powerful men who had been exiled to Siberia had not been filled. Thought languished: men's minds were working, but nothing was yet attained. To speak was dangerous, and indeed there was nothing to say; suddenly a mournful figure quietly rose and asked for a hearing in order calmly to utter his *lasciate ogni speranza*.

In the summer of 1836 I was sitting quietly at my writing-table in Vyatka when the postman brought me the latest number of the *Telescope*. One must have lived in exile and in the wilds to appreciate a new book. I abandoned everything, of course, and set to work to cut *The Telescope*. I saw 'Philosophical Letters', written to a lady, unsigned. In a footnote it was stated that these letters had been written by a Russian in French, that is, that it was a translation. This put me against them rather than for them, and I proceeded to read the 'criticism' and the 'miscellany'.

At last the turn came for the *Letter*; from the second or third page I was struck by the mournfully earnest tone. Every word breathed of prolonged suffering, which by now was calmer, but was still bitter. It was written as only men write who have been thinking for many years, who have thought much and learned much from life and not from theory. . . . I read further: the letter grew and developed, it turned into a dark denunciation of Russia, the protest of one who, in return for all he has endured, longs to utter some part of what is accumulated in his heart.

Twice I stopped to take breath and collect my thoughts and feelings, and then again I read on and on. And this was published in Russian by an unknown author. . . . I was afraid I had gone out of my mind. Afterwards I read the *Letter* aloud to Vitberg, then to Skvortsov, a young teacher in the Vyatka High School; then I read it again to myself.

It is most likely that exactly the same thing was happening in various provincial and district capitals, in Moscow and Petersburg and in country gentlemen's houses. I learned the author's name a few months later.

Long cut off from the people, part of Russia had been suffering in silence under the most incapable and prosaic yoke, which gave them nothing in return. Everyone felt the oppression of it, everyone had something weighing on his heart, and yet all were silent; at last a man had come who in his own way told them what it was. He spoke only of pain; there was no ray of light in his words, nor indeed in his view. Chaadayev's *Letter* was a merciless cry of pain and reproach against Petrine Russia, which deserved the indictment; had it shown pity or mercy to the author or anyone else?

Of course such an utterance was bound to provoke opposition, or Chaadayev would have been perfectly right in saying that Russia's past was empty, its present insufferable, and that there was no future for it at all; that it was 'a *lacuna* of the intellect, a stern lesson given to the nations of the plight to which people can be brought by alienation and slavery'. This was both penitence and accusation; to know beforehand the means of reconciliation is not the business of penitence, nor the business of protest – or consciousness of guilt becomes a jest, and expiation insincere.

But it did not pass unnoticed; for a minute everyone, even the drowsy and the stunned, recoiled in alarm at this ominous voice. All were astounded and most were offended, but a dozen men loudly and warmly applauded its author. Talk in the drawing-rooms anticipated government measures – provoked them. [...]

The review was at once prohibited; Boldyrev, the censor, an old man, and the Rector of Moscow University, was dismissed; Nadyezhdin the publisher was sent to Ust-Sysolsk; Nicholas ordered Chaadayev himself to be declared insane, and to be obliged to sign an undertaking to write nothing. Every Saturday he was visited by the doctor and the *politsmeyster*; they interviewed him and made a report, that is, gave out over their signature fifty-two false statements by the command of His Majesty – an intelligent and moral proceeding. It was they of course who were punished. Chaadayev looked with profound contempt on these tricks of the truly insane arbitrariness of power. Neither the doctor nor the *politsmeyster* ever hinted at what they had come for.

I had seen Chaadayev once before my exile. It was on the very day of Ogarëv's arrest. I have mentioned already that on that day there was a dinner-party at M. F. Orlov's.[3] All the guests were assembled when a man, bowing coldly, walked into the room. His unusual appearance, handsome, with a striking air of independence, was bound to attract everyone's attention. Orlov took me by the hand and introduced me: it was Chaadayev. I remember little of that first meeting; I had no thoughts to spare for him; he was as always cold, grave, clever, and malicious. After dinner Mme Rayevsky, Orlov's mother-in-law, said to me:

'How is it you are so sad? Oh you young people! I don't know what has come over you in these days.'

'Then you do think,' said Chaadayev, 'that there still are young people in these days?'

That is all that has remained in my memory.

On my return to Moscow I made friends with him and from that time until I went away we were on the best of terms.

Chaadayev's melancholy and peculiar figure stood out sharply like a mournful reproach against the faded and dreary background of Moscow 'high life'.[4] I liked looking at him among the tawdry aristocracy, feather-brained Senators, grey-headed scapegraces, and venerable nonentities. However dense the crowd, the eye found him at once. The years did not mar his graceful figure; he was very scrupulous in his dress, his pale, delicate face was completely motionless when he was silent, as though made of wax or of marble – 'a forehead like a bare skull,'[5] – his blue-grey eyes were melancholy and at the same time there was something kindly in them, though his thin lips smiled ironically. For ten years he stood with folded arms, by a column, by a tree on the boulevard, in drawing-rooms and theatres, at the club and, an embodied veto, a living protest, gazed at the vortex of faces senselessly whirling round him. He became whimsical and eccentric, held himself aloof from society, yet could not leave it altogether, then uttered his message, which he had quietly concealed, just as in his features he concealed passion under a skin of ice. Then he was silent again, again showed himself whimsical, dissatisfied, irritated; again he was an oppressive influence in Moscow society, and again he could not leave it. Old and young alike were awkward and ill at ease with him; they were abashed, God knows why, by his immobile face, his direct gaze, his mournful mockery, his malignant condescension. What made them receive him, invite him ... still more, visit him? It is a very difficult question.

Chaadeyev was not wealthy, particularly in his later years; he was not eminent – a retired captain of cavalry with the iron Kulm cross[6] on his breast. It is true, as Pushkin writes, that he would

> 'In Rome have been a Brutus,
> In Athens Pericles,
> But here, under the yoke of Tsars,
> Was only Captain of Hussars.'[7]

Acquaintance with him could only compromise a man in the eyes of the ruling police. To what did he owe his influence? Why did the 'swells' of the English Club, and the patricians of the Tverskoy Boulevard flock on Mondays to his modest little study in Old Basmannaya Street? Why did fashionable ladies gaze at the cell of the morose thinker? Why did generals who knew nothing about civilian affairs feel obliged to call upon the old man, to pretend awkwardly to be people of culture, and brag afterwards, garbling some phrase of Chaadayev's uttered at their expense? Why did I meet at Chaadayev's the savage Tolstoy 'the American',[8] and the savage Adjutant-General Shipov[9] who destroyed culture in Poland?

Chaadayev not only made no compromise with them, but worried them and made them feel very clearly the difference between himself and them. Of course these people went to see him and invited him to their gatherings from vanity, but that is not what matters; what is important is the involuntary recognition that thought had become a power, that it had its honoured place in spite of His Majesty's command. In so far as the authority of the 'insane' Captain Chaadayev was recognised, the 'insane' power of Nicholas Pavlovich was diminished.

. . .

Chaadayev and the Slavophils alike stood facing the unsolved Sphinx of Russian life, the Sphinx sleeping under the overcoat of the soldier and the watchful eye of the Tsar; they alike were asking: 'What will come of this? To live like this is impossible: the oppressiveness and absurdity of the present situation is obvious and unendurable – where is the way out?'

'There is none,' answered the man of the Petrine epoch of exclusively Western civilisation, who in Alexander's reign had believed in the European future of Russia. He sadly pointed to what the efforts of a whole age had led to. Culture had only given new methods of oppression, the church had become a mere shadow under which the police lay hidden; the people still tolerated and endured, the government still crushed and oppressed. 'The history of other nations is the story of their emancipation. Russian history is the development of serfdom and autocracy.' Peter's upheaval made us into the worst that men can be made into – *enlightened* slaves. We have suffered enough, in this oppressive, troubled moral condition, misunderstood by the people, struck down by the government – it is time to find rest, time to bring peace to one's soul, to find something to lean on . . . this almost meant 'time to die', and Chaadayev thought to find in the Catholic Church the rest promised to all that labour and are heavy laden.

From the point of view of Western civilisation in the form in which it found expression at the time of restorations, from the point of view of Petrine Russia, this attitude was completely justified. The Slavophils solved the question in a different way.

Their solution implied a true consciousness of the *living soul* in the people; their instinct was more penetrating than their reasoning. They saw that the existing condition of Russia, however oppressive, was not a *fatal disease*. And while Chaadayev had a faint glimmer of the possibility of saving individuals, but not the people, the Slavophils had a clear perception of the ruin of individuals in the grip of that epoch, and faith in the salvation of the people.

'The way out is with us,' said the Slavophils, 'the way out lies in renouncing the Petersburg period, in going back to the people from whom we have been separated by foreign education and foreign government; let us return to the old ways!'

But history does not turn back; life is rich in materials, and never needs old clothes. All reinstatements, all restorations have always been masquerades. We have seen two; the Legitimists did not go back to the days of Louis XIV nor the Republicans to the 8th of Thermidor. What has once happened is stronger than anything written; no axe can hew it away.

More than this, we have nothing to go back to. The political life of Russia before Peter was ugly, poor and savage, yet it was to this that the Slavophils wanted to return, though they did not admit the fact; how else are we to explain all their antiquarian revivals, their worship of the manners and customs of old days, and their very attempts to return, not to the existing (and excellent) dress of the peasants but to the clumsy, antiquated costumes?

In all Russia no one wears the *murmolka* but the Slavophils. K. S. Aksakov wore a dress so national that people in the street took him for a Persian, as Chaadayev used to tell for a joke.

They took the return to the people in a very crude sense too, as the majority of Western democrats did also, accepting the people as something complete and finished. They supposed that sharing the prejudices of the people meant being at one with them, that it was a great act of humility to sacrifice their own reason instead of developing reason in the people. This led to an affectation of devoutness, the observance of rites which are touching when there is a naïve faith in them and offensive when there is visible premeditation. The best proof of the lack of reality in the Slavophils' return to the people lies in the fact that they did not arouse in them the slightest sympathy. [...]

. . .

2

On my return from Novgorod to Moscow I found both parties at the barricades. The Slavophils were in full fighting order, with their light cavalry under the leadership of Khomyakov and extremely heavy infantry under that of Shevyrëv and Pogodin,[10] with their sharp-shooters, chasseurs, and ultra-Jacobins, who rejected everything that had existed later than the Kiev period, and the moderate Girondists who rejected only the Petersburg period;

they had their own chairs at the university and their own monthly review, which always came out two months late, but still did come out. The main body was composed of orthodox Hegelians, Byzantine theologians, mystical poets, a great number of women, and so on.

Our warfare greatly interested the literary *salons* of Moscow. Moscow in general was at that time entering the period of enthusiasm for intellectual subjects when, political questions being impossible, literary ones become the problems of life. The appearance of a remarkable book, *Dead Souls*,[11] for instance, was an event. Criticisms favourable and unfavourable were read and commented upon with the attention with which parliamentary debates *used to be* followed in England or France. The suppression of all other spheres of human activity threw the cultured part of society into the world of books, and only in it did there really occur, muffled and in undertones, the protest against the oppression of Nicholas, the protest which we heard more loudly and openly on the day after his death.

In the person of Granovsky Moscow society welcomed Western thought forcing itself towards freedom, the idea of intellectual independence and the struggle for it. In the persons of the Slavophils it protested against the outrage done to its feelings of nationalism by the Biron-like arrogance of the Petersburg government.

Here I must make a reservation.

I knew two circles in Moscow, the two opposite poles of its social life, and can speak only of them. At first I got lost in the society of old people, officers of the Guards in Catherine's time who were father's comrades, and other old gentlemen who had found a quiet refuge in the tolerant, hospitable Senate, who were his brother's. Afterwards I knew only *young* Moscow, literary, fashionable Moscow, and I speak only of it. I knew nothing and cared to know nothing of what lived and germinated in the space between the veterans of the pen and the sword, who were waiting for funerals suitable to their rank, and their sons and grandsons, who sought no rank and cared only for 'books and ideas'. That world that stood between them, the real Russia of Nicholas, was colourless and vulgar, without the idiosyncrasy of the age of Catherine, without the dash and daring of the men of 1812, without our aspirations and interests. It was a pitiful, crushed generation in which a few martyrs struggled, were suffocated and perished. When I speak of the Moscow *salons* and dining rooms, I speak of those in which A. S. Pushkin once reigned supreme; in which up to our own day the Decembrists set the tone; in which Griboyedov laughed; in which M. F. Orlov and A. P. Yermolov met a friendly welcome because they were under the ban; those, finally, in which Khomyakov started arguing at nine in the evening and went on till four in the morning; in which K. S. Aksakov with a *murmolka* in his hand furiously defended Moscow, though no one had attacked it, and never took a goblet of champagne in his hand without repeating secretly a prayer and a toast which everyone knew; in which Redkin logically deduced a personal God *ad majorem gloriam Hegeli*; in which Granovsky appeared with his firm and gentle speech; in which everyone remembered Bakunin and Stankevich; in which Chaadayev with his delicate, wax-like face, scrupulously dressed, enraged the panic-stricken aristocrats and Orthodox Slavophils with his biting remarks, always cast in an original mould and intentionally iced; in which A. I. Turgenev,[12] young in spite of his age, gossiped nicely about all the celebrities of Europe, from Chateaubriand and Récamier to Schelling and Rahel Varnhagen[13]; in which Botkin and Kryukov *pantheistically* enjoyed M. S. Shchepkin's stories; and into which, finally, Belinsky sometimes fell like one of Congreve's rockets, setting fire to anything handy.

Life in Moscow is on the whole more rustic than urban, only the gentlemen's houses are nearer each other. Not everything in it has the same denominator, but specimens of varying periods, cultures, social strata and latitudes and longitudes in Russia, live after their own fashion. In it the Larins[14] and Famusovs calmly live out their days; and not only they but Vladimir Lensky and our eccentric Chatsky – of Onegins there have been even too many. With little to do they all lived without haste, with no particular worries, their sleeves not rolled up. The easy-going ways of the Russian country gentleman are dear to our hearts, we must own; there is a spaciousness of their own about them which we do not find in the *petit-bourgeois* life of the West. The obsequious time-serving, of which Miss Wilmot speaks in the *Memoirs* of Princess Dashkov, and which I still came across myself, did not exist in the circles of which I am speaking. The rank and file of this society was composed of landowners not in the service, or serving not on their own account but to pacify their relations, and of young literary men and professors. This society had the fluidity of relationships not yet settled and of habits not reduced to a sluggish orderliness, a freedom which is not found in the more ancient life of Europe; and at the same time there is preserved in our society the tradition of Western politeness grafted on to us by our education and now vanishing in the West; this courtesy, blended with the Slav *laisser aller*, and sometimes dissipation, composed the special Russian nature of *Moscow* society, to its great grief because it was mortally keen to be *Parisian*, and probably still is.

We still only know of Europe from back numbers; we are still haunted by the days when Voltaire reigned over the Parisian *salons* and one was invited to hear Diderot arguing, as if to partake of a sturgeon. [...]

· · ·

They say that Moscow – young Moscow – has grown old, has not survived Nicholas; that even the university has degenerated, and that the landowning temper has come out in too strong relief in face of the question of emancipation; that its English Club has become less English than ever; that in it the Sobakeviches[15] are clamouring against emancipation and the Nozdrëvs noisily maintaining the natural and inalienable rights of the nobility. Perhaps! ... But the Moscow of the 'forties was not like that, and it was that Moscow that actively participated for and against the *murmolka*; girls and ladies read very boring articles, listened to very long discussions, and themselves argued in defence of Konstantin Aksakov or Granovsky, *only* regretting that Aksakov was too Slavophil and Granovsky insufficiently patriotic.

The arguments were renewed at every literary and non-literary evening at which we met, and that was two or three times a week. On Mondays we assembled at Chaadayev's, on Tuesdays at Sverbeyev's,[16] on Sundays at Mme A. P. Yelagin's.[17]

Besides those who took part in the arguments, besides the people who had opinions, *amateurs* and even *amatrices* would come to these evenings and sit until two o'clock in the morning to see which of the matadors would dispatch which, and how he would be dispatched himself; they came as in old days people used to go to prize-fights, and to the amphitheatre behind the Rogozhsky Gate.

The Ilya of Murom,[18] the *bogatyr* who, on the side of Orthodoxy and Slavophilism, struck down everyone, was Alexey Stepanovich Khomyakov, 'Gorgias the disputer of this world',[19] to use the expression of the half-crazy Moroshkin.[20] A man of powerful and mobile intelligence, rich in resources and indiscriminate in the use of them, fitted with a good memory, and the power of rapid reflection, he spent his whole life in heated and indefatigable argument. An unwearying and unresting fighter, he cut and thrust, attacked and pursued, pelted with witticisms and quotations, frightened his opponents and drove them into an enchanted forest from which there was no escape without saying a prayer – in short, whose conviction soever he attacked the conviction was done for, whose logic soever he attacked, the logic was done for.

Khomyakov really was a dangerous opponent; a hardened old duellist of dialectics, he took advantage of the slightest inadvertence, the slightest concession. An extraordinarily gifted man, with formidable stores of erudition at his disposal, he was like the medieval knights who guarded the Madonna and slept fully armed. At any hour of the day or the night he was ready for the most intricate argument, and to secure the triumph of his Slavophil views turned everything in the world to use, from the casuistry of Byzantine theologians to the subtleties of a shifty lawyer. His refutations, often only apparent, always dazzled and confounded his opponent.

Khomyakov was very well aware of his strength, and played with it; he lapidated people with words, intimidated them by his learning, mocked everything, made a man laugh at his own beliefs and convictions, leaving him in doubt whether he himself really had anything left which was sacred. In masterly fashion he caught those who had halted half-way and roasted them on the dialectical grid-iron, frightened the timid, reduced the dilettante to despair, and, with all this, laughed, *as it seemed*, simply and candidly. I say 'as it seemed' because there was in his somewhat Oriental features a look as of something concealed and a sort of artless Asiatic cunning together with the Russian canniness. On the whole he rather confused than convinced.

His philosophical contentions consisted in his rejecting the possibility of attaining truth by reason; he attributed to reason a formal faculty only, the faculty of developing embryos, or seeds, received in other ways and relatively complete (that is, imparted by revelation and accepted through faith). If reason is left to itself, then, wandering in empty space, and building category after category, it may reveal its own laws, but will never reach the conception of the spirit, nor the conception of immortality – and so on. On this basis Khomyakov knocked out people who halted between religion and science. However they struggled in the forms of the Hegelian method, whatever constructions they made, Khomyakov went with them step by step and just before the end blew down the house of cards built of logical formulas or tripped them up and sent them falling into 'materialism' which they shamefacedly renounced, or into 'atheism' of which they were simply afraid. Khomyakov triumphed!

Since I had several times been present while he was arguing, I noticed this trick, and the first time that it was my lot to try my strength with him I myself enticed him into these deductions. Khomyakov screwed up his slanting eyes, shook his pitch-black curls, and smiled in anticipation.

'Do you know,' he said suddenly, as though himself surprised by a new idea, 'it is not merely impossible by reason alone to arrive at a rational spirit developing in nature, but by reason alone you can reach no other conception of nature than that of a simple, uninterrupted fermentation which has no aim and may either go on or come to a stop. And if that is so, you also cannot prove that history will not break down tomorrow, will not perish together with the human race, together with the planet.'

'I didn't say,' I answered, 'that I undertook to prove this. I know very well that it is impossible.'

'What?' said Khomyakov, somewhat surprised, 'you can accept these frightening results of the cruel theory of immanence, and nothing in your soul revolts?'

'I can, because the deductions of reason are independent of whether I desire them or not.'

'Well, you are consistent, at any rate. But what violence a man must do to his soul to resign himself to these dismal deductions of your science, and to accustom himself to them.'

'Prove that your non-science is more true, and I will accept it as frankly and fearlessly, whatever it may lead me to, even to the Iverskaya Madonna.'

'For that you must have faith.'

'But Alexey Stepanovich, you know the saying: "There's no doing impossibilities." '

Many people thought – indeed I sometimes did myself – that Khomyakov argued from an artistic need for argument, that he had no deep convictions; and for this his manner, his everlasting laugh, and the superficiality of his critics were responsible. I do not think that any one of the Slavophils did more to spread their theories than Khomyakov. His whole life – and he was a very wealthy man and not in the service – was devoted to propaganda. Whether he laughed or wept depended on his nerves, on the cast of his mind, on how he had been formed by his environment and how he reflected it; this has nothing to do with depth of conviction.

Perhaps in continual preoccupation with the trivial activity of discussion, and the busy idleness of polemic, Khomyakov suppressed the feeling of emptiness which, for its part, suppressed everything luminous in his comrades and nearest friends, the Kireyevskys.

That these people were crushed and torn to pieces by the age of Nicholas was obvious. In the heat of polemical argument one might sometimes forget this – to do so now would be feeble and pitiful.

The two Kireyevsky brothers stand like mournful shades on the border of the national renaissance; not recognised by the living, not sharing their interests, they did not take off their shrouds.

The prematurely aged face of Ivan Kireyevsky bore deeply bitten traces of the suffering and conflict which had been followed by the mournful calm of the sea-swell above a foundered ship. His life was a failure. He threw himself with ardour – in 1833, if I remember rightly – into a monthly review, *The European*. The two numbers that appeared were excellent, but on the publication of the second *The European* was prohibited. [...] Kireyevsky, who had lost a great deal of his fortune over *The European*, dismally whiled away his time in the desert of Moscow life: nothing presented itself there; he could not stand it, and went away to the country, burying in his heart profound distress and yearning for activity. This man, too, firm and true as steel, was consumed by the rust of that terrible time. Ten years later he went back to Moscow from his hermit's life, a mystic and an Orthodox believer.

His situation in Moscow was a hard one. He found no complete intimacy or sympathy either in his friends or in us. Between him and us stood the church wall. A worshipper of liberty and of the great age of the French Revolution, he could not share the disdain of the new Old Believers for everything European. He once said with intense sadness to Granovsky:

'In heart I am closer to you, but I do not share many of your convictions; to our people I am nearer in belief, but just as far from them in other things.'

And really his life was going out, alone in his own family.[21] Beside him stood his brother and friend, Pëtr. Sadly, as though their tears were not yet dried, as though misfortune had visited them the day before, both the brothers appeared at parties and assemblies. I looked at Ivan Kireyevsky as at a widow, or a mother who had been bereft of her son; life had deceived him, and ahead all was emptiness, and the only consolation:

> 'Wait a little,
> Thou too shalt rest!'[22]

One was sorry to disturb his mysticism. I used to feel the same scruple in the old days with Vitberg. The mysticism of both was romantic; it was as though the truth had not disappeared behind it, but was hiding in fantastic outlines and monkish cassocks. One only feels a ruthless need to wake a man up when he invests his madness in a polemical uniform or when one's affection for him is so great that any dissonance rends the heart and gives one no peace.

And what argument could one use to a man who said things like this: 'I once stood in a chapel and gazed at a wonder-working icon of the Mother of God, thinking of the childlike faith of the people praying to it; some women, sick people and old men knelt, crossing themselves and bowing down to the earth. With burning hope I then gazed at the holy features, and little by little the secret of their marvellous power began to grow clear to me. Yes, this was not simply a painted board ... for whole ages it had absorbed these streams of passionate exaltation, the prayers of the afflicted and unhappy; it must have become filled with power that pours from it, that is reflected from it upon the faithful. It had become a living organism, a meeting place between the Creator and men. Thinking of this, I looked once more at the old men, at the women and children prostrate in the dust, and at the holy icon – then I myself saw the features of the Mother of God suffused with life; she looked with love and clemency at these simple folk ... and I sank on my knees and meekly prayed to her.'

Pëtr Kireyevsky was even more incorrigible and went even further in Orthodox Slavophilism; his was perhaps a less gifted nature, but he was single-minded and strictly consistent. He did not, like his brother Ivan or the Slavophil Hegelians, try to reconcile religion with science, and Western civilisation with Muscovite nationalism; quite the contrary: he rejected all compromises. Firmly and independently he stood his ground, neither seeking quarrels nor avoiding them. He had nothing to fear: he was so irrevocably devoted to his opinion and so welded to it in sorrowful sympathy for the Russia of his day that his position was easy. It was as impossible to agree with him as with his brother; but it was easier to understand him, as it is easier to understand every ruthless extreme. He had discerned (and this I appreciated only long afterwards) some part of the bitter, crushing truths concerning the social condition of Western Europe which we only came to see after the upheavals of 1848. He perceived them with melancholy clear-sightedness, divined them through hatred and resentment for the evil wrought by Peter in the name of Western civilisation. That is why Pëtr Kireyevsky had not, as his brother had, together with his Orthodoxy and Slavophilism, yearnings towards some humane and religious philosophy in which his lack of faith in the present would be resolved. No, his austere nationalism involved complete, final estrangement from all that was Western.

It was their common misfortune that they had been born either too early or too late; the 14th of December came upon us as children, but them as young men. This made a great difference. At that time we were at our lessons, knowing nothing at all of what was really being done in the world of action. We were full of theoretical dreams, we were Gracchi and Rienzi in the nursery; afterwards, confined to a small circle, we spent our academic years in friendship together; as we passed out of the gates of the university we encountered the gates of a prison. Prison and exile in youth, in the grey and stifling days of persecution, are extremely beneficial; they are a hardening process; only feeble organisms are subdued by prison, those in whom struggle has been the passing impulse of youth and not a talent, not an internal need. To be the object of open persecution strengthens the desire for resistance;

danger when doubled trains for endurance and moulds the conduct. All this provides an interest, a distraction; it goads and angers; and the convict or the exile is more often prone to moments of rage than to the exhausting hours of listless, impotent despair of men who are at complete liberty but in a vulgar and oppressive environment.

When we came back from exile a new spirit was already stirring in the university, in literature, in society itself. Those were the days of Gogol and Lermontov, of Belinsky's articles and of the lectures of Granovsky and the young professors.

It had been very different with our predecessors; their premature coming of age was marked by the bell which tolled for the execution of Pestel[23] and pealed for the coronation of Nicholas; they were too young to take part in the conspiracy of the 14th of December, and not young enough to be at school after it. They were faced with the ten years which ended in Chaadayev's gloomy *Letter*. Of course they could not have grown much older in those ten years, but they were reined back and thrown down, surrounded by a society with no living interests, pitiful, cowardly, cringing. And those were the first ten years of manhood! Inevitably a man was driven, like Onegin, to envy the paralysis of the Tula assessor, to go to Persia like Lermontov's Pechorin, to become a Catholic like the real Pechërin, or to throw himself into desperate Orthodoxy or raging Slavophilism, if he had no desire to get drunk, to flog peasants, or to play cards.

At the first moment when Khomyakov was conscious of this emptiness he went for a tour in Europe, during the somnolent, boring reign of Charles X; after finishing in Paris his forgotten tragedy, *Yermak*, and talking to various Czechs and Dalmatians on the way home, he returned. Everything was boring! Fortunately the Turkish war broke out; he joined a regiment, quite unnecessarily, quite aimlessly, and went to Turkey. The war ended, and another forgotten tragedy, *Dmitry the Pretender*, was finished. Boredom again!

In this boredom, in this depression, in the midst of this fearful environment and a fearful emptiness a new thought flashed upon him: it was greeted with derision as soon as it was uttered; that only made Khomyakov fly the more furiously to the defence of it, and drove it more deeply into the very flesh and blood of the Kireyevskys.

The seed was scattered; their energies all went into the sowing and the guarding of the young crops. Men were needed of another generation, not whipped and distorted, by whom their thought would be accepted not through suffering and sickness, as the teachers had reached it, but by transmission, by inheritance. Young men responded to their summons, men of Stankevich's circle joined them, and among them were such powerful personalities as Konstantin Aksakov and Yury Samarin.

Konstantin Aksakov did not laugh like Khomyakov and was not wrapped in hopeless complaining like the Kireyevskys. Being a youth on the threshold of manhood, he threw himself with energy into the work. In his convictions there was no uncertain testing of the ground, no melancholy sense of being a voice crying in the wilderness, no dark aspiration,[24] no remote hopes, but a fanatical faith, intolerant, elbowing its way in, biased, that faith which goes before victory. Aksakov was partial, like every fighter; a calmly balanced eclecticism is no equipment for battle. He was surrounded by a hostile environment, a powerful environment that had great advantages over him; he had to fight his way through all sorts of enemies one after another, and to set up his standard. How could he be tolerant!

His whole life was an uncompromising protest against Petrine Russia, against the Petersburg epoch, in the name of the unrecognised, oppressed life of the Russian people. His dialectical powers were inferior to those of Khomyakov, and he was not a poet and thinker like Ivan Kireyevsky, but he was ready to go out into the market-place for his faith, he would have gone to the scaffold; and when that is felt behind a man's words they become fearfully convincing. Early in the 'forties he was preaching the village commune, the *mir*, and the workmen's guild. He taught Haxthausen[25] to understand them and, consistent to the point of childishness, was the first to put his trousers inside his high boots, and to wear a shirt with a collar fastened at the side.

'Moscow is the capital of the Russian people,' he used to say, 'but Petersburg is only the residence of the Emperor.'

'And observe,' I answered, 'to what lengths the difference goes – in Moscow they will infallibly put you in the lock-up, but in Petersburg they will take you to the *Hauptwache*.'

'To the end of his days Aksakov remained an everlastingly enthusiastic and boundlessly generous youth; he carried away and was carried away, but was always perfectly single-hearted. In 1844, when our differences had reached such a point that neither the Slavophils nor we wanted to go on meeting, I was walking along the street one day when Konstantin Aksakov drove up in a sledge. I bowed to him in a friendly way. He was on the point of driving by, but he suddenly stopped the coachman, got out of his sledge and came towards me.

"It hurt me too much," he said, "to pass you and not say goodbye. You understand that after all that has happened between your friends and mine I shan't be coming to see you; it's a pity, a pity; but there is no help for it. I wanted to shake you by the hand and say goodbye." He went quickly towards his sledge, but suddenly turned round. I was standing in the same place, and I was sad; he rushed up to me, embraced me and kissed me warmly. I had tears in my eyes. How I loved him at that moment of our quarrel!²⁶'

The quarrel in question was the result of the polemics of which I have spoken.

Granovsky and I still managed to get on with them somehow, without giving up our principles; we did not make a personal question of our difference of opinion. Belinsky, passionate in his intolerance, went further and bitterly reproached us. 'I am a Jew by nature,' he wrote to me from Petersburg, 'and cannot eat at the same table with the Philistines. ... Granovsky wants to know whether I have read his article in *The Muscovite*. No, and I am not going to read it; tell him I am not fond of meeting my friends in improper places, and I don't make appointments with them there.'

In return for this the Slavophils too abused him. *The Muscovite*, irritated by Belinsky, by the success of the *Notes of the Fatherland* and of Granovsky's lectures, used any weapon that came to hand in self-defence, and spared Belinsky least of all, speaking of him in so many words as a dangerous man who thirsted for destruction and 'rejoiced at the sight of a conflagration'.

The Muscovite, however, was pre-eminently the organ of the university doctrinaire set of the Slavophils. This set might be described not merely as the university, but to some extent as the *government* party. This was a great novelty in Russian literature. Among us servility either keeps quiet, takes bribes, and can barely read or write or, disdainful of prose, strikes chords on the lyre of a loyal subject.

Bulgarin and Grech[27] are in no way typical: no one was deceived by them and no one mistook the cockade of their livery for the badge of an opinion.

Pogodin and Shevyrëv, the editors of *The Muscovite*, quite on the contrary, were conscientiously servile: Shevyrëv I do not know why: possibly influenced by the examples of his ancestor, who, in the midst of the tortures and agonies of the reign of Ivan the Terrible, sang psalms and almost prayed for the ferocious old man's days to be prolonged; Pogodin – from hatred of the aristocracy.

There are periods at which thinkers are on the side of authority, but that is only when authority is progressive, as in the days of Peter I, is defending its country as in 1812, or is treating its wounds and letting it rest as in the reign of Henry IV of France and *perhaps* of Alexander II.[28] But to select the most arid and narrow epoch of Russian autocracy and, leaning upon Father Tsar, to take up arms against the individual misdeeds of the aristocracy, which is developed and supported by the authority of that same Tsar, is absurd and pernicious.

People say that under the aegis of devotion to the Imperial power the truth can be told more boldly. Why then did they not tell it?

Pogodin was a useful professor who appeared, with energy that was new and a Heeren[29] that was not, on the ash-heap of Russian history, which had been corroded and reduced to smoke and dust by Kachenovsky.[30] But as a writer he was of little importance, in spite of the fact that he wrote everything, even *Götz von Berlichingen*, in Russian. His rugged, unpolished style, his coarse way of throwing out gnawed, crop-eared remarks and unmasticated thoughts, inspired me somehow as in old days, and I wrote a parody of him, a little fragment of *Vëdrin's Notes of Travel*. Strogonov (the Director of Moscow University), said after reading it: 'I'm sure Pogodin imagines he really wrote it himself.'

It is doubtful whether Shevyrëv ever did anything at all as a professor. As for his literary articles, I do not remember a single original idea or a single independent opinion in anything he wrote. On the other hand his style was quite the opposite of Pogodin's, being windy, spongy, rather like a blancmange that has not set, in which the bitter almond flavouring has been forgotten, although under its treacle a mass of jaundiced, conceited irritability was fermenting. As one reads Pogodin one feels as though he were swearing, and one looks round to see whether there are ladies in the room. Reading Shevyrëv one slumbers and keeps dreaming of something quite different.

Speaking of the style of these Siamese twins of Moscow journalism inevitably reminds one of Georg Forster,[31] the celebrated companion of Captain Cook in the Sandwich Islands, and of Robespierre at the Convention of the one and indivisible Republic. Being professor of botany in Vilna and listening to Polish, so rich in consonants, Forster remembered his friends in Otaheite [Tahiti] who spoke almost entirely in vowel sounds and observed: 'If those two languages were mixed what a smooth and sonorous tongue it would make!'

However, badly as they wrote, the twins of *The Muscovite* began attacking not only Belinsky, but also Granovsky for his lectures, and always with the same unhappy lack of tact that set all decent people against them. They accused Granovsky of partiality for Western culture, for a certain 'order of ideas' for which Nicholas, from 'an idea of order', clapped men in fetters and sent them to Nerchinsk.

Granovsky took up their gauntlet, and his bold and noble reply made them blush. He asked his accusers publicly from the lecturer's platform why he ought to hate Western Europe, and if he did hate Western culture what inducement would he have to lecture on its history.

'I am accused,' said Granovsky, 'of using history merely as a means of expressing my own views. That is partly true; I have convictions and I bring them forward in my lectures. If I had none I should not appear before you in public simply in order, more or less interestingly, to describe a succession of events.'

Granovsky's answers were so simple and manly, and his lectures so attractive, that the Slavophil doctrinaires subsided, and the young people applauded no less than we. At the end of the course an effort was even made at reconciliation. We gave Granovsky a dinner after his final lecture. The Slavophils wanted to join us in it, and Yury Samarin was chosen by them (as I was by our side) as steward. The banquet was a success; at the end of it, after many toasts, drunk not only unanimously but as bumpers, we embraced the Slavophils and kissed them in the Russian style. Ivan Kireyevsky only begged of me one thing, that I would alter the spelling of my name, and by changing the *e* into a Slavonic vowel make it

more Russian to the ear. But Shevyrëv did not even insist on that;
on the contrary as he embraced me he repeated in his soprano: 'He
is a good man even with an *e*! he is a Russian even with an *e*!' On
both sides the reconciliation was genuine and without reserva-
tions, which, of course, did not prevent us from disagreeing more
than ever a week later.

Reconciliations as a rule are only possible when they are unnecessary, that is when personal exasperation is over, or when
opinions have drawn closer and when people see themselves that
they have nothing to quarrel about. Otherwise every reconciliation involves a reciprocal weakening; both sides fade: that is,
surrender their bright colours. Our attempt at a Kuchuk-
Kaynardzhi[32] very soon turned out to be impracticable, and the
conflict raged with fresh obduracy.

On our side it was impossible to rope in Belinsky; he sent us
threatening epistles from Petersburg, excommunicated and anathe-
matised us, and wrote more angrily than ever in the *Notes of the
Fatherland*. At last he pointed a triumphant finger at the 'dodges'
of Slavophilism and repeated reproachfully, 'there you have
them,' while we hung our heads in contrition. Belinsky was
right!

A poet,[33] at one time a favourite, who became a Slavophil
through family connections and a sanctimonious bigot through
illness, tried with his dying hand to have a lash at us; unluckily
the police whip was again the means he chose for the purpose. In a
play entitled *Our Opponents*, he called Chaadayev a renegade
from Orthodoxy, Granovsky a false teacher corrupting the young,
me a servant wearing the gorgeous livery of Western culture, and
all three of us traitors to our country. Of course, he did not
mention our names; those were put in by the readers who
enthusiastically carried this spy's report in verse from drawing-
room to drawing-room. K. Aksakov indignantly answered him,
also in verse, branding with biting words his spiteful attacks, and
saying that their real opponents were various Slavophils who
played the gendarmes as if in the name of Christ.

This incident added much bitterness to our relationship. The
poet's name, the name of the man who recited the poem,[34] the
circle in which he lived, the circle which was enthusiastic over it –
all helped to increase the irritation caused by it.

Our quarrels very nearly led to an enormous calamity, to the ruin of the two purest and best representatives of the two parties. All the efforts of their friends only just managed to suppress Granovsky's quarrel with Pëtr Kireyevsky, which was quickly coming to a duel.

In the midst of these circumstances Shevyrëv, who could never resign himself to the colossal success of Granovsky's lectures, had the happy thought of trying to beat him in his own field, and announced a course of public lectures. He lectured on Dante, on Nationalism in Art, on Orthodoxy in Science, and so on; his audience was numerous, but it remained cold. He displayed boldness at times and this was very much appreciated, but the general effect was negligible. One lecture has remained in my memory, the one in which he talked of Michelet's *Le Peuple* and George Sand's novel *La Mare au Diable*, because in it he touched vividly on a subject of living and contemporary interest. It was difficult to arouse sympathy when talking of the charms of the ecclesiastical writers of the Eastern Church and lauding the Greco-Russian Church. Only Fëdor Glinka[35] and his wife Yevdokia, who wrote of 'the milk of the Most Pure Virgin', usually sat side by side in the front row, modestly casting down their eyes when Shevyrëv was particularly immoderate in his praises of the Orthodox Church.

Shevyrëv spoilt his lectures, just as he spoilt his articles, by sallies against ideas, books, and persons, whom one could hardly have defended in our country without being clapped in prison.

Meanwhile, in spite of all the devices invented in order to contrive to make a success of *The Muscovite*, it was definitely a failure. To make a polemical journal lively one must have without fail the instinct of modernity, one must have that delicate ticklishness of the nerves which is at once stimulated by all that stimulates society. The editors of *The Muscovite* were entirely destitute of this intuitive vision and, however they flourished poor Nestor and poor Dante, they were at last themselves convinced that in our depraved age you could have no success, either with the roughly chopped chaff of Pogodin's phrases or the sing-song suavity of Shevyrëv's eloquence. After much consideration they determined to offer the chief editorship to Ivan Kireyevsky. The choice of Kireyevsky was a particularly happy one, not only because of his intelligence and talents, but also on the financial side. There is no one in the world with whom I myself should so much like to transact business as with Kireyevsky.

To give an idea of his commercial philosophy I shall relate the following anecdote. He had a stud-farm from which horses were brought to Moscow, a price was put on them and they were sold. On one occasion a young officer came to him to buy a horse to which he had taken a great fancy; the groom, seeing this, put up the price. After some bargaining the officer agreed to his terms and went to Kireyevsky. The latter, as he took the money, looked in the list and observed to the officer that the horse was priced at eight hundred roubles, not at a thousand, and that the groom must have made a mistake. The officer was so puzzled by this that he asked permission to look at the horse again, and after examining it refused to buy it, saying: 'It must be a nice sort of horse, if the owner is ashamed to take the money. . . .' Where could one find a better editor?

He set to work zealously, wasted a great deal of time and moved to Moscow with this object, but for all his talent he could do nothing with the magazine. The Muscovite did not respond to any brisk, widespread demand in society, and therefore could not have any circulation except in its own coterie. Its failure must have been a great grief to Kireyevsky.

The Muscovite did not recover after its second breakdown, and the Slavophils themselves perceived that no one could make much headway in that boat. They began to think of another magazine.

This time it was not they who came off victorious. Public opinion clamorously decided in our favour. In the dark night when The Muscovite was sinking and the Lighthouse was no longer lighting it up from Petersburg, Belinsky, who had nourished the Notes of the Fatherland with his own blood, set their illegitimate offspring[36] on its feet and gave them both such a shove that they were able for some years to continue on their course with no staff but proof-correctors, printers, and the publicans and sinners of literature. Belinsky's name was enough to make the fortune of two shops and to concentrate all that was best in Russian literature in the publications in which he took part, while Kireyevsky's talent and Khomyakov's contributions could bring neither circulation nor readers to The Muscovite.

Such was the field of battle when I left it and went away from

Russia. Both sides spoke out once more,* and all the questions have been displaced by the great events of 1848.

Nicholas has died, a new life drew the Slavophils and ourselves beyond the boundaries of our feud. We stretched out our hands to them, but where are they? Gone! And K. Aksakov is gone, and those 'opponents who were dearer to us than many of our own side' are no more.

It was a hard life that burnt men away like a candle left to the wind of autumn.

They were all living when I wrote this chapter the first time. This time let it end with the following lines spoken on the death of Aksakov:[37]

'The Kireyevskys, Khomyakov, and Aksakov *have done their work*; whether their lives were short or long, they could, as they closed their eyes, say to themselves with full conviction that they had done what they meant to do, and, if they could not stop the courier's troika which Peter had sent on its way and in which Biron sat goading the driver to gallop over cornfields and crush the people, they did bring too impassioned public opinion to a halt and made all earnest people reconsider their position.

'With them a sudden change begins in Russian thought and, when we say that, it seems impossible to suspect us of partiality.

'Yes, we were their opponents, but very strange ones. We had the same love, but not the same way of loving.

'Both they and we had been from earliest years possessed by one powerful, unaccountable, physiological, passionate feeling, which they took for memory and we for prophecy – a feeling of boundless love, that embraced the whole of one's existence, for the Russian people, the Russian way of living, the Russian cast of mind. And like Janus, or the two-headed eagle, they and we looked in different directions while one heart throbbed within us.

'They transferred all their love, all their tenderness to their oppressed mother. In us, brought up away from home, this tie was weaker. We had been in the hands of a French governess, and learned later on that our mother was not she but an over-driven

* K. Kavelin's article (in *Sovremennik*, 1847, no. 1), and Yury Samarin's reply to it (in *Moskvityanin*, 1847, no. 2). They are dealt with in *Du Développement des idées révolutionnaires en Russie* [ch. VI.].

peasant woman, and this we ourselves divined from the likeness in our features and because her songs were more akin to us than the *vaudevilles*. We loved her dearly, but her life was too narrow. We were stifled in her little room which was all blackened faces looking out from silver settings, all priests and church servitors to frighten the unfortunate woman, knocked silly by soldiers and clerks. Even her everlasting wailing for her lost happiness rent our hearts: we knew she had no bright memories; we knew something else too: that her happiness lay in the future, that the new life was stirring under her heart; this was our younger brother, to whom without the mess of pottage we should yield our seniority. And meanwhile:

> "Mutter, Mutter, lass mich gehen
> Schweifen auf den wilden Höhen!"[38]

'Such were our family dissensions fifteen years ago. Much water has flowed away since then, and we have met the *mountain spirit* that has checked our flight, while they have stumbled out of a world of holy relics on to living Russian problems. It would be strange for us to adjust accounts, for there is no monopoly of understanding; time, history, and experience have brought us closer to each other, not because they have pulled us towards them, or we them towards us, but because both they and we are nearer to a true outlook now than we were then, when we pitilessly tore each other to pieces in articles in the magazines though even then I do not remember that we ever had any doubt of the warmth of their love for Russia, or they of ours.

'This faith in one another, this common love gives us, too, the right to do homage at their tombs and to throw our handful of earth upon their dead, in the sacred hope that on their graves and ours young Russia may blossom in strength and grandeur.'

*

CHAPTER VII

RIFTS IN THE MOSCOW CIRCLE

AFTER the reconciliation with Belinsky in 1840 our little group of friends went ahead without any important disagreement: there were shades of opinion and personal views, but what was of most importance and common to all was based on the same principles. I do not think it could have gone on like that for ever. We were bound to reach a limit, a barrier which some would pass over and others would get caught up on.

Three or four years later I began with profound regret to notice that though we started from the same first principles we were reaching different conclusions – and not because we interpreted them differently but because not all of us *liked* them.

At first these disputes were half jocular. We used to laugh, for instance, at the Little Russian obstinacy with which Redkin tried to deduce a logical proof of a personal soul. This reminds me of one of the last jests of dear, kind-hearted Kryukov. He was already very ill and Redkin and I were sitting by his bedside. It had been a foul day, and all at once there was a flash of lightning followed by a loud clap of thunder. Redkin went to the window and let down the blind.

'Will that do any good?' I asked him.

'Why,' Kryukov answered for him, 'Redkin believes in *die Persönlichkeit des absoluten Geistes*, and so covers the window that He may not see where to aim if He should think fit to shoot an arrow at him.'

But it might well have been imagined that such an essential difference in outlook would not remain a jesting matter for long.

On one sheet in a notebook of that time occurs the following sentence, written with evident *arrière-pensée*: 'Personal relationships are very bad for straightforward thinking. Out of respect for the excellent qualities of individuals we sacrifice for their sakes the acuteness of our minds. It needs great strength to weep and yet be able to sign the death-warrant of Camille Desmoulins.'

The germs of the angry dissensions of 1846 were already latent in this envy of Robespierre's strength.

The questions upon which we came in collision were not casual ones; like fate, there was no riding round them. They are the granite stumbling-blocks on the road of knowledge which have been the same in all ages, frightening men away and beckoning them on. And just as liberalism carried out consistently inevitably brings a man face to face with the social question, so learning – if only a man entrusts himself to it without an anchor – will without fail beat him with its waves upon the grey cliffs on to which all who have had the temerity to think have been cast – from the Seven Sages of Greece down to Kant and Hegel. Instead of simple explanations almost everyone has tried to skirt them, and has only covered them with fresh layers of symbols and allegories; and that is how it is that even now they stand up as menacingly, and navigators are afraid to make straight for them and to convince themselves that they are not cliffs at all but only fog seen in a fantastic light.

This step is not easy, but I believed both in the strength and in the will of our friends; they had not to seek the fairway anew as Belinsky and I had had to do. He and I had struggled long and desperately in the squirrel's wheel of dialectic repetition and had leapt out of it in the end at our own risk. They had our example before their eyes and Feuerbach in their hands. For a long time I did not believe it, but at last I became convinced that, though our friends did not share Redkin's form of proof, yet in essence they were more in agreement with him than with me, and that for all the independence of their minds there were still truths of which they were frightened. I differed from everyone except Belinsky, even from Granovsky and Yevgeny Korsh.

This discovery filled me with deep regret; the threshold at which they hesitated, once stated in words, could no longer be merely implied. Discussions arose from the inner need to reach the same level once more; to do this we had, so to speak, to call to each other to find out where each one stood.

Before we ourselves brought our theoretical rift into the light of day it had been noticed by the younger generation, who stood incomparably nearer to my point of view. Not only in the university and the Lycée, but even in the clerical colleges, young people were eagerly reading my articles on 'Dilettantism in Science' and my 'Letters on the Study of Nature'. This last fact I learned from Count S. Strogonov to whom Filaret[1] complained of it, threatening to take defensive measures against such pernicious spiritual fare.

About the same time I learned from a different source of their success among seminarists. This incident gives me so much pleasure that I cannot help telling of it.

The son of a priest of our acquaintance who lived near Moscow, a young man of seventeen, came several times to me for the *Notes of the Fatherland*. He was shy, scarcely spoke, blushed, was confused and in a hurry to get away. His open and intelligent face was eloquent in his favour, and at last I broke down his youthful diffidence and began talking to him about the *Notes of the Fatherland*. He read the philosophical articles with great attention and judiciousness. He told me how eagerly the seminary students taking the higher course read my historical exposition of philosophical systems and how this astonished them after the philosophy of Burmeister[2] and Wolf.[3]

The young man began to visit me sometimes, and I had plenty of time to convince myself of his abilities and capacity for work.

'What do you intend doing when you have finished your studies?' I once asked him.

'Enter the priesthood,' he answered, blushing.

'Have you thought seriously of the lot that awaits you if you go into the priesthood?'

'I have no choice: my father definitely objects to my taking up any secular calling. I shall have leisure enough for my occupations.'

'Don't be angry with me,' I replied, 'but I cannot help telling you my opinion frankly. Your conversation, your way of thinking, which you have not concealed from me, and the sympathy you feel for my work – all this and, what is more, my sincere interest in your future, together with my age, gives me certain rights. Think again a hundred times before you put on the cassock. It will be far more difficult to take it off, and perhaps it will be hard for you to breathe in it. I shall ask you one very simple question: Tell me, is there in your soul a belief in even one dogma of the theology that you are being taught?'

The young man dropped his eyes and said after a pause:

'I am not going to lie to you – no!'

'I knew that. Only think now of your future fate. You will have every day for the whole of your life to lie loudly before all the world, to be false to the truth; why, that is the sin against the

Holy Spirit, conscious, premeditated sin. Would there be enough in you to manage such duplicity? Your whole station in life will be a falsehood. How will you meet the eye of one who is praying in earnest; how will you comfort the dying with heaven and eternal life; how will you absolve men's sins? And you will be forced to try to convert heretics,[4] too, and to condemn them for their heresy.'

'This is awful! awful!' said the young man, and he went away shaken and disconcerted.

He came back the next evening.

'I have come to tell you,' he said, 'that I have thought a great deal about what you said. You are perfectly right: the priestly calling is impossible for me and I assure you that I would sooner go for a soldier than allow myself to be made a priest.'

I pressed his hand warmly and promised that when the time came I would do my utmost to persuade his father to agree to his wishes.

So it has been my lot to save a soul alive, or at least to contribute to its salvation.

I was able to get a closer view of the bent of the students for philosophy. Through the whole academic year of 1845 I attended the lectures on comparative anatomy. In the lecture-room and the dissecting theatre I became acquainted with a new generation of young people.

Their prevailing tendency was absolutely realistic, that is, positively scientific. It is remarkable that this was the tendency of almost all the students who came from the Tsarskoye Selo Lycée. The Lycée, turned by the suspicious and deadening despotism of Nicholas out of its beautiful park,[5] still remained the same great nursery of talent; Pushkin's bequest, the poet's blessing, survived the churlish blows of ignorant authority.

It was with joy that I welcomed these students from the Lycée who were at Moscow University, a new, vigorous generation.

It was these young university students, devoted with all the impatience and fire of youth to the world of realism that was opening before them with its ruddy flush of health, who discerned, as I have said, where it was that we differed from Granovsky. Passionately as they loved him, they were beginning to revolt against his 'romanticism'. They urgently desired that I should bring him over to our side, regarding Belinsky and me as the representatives of their philosophical opinions.

This was the position at the begining of 1846. Granovsky began a new course of public lectures. Again all Moscow gathered round his platform, again his formative, musing eloquence set their hearts quivering; but the completeness, the enthusiasm that there had been in his first course was lacking, as though he were tired or as though some idea with which he could not cope were absorbing and hindering him. That was just how it was, as we shall see much later.

At one of these lectures in March a common acquaintance of ours ran in headlong to tell us that Ogarëv and Satin[6] had arrived from foreign parts.

We had not met for several years and very rarely corresponded. ... What would they be like? ... How would they stand? ... With beating hearts Granovsky and I dashed off to the Yar where they were staying. And here they were at last – and how changed, and what a beard, and we had not seen each other for some years – we fell to looking at trifles and talking of trifles though we felt that we wanted to talk of something else.

At last our little circle was almost all assembled – now we should begin to live!

We had spent the summer of 1845 at a villa in Sokolovo. It is a beautiful corner of the Moscow district, some fifteen miles from the town on the road to Tver. There we took a little country house which stood almost wholly in the park which sloped away downhill to a little river. On the one side stretched our Great Russian ocean of cornfields; on the other there was a wide view into the distance, for which reason the owner of the house had not failed to call the summer house which was built there 'Belle Vue'.

Sokolovo had belonged at one time to the Rumyantsevs. The wealthy landowners and aristocrats of the eighteenth century, with all their faults, were possessed of a breadth of taste which they have not transmitted to their heirs. The old-fashioned manors and homesteads along the Moskva river are unusually fine, especially those in which the last two generations have made no reforms and no changes.

We had a splendid time there. No serious cloud darkened the summer sky; we lived in our park, working hard and going for long walks. Ketscher grumbled less, though he did sometimes lift his eyebrows very high and utter weighty sayings with vivid mimicry. Granovsky and Yevgeny Korsh used to come for the night almost every Saturday and sometimes stayed till Monday. Shchepkin had taken another villa a little way off. He often walked over, wearing a broad-brimmed hat and a white coat like Napoleon at Longwood, with a basket of mushrooms he had gathered; he made jokes, sang Little Russian songs, and was almost the death of us with his stories, which I do believe would have made Ioann the Sorrowful, who spent his life shedding tears over the sins of this world, shed them from laughing. . . .

Sitting in a friendly group in a corner of the park under a big lime tree, we had had no regret, except for Ogarëv's absence. Well, here he was, and in 1846 we went to Sokolovo once more and he went with us; Granovsky took a little lodge for the whole summer, and Ogarëv was installed in the *entresol* above the steward, a naval officer who had lost one ear.

And with all this, an undefined feeling was whispering to me two or three weeks later that our *villeggiatura* had not been a success and that there was no putting it right. Who has not had the experience of preparing some festivity, rejoicing in anticipation at the coming gaiety of his friends; and when they arrive everything goes well, there is nothing amiss, yet the expected gaiety does not come off. Life goes on well and briskly only when one does not feel the blood circulating in one's veins and does not think how the lungs rise and fall. If you feel the recoil of every shock, you may be sure there will be pain, a disharmony which cannot always be overcome.

The first days after our friends' arrival were spent in the inspiriting fumes of festivities; before they were over my father was taken ill. His death and all the worries and business that followed had distracted us from theoretical questions. In the peace of our life at Sokolovo our divergences were bound to be put in words.

Ogarëv, though he had not seen me for four years, was following exactly the same line as I was. We had moved over the same ground by different paths and found ourselves together. Natalie was with us, too. The serious and at first sight overwhelming conclusions that we had come to did not alarm her; she imparted to them a special, poetical *nuance*.

Arguments became more frequent and came back in a thousand

variations. One day we were dining in the garden. Granovsky was reading in the *Notes of the Fatherland* a letter of mine on the study of nature (it was one on the Encyclopaedists, I remember) and was extremely pleased with it.

'But what is it you like?' I asked him. 'Can it be only the method of exposition? You cannot possibly agree with the underlying implications of it.'

'Your opinions,' answered Granovsky, 'are just as much an historical moment in the science of thought as the writings of the Encyclopaedists themselves. I like in your articles just what I like in Voltaire or Diderot; they touch vividly and sharply upon questions which rouse a man and urge him forward, and as for the one-sidedness of your views I don't want to go into that. Does anyone talk of Voltaire's theories nowadays?'

'Do you mean to say that there is no standard of truth and that we rouse men only to talk trifles to them?'

The conversation continued for some time on these lines. At last I observed that the development of science, its contemporary condition, *obliges* us to accept certain truths, apart from whether we want to or not; that, once recognised, they cease to be historical problems and become simply irrefutable facts of consciousness like the theories of Euclid, like the laws of Kepler, like the inseparability of cause and effect, of spirit and matter.

'All that is so far from being obligatory,' Granovsky objected, with a slight change in his expression, 'that I shall never accept your dry, cold idea of the unity of soul and body, for with it the immortality of the soul disappears. You may not need it, but I have buried too much to give up this belief. Personal immortality is essential for me.'

'Life in this world would be a splendid affair,' I said, 'if anything anyone needed were always true here and now, as in fairy tales.'

'Only think, Granovsky:' added Ogarëv, 'why, it's a sort of running away from unhappiness.'

'Listen,' answered Granovsky, turning pale and assuming the air of a disinterested outsider: 'you will truly oblige me if you never speak to me on these subjects again; there are plenty of interesting things to talk about with far more profit and pleasure.'

'Certainly. With the greatest pleasure,' I said, feeling a chill on my face. Ogarëv said nothing; we all looked at one another and that glance was quite enough; we all loved one another too much not to gauge to the full by our expressions what had happened. Not a word more was said. The argument was not continued. Natalie tried to cover up the incident and set things right, and we helped her. Children, who always come to the rescue in these cases, served as a subject of conversation, and the dinner ended so peacefully that no stranger who had arrived after the conversation would have noticed anything wrong. . . .

After dinner Ogarëv jumped on his horse Kortik, I mounted a gendarme's cast nag and we rode out into the open country. We were as sad as though someone near and dear had died; for till then Ogarëv and I had expected that we should come to an agreement, that our friendship would blow away our differences like dust; but the tone and meaning of Granovsky's last words had revealed a distance between us such as we had never imagined. So here was the boundary line, the limit, and with it the censorship! Neither he nor I spoke all the way. As we were riding home we shook our heads sadly and both said with one voice: 'So it seems we are alone again.'

. . .

I met Granovsky the next day as though nothing had happened, a bad sign on both sides. The pain was still so keen that it could find no words; and dumb pain that has no outlet gnaws away thread after thread like a mouse in the stillness. . . .

Two days later I was in Moscow. Ogarëv and I went to see Korsh. He was as solicitously kind and sorrowfully nice to us as though he was sorry for us, but, hang it all, had we committed a crime? I asked Korsh straight out, had he heard of our dispute? He had; he said that we had all been too heated over abstract subjects; pointed out that the perfect identity between people and opinions of which we dreamed did not exist, that people's sympathies, like a chemical affinity, have their limit of saturation which cannot be exceeded without stumbling upon aspects at which men once more became strangers. He jested at our being so young when over thirty, and he said all this with friendliness and tact; one could see that he did not feel easy either.

We parted peacefully. Blushing a little I thought of my 'naïveté', and afterwards, when I was alone and went to bed, it seemed to me that another bit of my heart had been taken away – skilfully and painlessly, but it was gone!

Nothing further happened ... only everything was clouded over with something dark and dull; the freedom from constraint, the complete *abandon* had vanished from our circle. We became more careful, we edged round certain questions, that is, we really did step back at 'the limit of chemical affinity' – and all this gave us the more pain and grief because we had a great and genuine love for one another.

I may have been too intolerant, may have argued conceitedly and answered sarcastically ... perhaps so ... but in reality I am convinced even now that in really intimate relationships it is essential to have the same religion, to be at one in the chief theoretical convictions. Of course theoretical agreement alone is not enough for intimacy between men; I was nearer in sympathy, for instance, to Ivan Kireyevsky than to many of my own set. What is more, one may be a good and faithful *ally*, agreeing in some definite cause and differing in opinions. I was on such terms with men for whom I had infinite respect, though I differed from them on many subjects – for instance, with Mazzini and with Worcell.[7] I did not try to convince them nor they me; we had enough in common to go the same way together without quarrelling. But between us brothers of one family, twins, as it were, who had lived one life together, it was impossible to differ so profoundly.

If only we had had some unavoidable business which would have absorbed us completely; but as it was, all our activity lay precisely in the sphere of thought and the propaganda of our convictions ... how was compromise possible in that field? ...

The rift in one of the walls of our temple of friendship grew wider, as always happens, through trifles, misunderstandings, unnecessary openness where it would have been better to be silent and harmful silence where it was essential to speak; these things are decided only by the tact of the heart: there are no rules.

Soon afterwards everything was at sixes and sevens among the ladies too. ...

There was nothing to be done at the moment.

To go away, far away, for a long time, to go without fail! But it was not so easy to go. The rope of police supervision was round my legs, and without permission from Nicholas a foreign passport could not be got.

CHAPTER VIII

TO PETERSBURG FOR A PASSPORT

A FEW months before my father's death Count Orlov was appointed to succeed Benckendorf.[1] I then wrote to Olga Alexandrovna to ask whether she could manage to procure me a passport for abroad or permission on some pretext or other to go to Petersburg to get one for myself. My old friend answered that the latter was easier to arrange and a few days later I received from Orlov His Majesty's permission to go to Petersburg for a short time to arrange my affairs. My father's illness, his death, the actual arrangement of my affairs, and some months spent in the country, delayed me till winter. At the end of November I set off for Petersburg, having first sent a request for a passport to the Governor-General. I knew that he could not grant it because I was still under *strict* police supervision: all I wanted was that he should send on the request to Petersburg.

. . .

The second day after my arrival in Petersburg the house porter came to ask me from the local police: 'With what papers had I come to Petersburg?' The only paper I had, the decree concerning my retirement from the service, I had sent to the Governor-General with my request for a passport. I gave the house-porter my permit, but he came back to say that it was valid for leaving Moscow but not for entering Petersburg. A police-officer came too, with an invitation to the *oberpolitsmeyster*'s office. I went to Kokoshkin's office, which was lit by lamps although it was daytime, and after an hour he arrived. Kokoshkin more than other persons of the same selection was the picture of a servant of the Tsar with no ulterior designs, a man in favour, ready to do any dirty job, a favourite with no conscience and no bent for reflection. He served and made his pile as naturally as birds sing.

Perovsky told Nicholas that Kokoshkin was a great bribe-taker.

'Yes,' answered Nicholas, 'but I sleep peacefully at night knowing that he is *politsmeyster* in Petersburg.'

I looked at him while he was dealing with other people. . . .

What a battered, senile, depraved face he had; he was wearing a curled wig which was glaringly incongruous with his sunken features and wrinkles.

After conversing with some German women in German and with a familiarity that showed they were old acquaintances, which was evident also from the way the women laughed and whispered, Kokoshkin came up to me, and looking down asked in a rather rude voice:

'Why, are you not forbidden by His Majesty to enter Petersburg?'

'Yes, but I have permission.'

'Where is it?'

'I have it here.'

'Show it. How's this? You are using the same permit twice.'

'Twice?'

'I remember that you came here before.'

'I didn't.'

'And what is your business here?'

'I have business with Count Orlov.'

'Have you been at the Count's, then?'

'No, but I have been at the Third Division.'

'Have you seen Dubelt?'

'Yes.'

'Well, I saw Orlov himself yesterday and he says that he has sent you no permit.'

'It's in your hands.'

'God knows when this was written, and the time has expired.'

'It would be an odd thing for me to do, wouldn't it? to come without permission and begin with a visit to General Dubelt.'

'If you don't want any trouble, be so good as to go back, and not later than the next twenty-four hours.'

'I was not proposing to remain here long, but I must wait for Count Orlov's answer.'

'I cannot give you leave to do so; besides, Count Orlov is much displeased at your coming without permission.'

'Kindly give me my permit and I will go to the Count at once.'

'It must remain with me.'

'But it is a letter to me, addressed to me personally, the only document on the strength of which I am here.'

'The document will remain with me as a proof that you have

been in Petersburg. I earnestly advise you to go tomorrow in order
that nothing worse may befall you.'

He nodded and went out. Much good it is talking to them!

. . .

I went to the Third Division and told Dubelt what had happened. He burst out laughing. 'What a muddle they everlastingly
make of everything! Kokoshkin reported to the Count you had
come without permission and the Count said you were to be sent
away, but I explained the position to him afterwards; you can stay
as long as you like. I'll have the police written to at once. But now
about your petition: the Count does not think it would be of any
use to ask permission for you to go abroad. The Tsar has refused
you twice, the last time when Count Strogonov interceded for
you; if he refuses a third time, you won't get to *the waters* during
this reign, for certain.'

'What am I to do?' I asked in horror, for the idea of travel and
freedom had taken such deep root in my heart.

'Go to Moscow: the Count will write a private letter to the
Governor-General telling him that you want to go abroad for the
sake of your wife's health, assuring him that he knows nothing of
you but what is good, and asking him whether he thinks it would
be possible to relieve you from police supervision. He can make
no answer but "yes" to such a question. We shall report to the
Tsar the removal of police supervision, then you take out a passport for yourself like anybody else, and you can go to any
watering-place you like, and good luck to you.'

All this seemed to me extraordinarily complicated, and indeed I
fancied it was a device simply to get rid of me. They could not
refuse me point-blank, for it would have brought down upon them
the wrath of Olga Alexandrovna, whom I visited every day. When
once I had left Petersburg I could not come back again, cor-
responding with these gentry is a difficult business. I communicated some of my doubts to Dubelt; he began frowning, that is,
grinning more than ever with his lips and screwing up his eyes.

'General,' I said in conclusion, 'I do not know, but the fact is
I do not even feel certain that Strogonov's representation reached
the Tsar.'

Dubelt rang the bell and ordered the file about me to be
brought, and while waiting for it he said to me good-naturedly:

'The Count and I are suggesting to you the course of proceeding by which we think you most likely to get your passport; if you have more certain means at your disposal, make use of them; you may be sure that we shall not hinder you.'

'Leonty Vasilevich is perfectly right,' observed a sepulchral voice. I turned round; beside me, looking older and more grey-headed than ever, stood Sakhtynsky, who had received me five years before at the same Third Division. 'I *advise* you to be guided by his opinion if you want to go.'

I thanked him.

'And here's the file,' said Dubelt, taking a thick writing book from the hands of a clerk (what would I not have given to read the whole of it! In 1850 I saw my *dossier* in Carlier's office in Paris;[2] it would have been interesting to compare them). After rummaging in it he handed it to me open; there was Benckendorf's report after Strogonov's letter petitioning for permission for me to go for six months to a watering-place in Germany. In the margin was written in big letters in pencil: 'Too soon.' The pencil marks were glazed over with varnish, and below was written in ink ' "Too soon" written by the hand of his Imperial Majesty. – Count A. Benckendorf.'[3]

'Do you believe now?' asked Dubelt.

'Yes, I do,' I answered, 'and I am so sure of your words that I shall go to Moscow tomorrow.'

'Well, you can stay and amuse yourself here a little; the police will not worry you now, and before you go away look in, and I'll tell them to show you the letter to Shcherbatov. Goodbye. *Bon voyage*, if we don't meet again.'

'A pleasant journey,' added Sakhtynsky.

We parted, as you see, on friendly terms.

On reaching home I found an invitation, from the superintendent of the Second Admiralty Police-Station I believe it was. He asked me when I was going.

'Tomorrow evening.'

'Upon my word, but I believe, I thought ... the general said today. His Excellency will put it off, of course. But will you allow me to make certain of it?'

'Oh yes, oh yes; by the way, give me a permit.'

'I will write it in the police-station and send it to you in two hours' time. By what convenience are you thinking of going?'

'The Serapinsky, if I can get a seat.'

'Very good, and if you do not succeed in getting a seat kindly let us know.'

'With pleasure.'

In the evening a policeman turned up again; the superintendent sent to tell me that he *could* not give me the permit, and that I must go at *eight o'clock* next morning to the *oberpolitsmeyster*'s.

What a plague and what a bore! I did not go at eight o'clock, but in the course of the morning I looked in at the office of the *oberpolitsmeyster*. The police-station superintendent was there; he said to me:

'You cannot go away: there is a paper from the Third Division.'

'What has happened?'

'I don't know. The general gave orders you were not to be given a permit.'

'Does the director know?'

'Of course he knows,' and he pointed out to me a colonel in uniform and wearing a sword sitting at a big table in another room; I asked him what was the matter.

'To be sure,' he said, 'there was a paper, and here it is.' He read it through and handed it to me. Dubelt wrote that I had a perfect right to come to Petersburg and could remain *as long as I liked*.

'And is that why you won't let me go? Excuse me, I can't help laughing; yesterday the *oberpolitsmeyster* was chasing me away against my will, today he is keeping me against my will, and all this on the ground that the document gives me leave to remain *as long as I like*.'

The absurdity was so evident that even the colonel-secretary laughed.

'But why should I throw money away, paying for a place in the *diligence* twice over? Please tell them to write me a permit.'

'I cannot, but I will go and inform the general.'

Kokoshkin ordered them to write me a permit, and as he walked through the office said to me reproachfully: 'It's beyond anything. First you want to stay, then you want to go; why, you have been told that you can stay.'

I made no answer.

When we had driven out of the city gates in the evening and I saw once more the endless plain stretching away towards the Four Hands,[4] I looked at the sky and vowed with all my heart never to return to that city of the despotism of blue, green, and variegated police, of official muddle, of flunkeyish insolence, of gendarme romance, in which the only civil man was Dubelt, and he chief of the Third Division.

Shcherbatov answered Orlov reluctantly. He had at that time a secretary who was not a colonel but a pietist, who because of my articles hated me as an 'atheist and Hegelian'. I went myself to deal with him. The pious secretary, in an oily voice and with Christian unction, told me that the Governor-General knew nothing about me, that he did not doubt my lofty moral qualities, but that he would have to make inquiries of the *oberpolitsmeyster*. He wanted to drag the business out; moreover, this gentleman did not take bribes. In the Russian service disinterested men are the most frightful of all; the only ones who do not take bribes in all simplicity are Germans; if a Russian does not take money he will take it out in something else, and from such villains God spare us. Fortunately *oberpolitsmeyster* Luzhin gave me a good character.

On returning home ten days later I bumped into a gendarme at my door The appearance of a police-officer in Russia is as bad as a tile falling on one's head, and therefore it was not without a particularly unpleasant feeling that I waited to hear what he had to say to me; he handed me an envelope. Count Orlov informed me of his Imperial Majesty's command that I should be relieved from police supervision. Together with this I received the right to a foreign passport.

...

'Six or seven sledges accompanied us as far as Chërnaya Gryaz. There for the last time we clinked glasses and parted, sobbing.

'It was evening, the covered sledge crunched through the snow ... you looked sadly after us but did not guess that it meant a funeral and eternal separation. All were there, only one was missing, the nearest of the near: he alone was ill,[5] and by his absence, as it were, washed his hands of my departure.

'It was the 21st of January, 1847 ...'[6]

The sergeant gave me back the passports: a little old soldier in a clumsy shako covered with oilskin, carrying a rifle of incredible size and weight, lifted the barrier; a Ural Cossack with narrow little eyes and broad cheekbones, holding the reins of his little, shaggy, dishevelled nag, which was covered all over with little icicles, rode up to wish me a happy journey; the pale, thin, dirty little Jewish driver with rags twisted four times round his neck clambered on the box.

'Goodbye! Goodbye!' said our old acquaintance, Karl Ivanovich, who was seeing us as far as Taurogen, and Tata's wet-nurse, a handsome peasant woman, dissolved in tears as she said farewell.

The little Jew whipped up his horses, the sledge moved off. I looked back, the barrier had been lowered, the wind swept the snow from Russia on to the road and blew to one side the tail and mane of the Cossack's horse.

The nurse in a sarafan and a warm jacket was still looking after us and weeping; Sonnenberg, that symbol of the parental home, that comic figure from the days of childhood, waved his silk hand-kerchief – all round us was the endless steppe of snow.

'Goodbye, Tatyana! Goodbye, Karl Ivanovich!'

Here was a milestone and on it, covered with snow, a thin, *single-headed* eagle with outspread wings ... and that's a good thing: one head less.

*

PARIS – ITALY – PARIS

1847–1852

WHEN I began to publish yet another part of *My Past and Thoughts*, I paused in hesitation before the discontinuity of the narratives, the pictures and of my, so to speak, interlinear comments on them. There is less external unity in them than in the earlier parts. I cannot weld them into one. In filling in the gaps it is very easy to give the whole thing a different background and a different lighting – the truth of *that time* would be lost. *My Past and Thoughts* is not an historical monograph, but the reflection of historical events on a man who has accidentally found himself in their path. That is why I have decided to leave my disconnected chapters as they were, stringing them together like the mosaic pictures in Italian bracelets – all of which refer to one subject but are only held together by the setting and the chain.

My *Letters from France and Italy*[1] are essential for completing this part, especially in regard to the year 1848; I had meant to make extracts from them, but that would have involved so much reprinting that I could not make up my mind to it.

Many things that have not appeared in *The Pole Star* have been put into this edition, but I cannot give everything to my readers yet, for reasons both personal and public. The time is not far off when not only the pages and chapters here omitted, but the whole volume, which is the most dear to me, will be published.

GENEVA, 29 July 1866.

CHAPTER I

THE JOURNEY

At Lautzagen the Prussian gendarmes invited me into the guard-room. An old sergeant took the passports, put on his spectacles, and with extraordinary precision began reading aloud all that was unnecessary:

Auf Befehl s.k. M. Nikolai des Ersten . . . allen und jeden denen daran gelegen, etc. etc. . . . Unterzeichner Peroffski, Minister des Innern, Kammerherr, Senator und Ritter des Ordens St. Wladimir . . . Inhaber eines goldenen Degens mit der Inschrift für Tapferkeit . . .

. . .

[. . .] after reading three times in the three passports all General Perovsky's decorations, including his clasp for an unblemished record, he asked me:

'But who are you, *Euer Hochwohlgeboren?*'

I stared, not understanding what he wanted of me.

'*Fräulein Maria E., Fräulein Maria K., Frau H.*² – they are all women, there is not one man's passport here.'

I looked: there really were only the passes of my mother and two ladies we knew who were travelling with us; a cold shudder ran down my back.

'They would not have let me through at Taurogen without a passport.'

'*Bereits so*, but you can't go further.'

'What am I to do?'

'Perhaps you have forgotten it at the guardroom. I'll tell them to harness a sledge for you; you can go yourself, and your people can warm themselves here meanwhile. Heh! *Kerl! Lass er mal den Braunen anspannen.*'

I cannot remember this stupid incident without laughing, just because I was so utterly disconcerted by it. I was overwhelmed by losing that passport of which I had been dreaming for several years, which I had been trying to obtain for two years, and losing it the minute after crossing the frontier. I was certain I had put it in my pocket, so I must have dropped it – where could I look for it? It would be covered by snow. . . . I should have to ask for a new one, to write to Riga, perhaps to go myself: and then they would send in a report, would notice that I was going to the mineral waters in January. In short, I felt as though I were in Petersburg again; visions of Kokoshkin and Sakhtynsky, Dubelt and Nicholas, passed through my mind. Goodbye to my journey, goodbye to Paris, to freedom of the press, to concerts and theatres . . . once more I should see the clerks in the ministry, police – and every other sort of watcher, town constables with the two bright buttons on their backs that they use for looking behind them . . . and first of all I should see again the little scowling soldier in a heavy shako with the mysterious number '4' inscribed on it, the frozen Cossack horse . . . I might even see the nurse[3] again at 'Tavroga', as she had called it.

Meanwhile they put a big, melancholy, angular horse into a tiny sledge. I got in beside the driver in a military overcoat and high boots; he gave the traditional crack of the traditional whip – and suddenly the learned sergeant ran out into the porch wearing only his breeches, and shouted: 'Halt! Halt! Da ist der vermale-deite Pass,' and he held it unfolded in his hands.

I was overtaken by hysterical laughter.

'What's this you're doing to me? Where did you find it?'

'Look,' he said, 'your Russian sergeant folded them one inside the other: who could tell it was there? I never thought of unfolding them.'

And yet he had read three times over: Es ergehet deshalb an alle hohen Mächte und an alle und jede, welchen Standes und welcher Würde sie auch sein mögen. . . .

'I reached Königsberg[4] tired out by the journey, by anxiety, by many things. After a good sleep in an abyss of feathers, I went out next day to look at the town. It was a warm winter's day: the hotel-keeper suggested that we should take a sledge. There were bells on the horses and ostrich feathers on their heads . . . and we were gay; a load was lifted from our hearts: the unpleasant sensation of fear, the gnawing feeling of suspicion, had flown away. Caricatures of Nicholas were exhibited in the window of a book-shop, and I rushed in at once to buy a whole stock of them. In the evening I went to a small, dirty, inferior theatre, but came back from it excited, not by the actors but by the audience, which consisted mostly of workmen and young people; in the intervals everyone talked freely and loudly, and all put on their hats (an extremely important thing, as important as the right to wear a beard, etc.). This ease and freedom, this element of greater serenity and liveliness impresses the Russian when he arrives abroad. The Petersburg government is still so coarse and unpolished, so absolutely nothing but despotism, that it positively likes to inspire fear; it wants everything to tremble before it – in short, it desires not only power but the theatrical display of it. To the Petersburg Tsars the ideal of public order is the ante-room and the barracks.'

. . .

Berlin, Cologne, Belgium – all flashed past before our eyes; we looked at everything half absent-mindedly, in passing; we were in haste to arrive, and at last we did arrive.

. . . I opened the heavy, old-fashioned window in the Hôtel du Rhin; before me stood a column:

> '. . . with a cast-iron doll,
> With scowling face and hat on head,
> And arms crossed tightly on his breast.'[5]

And so I was really in Paris, not in a dream but in reality: this was the Vendôme column and the Rue de la Paix.

In Paris – the word meant scarcely less to me than the word 'Moscow!' Of that minute I had been dreaming since my childhood. If I might only see the Hôtel de Ville, the Café Foy in the Palais Royal, where Camille Desmoulins picked a green leaf, stuck it on his hat for a cockade and shouted '*à la Bastille!*'

I could not stay indoors; I dressed and went out to stroll about at random . . . to look up Bakunin, Sazonov:[6] here was Rue St-Honoré, the Champs-Élysées – all those names to which I had felt akin for long years . . . and here was Bakunin himself. . . .

I met him at a street corner; he was walking with three friends and, just as in Moscow, discoursing to them, continually stopping and waving his cigarette. On this occasion the discourse remained unfinished; I interrupted it and took him with me to find Sazonov and surprise him with my arrival.

I was beside myself with happiness!

And on that happiness I shall stop.

I am not going to describe Paris once more. My first acquaintance with European life, the triumphant tour of an Italy that had just leapt up from sleep, the revolution at the foot of Vesuvius, the revolution before St Peter's, and finally the news – like a flash of lightning – of the 24th of February – all that I have described in my *Letters from France and Italy*. I could not now with the same vividness reproduce impressions half effaced and overlaid by others. They make an essential part of my *Notes* – for what are letters but notes of a brief period?

CHAPTER II

...

WESTERN-EUROPEAN ARABESQUES
Notebook the First

1

THE DREAM

Do you remember, friends, how lovely was that winter day, bright and sunny, when six or seven sledges accompanied us to Chërnaya Gryaz, when for the last time we clinked glasses and parted, sobbing?

...

That was the 21st of January, 1847.[1]

Seven years[2] have passed since then, and what years! Among them were 1848 and 1852.

All sorts of things happened in those years, and everything was shattered – public and private: the European revolution and my home, the freedom of the world and my personal happiness.

Of the old life not one stone was left upon another. *Then* my powers had reached their fullest development; the previous years had given me pledges for the future. I left you boldly, with head-long self-reliance, with haughty confidence in life. I was in haste to tear myself away from the little group of people who were so thoroughly accustomed to each other and had come so close, bound by a deep love and a common grief. I was beckoned to by distance, space, open conflict, and free speech. I was seeking an independent arena, I longed to try my powers in freedom. ...

Now I no longer expect anything: after what I have seen and experienced nothing will move me to any particular wonder or to deep joy; joy and wonder are curbed by memories of the past and fear of the future. Almost everything has become a matter of indifference to me, and I desire as little to die tomorrow as to live long; let the end come as casually and senselessly as the beginning.

And yet I have found all that I sought, even recognition from this old, complacent world – and along with this I found the loss of all my beliefs, all that was precious to me, have met with betrayal, treacherous blows from behind, and in general a moral corruption of which you have no conception.

It is hard for me, very hard, to begin this part of my story; I have avoided it while I wrote the preceding parts, but at last I am face to face with it. But away with weakness: he who could live through it must have the strength to remember.

From the middle of the year 1848 I have nothing to tell of but agonising experiences, unavenged offences, undeserved blows. My memory holds nothing but melancholy images, my own mistakes and other people's: mistakes of individuals, mistakes of whole peoples. Where there was a possibility of salvation, death crossed the path. . . .

. . . The last days of our life in Rome conclude the bright part of my memories, that begin with the awakening of thought in child-hood and our youthful vow on the Sparrow Hills.

Alarmed by the Paris of 1847, I had opened my eyes to the truth for a moment, but was carried away again by the events that seethed about me.[3] All Italy was 'awakening' before my eyes! I saw the King of Naples tamed and the Pope humbly asking the alms of the people's love – the whirlwind which set everything in movement carried me, too, off my feet; all Europe took up its bed and walked – in a fit of somnambulism which we took for awaken-ing. When I came to myself, it had all vanished; la Sonnambula, frightened by the police, had fallen from the roof; friends were scattered or were furiously slaughtering one another. . . . And I found myself alone, utterly alone, among graves and cradles – their guardian, defender, avenger, and I could do nothing because I tried to do more than was usual.

And now I sit in London where chance has flung me – and I stay here because I do not know what to make of myself. An alien race swarms confusedly about me, wrapped in the heavy breath of ocean; a world dissolving into chaos, lost in a fog in which outlines are blurred, in which a lamp gives only murky glimmers of light.

. . . And that other land – washed by the dark-blue sea under the canopy of dark-blue sky . . . it is the one shining region left until the far side of the grave.

O Rome, how I love to return to your deceptions, how eagerly I run over day by day the time when I was intoxicated with you!

... A dark night. The Corso is filled with people, and here and there are torches. It is a month since a republic was proclaimed in Paris. News has come from Milan — there they are fighting, the people demand war,[4] there is a rumour that Charles Albert is on the way with troops. The talk of the angry crowd is like the intermittent roar of a wave, which alternately comes noisily up the beach and then pauses to draw breath.

The crowds form into ranks. They go to the Piedmontese ambassador to find out whether war has been declared.

'Fall in, fall in with us,' shout dozens of voices.

'We are foreigners.'

'All the better; *Santo Dio*, you are our guests.'

We joined the ranks.

'The front place for the guests, the front place for the ladies, *le donne forestiere!*'

And with passionate shouts of approval the crowd parted to make way. Ciceruacchio and with him a young Russian poet, a poet[5] of popular songs, pushed their way forward with a flag, the tribune shook hands with the ladies and with them stood at the head of ten or twelve thousand people — and all moved forward in that majestic and harmonious order which is peculiar to the Roman people.

The leaders went into the Palazzo, and a few minutes later the drawing-room doors opened on the balcony. The ambassador came out to appease the people and to confirm the news of the war; his words were received with frantic joy. Ciceruacchio was on the balcony in the glaring light of torches and candelabra, and beside him under the Italian flag stood four young women, all four Russians — was it not strange? I can see them now on that stone platform, and below them the swaying, innumerable multitude, mingling with shouts for war and curses for the Jesuits, loud cries of '*Evviva le donne forestiere!*'

In England they and we should have been greeted with hisses, abuse, and perhaps stones. In France we should have been taken for venal agents. But here the aristocratic proletariat, the descendants of Marius and the ancient tribunes, gave us a warm and genuine welcome. We were received by them into the European struggle ... and with Italy alone the bond of love, or at least of warm memory, is still unbroken.

And was all that ... intoxication, delirium? Perhaps — but I do not envy those who were not carried away by that exquisite dream. The sleep could not last long in any case: the inexorable Macbeth of real life had already raised his hand to murder sleep and ...

My dream was past – it has no further change.[6]

2

INTO THE STORM

On the evening of the 24th of June, coming back from the Place Maubert, I went into a café on the Quai d'Orsay. A few minutes later I heard discordant shouting, which came nearer and nearer. I went to the window: a grotesque comic *banlieue* was coming in from the surrounding districts to the support of order; clumsy, rascally fellows, half peasants, half shopkeepers, somewhat drunk, in wretched uniforms and old-fashioned shakos, they moved rapidly but in disorder, with shouts of 'Vive Louis-Napoléon!'

That ominous shout I now heard for the first time. I could not restrain myself, and when they reached the café I shouted at the top of my voice: 'Vive la République!' Those who were near the windows shook their fists at me and an officer muttered some abuse, threatening me with his sword; and for a long time afterwards I could hear their shouts of greeting to the man who had come to destroy half the revolution, to kill half the republic, to inflict himself upon France, as a punishment for forgetting in her arrogance both other nations and her own proletariat.

At eight o'clock in the morning of the 25th or 26th of June Annenkov and I went out to the Champs-Élysées. The cannonade we had heard in the night was now silent; only from time to time there was the crackle of rifle-fire and the beating of drums. The streets were empty, but the National Guards stood on either side of them. On the Place de la Concorde there was a detachment of the *Garde mobile*; near them were standing several poor women with brooms and some ragpickers and *concierges* from the houses near by. All their faces were gloomy and shocked. A lad of seventeen was leaning on a rifle and telling them something; we went up to them. He and all his comrades, boys like himself, were half drunk, their faces blackened with gunpowder and their eyes bloodshot from sleepless nights and drink; many were dozing with their chins resting on the muzzles of their rifles.

'And what happened then can't be described.' He paused, and then went on: 'Yes, and they fought well, too, but we paid them out for our comrades! What lots of them caught it! I stuck my bayonet up to my rifle-muzzle in five or six of them; they'll remember us,' he added, trying to assume the air of a hardened malefactor. The women were pale and silent; a man who looked like a *concierge* observed: 'Serve them right, the blackguards!' ... but this savage comment evoked not the slightest response. They were all of too ignorant a class to sympathise with the massacre and with the unfortunate boy who had been made into a murderer.

Silent and sad, we went to the Madeleine. Here we were stopped by a cordon of the National Guard. At first, after searching our pockets, they asked where we were going, and let us through; but the next cordon, beyond the Madeleine, refused to let us through and sent us back; when we went back to the first cordon we were stopped once more.

'But you saw us pass here just now!'

'Don't let them pass,' shouted an officer.

'Are you making fools of us, or what?' I asked.

'It's no use talking,' a shopman in uniform answered rudely. 'Take them up – and to the police: I know one of them' (he pointed at me); 'I have seen him more than once at meetings. The other must be the same sort too; they are neither of them Frenchmen. I'll answer for everything – march.'

We were taken away by two soldiers with rifles in front, two behind, and one on each side. The first man we met was a *représentant du peuple* with a silly badge in his button-hole; it was Tocqueville, who had written about America. I addressed myself to him and told him what had happened: it was not a joking matter; they kept people in prison without any sort of trial, threw them into the cellars of the Tuileries, and shot them. Tocqueville did not even ask who we were; he very politely bowed himself off, delivering himself of the following banality: 'The legislative authority has no right to interfere with the executive.' How could he have helped being a minister under Napoleon III!

The 'executive authority' led us along the boulevard to the Chaussée d'Antin to the *commissaire de police*. By the way, it will do no harm to mention that neither when we were arrested, nor when we were searched, nor when we were on our way, did I see a single policeman; all was done by the *bourgeois*-warriors. The boulevard was completely empty, all the shops were closed and the inmates rushed to their doors and windows when they heard our footsteps, and kept asking who we were: '*Des émeutiers étrangers,*' answered our escort, and the worthy *bourgeois* looked at us and gnashed their teeth.

From the police-station we were sent to the Hôtel des Capucines; the Ministry of Foreign Affairs had its quarters there, but at that time there was some temporary police committee there. We went with our escort into a large study. A bald old gentleman in spectacles, dressed entirely in black, was sitting alone at a table; he asked us over again all the questions that the *commissaire* had asked us.

'Where are your passports?'

'We never carry them with us when we are out for a walk.'

He took up a manuscript book, looked through it for a long time, apparently found nothing, and asked one of our escort:

'Why did you arrest them?'

'The officer gave the order; he says they are very suspicious characters.'

'Very well,' said the old gentleman; 'I will inquire into the case; you may go.'

When the escort had gone the old gentleman asked us to explain the cause of our arrest. I put the facts before him, adding that the officer might perhaps have seen me on the fifteenth of May at the Assembly; and then I told him of an incident of the previous day. I had been sitting in the Café Caumartin when suddenly there was a false alarm, a squadron of dragoons rode by at full gallop and the National Guard began to form ranks. Together with some five people who were in the café, I went up to a window; a National Guardsman standing below shouted rudely,

'Didn't you hear that windows were to be shut?'

His tone justified me in supposing that he was not addressing me, and I did not take the slightest notice of his words; besides, I was not alone, though I happened to be standing in front. Then the defender of order raised his rifle and, since this was taking place on the *rez-de-chaussée*, tried to thrust at me with his bayonet, but I saw his movement and stepped back and said to the others:

'Gentlemen, you are witnesses that I have done nothing to him — or is it the habit of the National Guard to bayonet foreigners?'

'*Mais c'est indigne, mais cela n'a pas de nom!*' my neighbours chimed in.

The frightened café-keeper rushed to shut the windows; a vile-looking sergeant appeared with an order to turn everyone out of the café. I fancied he was the same gentleman who had ordered us to be stopped. Moreover, the Café Caumartin was a couple of steps from the Madeleine.

'So that's how it is, gentlemen: you see what imprudence leads to. Why walk out at such a time? — minds are exasperated, blood is flowing....'

At that moment a National Guardsman brought in a maid-servant, saying that an officer had caught her in the very act of trying to post a letter addressed to Berlin. The old gentleman took the envelope and told the soldier to go.

'You can go home,' he said to us; 'only, please do not go by the same streets as before, and especially not by the cordon which arrested you. But stay, I'll send someone to escort you! he'll take you to the Champs-Élysées — you can get through that way.'

'And you,' he said, addressing the servant, giving her back the letter which he had not touched, 'post it in another letter-box, further away.'

And so the police gave protection from the armed *bourgeois*!

On the night of the 26th–27th of June, so Pierre Leroux relates, he went to Sénart to beg him to do something for the prisoners who were being suffocated in the cellars of the Tuileries. Sénart, a man well known as a desperate conservative, said to Pierre Leroux:

'And who will answer for their lives on the way? The National Guard will kill them. If you had come an hour earlier you would have found two colonels here: I had the greatest difficulty in bringing them to reason, and ended by telling them if these horrors went on I should give up the president's chair in the Assembly and take my place behind the barricades.'

Two hours later, on our returning home, the *concierge* made his appearance accompanied by a stranger in a dress coat and four men in workmen's blouses which badly disguised the moustaches of *municipales* and the deportment of gendarmes. The stranger un-buttoned his coat and waistcoat and, pointing with dignity to a tri-coloured scarf, said that he was Barlet, the *commissaire* of police (the man who on the 2nd of December, in the National Assembly, took by the collar the man who in his time had taken Rome – General Oudinot), and that he had orders to search my quarters. I gave him my key, and he set to work exactly as *politsmeyster* Miller had in 1834.[7]

My wife came in: the *commissaire*, like the officer of gendarmes who once came to us from Dubelt, began apologising. My wife looked at him calmly and directly and, when at the end of his speech he begged her indulgence, said:

'It would be cruelty on my part not to imagine myself in your place; you are sufficiently punished already by being obliged to do what you are doing.'

The *commissaire* blushed, but did not say a word. Rummaging among the papers and laying aside a whole heap of them, he suddenly went up to the fireplace, sniffed, touched the ashes and, turning to me with an important air, asked:

'What was your object in burning papers?'

'I haven't been burning papers.'

'Upon my word, the ash is still warm.'

'No, it is not warm.'

'*Monsieur, vous parlez à un magistrat!*'

'The ash is cold, all the same, though,' I said, flaring up and raising my voice.

'Why, am I lying?'

'What right have you to doubt my word? ... here are some *honest workmen* with you, let them test it. Besides, even if I had burnt papers: in the first place, I have a right to burn them; and in the second, what are you going to do?'

'Have you no other papers?'

'No.'

'I have a few letters besides, and very interesting ones; come into my room,' said my wife.

'Oh, your letters ...'

'Please don't stand on ceremony ... why, you are only doing your duty; come along.'

The *commissaire* went in, glanced very slightly at the letters, which were for the most part from Italy, and was about to go. ...

'But you haven't seen what is underneath here – a letter from the Conciergerie, from a prisoner, you see; don't you want to take it with you?'

'Really, Madame,' answered the policeman of the Republic, 'you are so prejudiced; I don't want that letter at all.'

'What do you intend to do with the Russian papers?' I asked.

'They will be translated.'

'The point is, where you will take your translator from. If he is from the Russian Embassy, it will be as good as laying information; you will destroy five or six people. You will greatly oblige me if you will mention at the *procés-verbal* that I beg most urgently that a Polish *émigré* shall be chosen as a translator.'

'I believe that can be done.'

'I thank you; and I have another request: do you know Italian at all?'

'A little.'

'I will show you two letters; in them the word France is not mentioned. The man who wrote them is in the hands of the Sardinian police; you will see by the contents that it will go badly with him if they get hold of the letters.'

'*Mais, ah ça!*' observed the commissaire, his dignity as a man beginning to be aroused; 'you seem to imagine that we are connected with the police of all the *despotic* powers. We have nothing to do with other countries. We are unwillingly compelled to take measures at home when blood is flowing in the streets and when foreigners interfere in our affairs.'

'Very well: then you can leave the letters here.'

The *commissaire* had not lied; he really did know *a little* Italian, and so, after turning the letters over, he put them in his pocket, promising to return them.

With that his visit ended. The letters from the Italian he gave back next day, but my papers vanished completely. A month passed; I wrote a letter to Cavaignac,[8] inquiring why the police did not return my papers nor say what they had found in them — a matter of very little consequence to them, perhaps, but of the greatest importance for my honour.

What gave rise to this last phrase was as follows. Several persons of my acquaintance had intervened on my behalf, considering the visit of the *commissaire* and the retention of my papers outrageous.

'We wanted to make certain,' Lamoricière[9] told them, 'whether he was not *an agent of the Russian Government.*'

This was the first time I heard of this abominable suspicion; it was something quite new for me. My life had been as open, as public, as though it had been lived in a glass hive, and now all at once this filthy accusation, and from whom? – from a republican government!

A week later I was summoned to the prefecture. Barlet was with me. We were received in Ducoux's room by a young official very like some free and easy Petersburg head-clerk.

'General Cavaignac,' he told me, 'has charged the Prefect to return your papers without any examination. The information collected concerning you renders it quite superfluous; no suspicion rests upon you; here is your portfolio. Will you be good enough to sign this paper first?'

It was a receipt stating that *all* the papers had been returned to me complete.

I stopped and asked whether it would not be more in order for me to look the papers through.

'They have not been touched. Besides, here is the seal.'

'The seal has not been broken,' observed Barlet soothingly.

'My seal is not here. Indeed, it was not put on them.'

'It is my seal, but you know you had the key.'

Not wishing to reply with rudeness, I smiled. This enraged them both: the head-clerk became the head of a department; he snatched up a penknife and, cutting the seal, said rudely enough: 'Pray look, if you don't believe, but I have not so much time to waste,' and walked out with a dignified bow. Their resentment convinced me that they really had not looked at the papers, and so, after a cursory glance at them, I signed the receipt and went home.

CHAPTER III

THE REVOLUTION OF 1848 IN FRANCE

I LEFT Paris in the autumn of 1847, without having formed any ties there; I remained completely outside the literary and political circles. There were many reasons for that. No immediate occasion of contact with them presented itself, and I did not care to seek one. To visit them simply in order to look at celebrities, I thought unseemly. Moreover, I particularly disliked the tone of condescending superiority which Frenchmen assume with Russians: they approve of us, encourage us, commend our pronunciation and our wealth; we put up with it all, and behave as though we were asking them a favour, or were even partly guilty, delighted when, from politeness, they take us for Frenchmen. The French overwhelm us with a flood of words, we cannot keep pace with them; we think of an answer, but they do not care to hear it; we are ashamed to show that we notice their blunders and their ignorance – they take advantage of all that with hopeless complacency.

To get on to a different footing with them one would have to impress them with one's consequence; to do this one must possess various rights, which I had not at that time, and of which I took advantage at once when they came to be at my disposal.

Moreover, it must be remembered that there are no people in the world with whom it is easier to strike up a nodding acquaintance than the French – and no people with whom it is more difficult to get on to really intimate terms. A Frenchman likes to live in company, in order to display himself, to have an audience, and in that respect he is as much a contrast to the Englishman as in everything else. An Englishman looks at people because he is bored; he looks at men as though from a stall in a theatre; he makes use of people as an entertainment, or as a means of obtaining information. The Englishman is always asking questions, the Frenchman is always giving answers. The Englishman is always wondering, always thinking things over; the Frenchman knows everything for certain, he is finished and complete, he will go no further: he is fond of preaching, talking, holding forth – about what, to whom, he does not care. He feels no need for personal intimacy; the café satisfies him completely. Like Repetilov in *Woe from Wit*, he does not notice that Chatsky is gone and Skalozub is in his place, that Skalozub is gone and Zagoretsky is in his place – and goes on holding forth about the jury-room, about Byron (whom he calls 'Biron'), and other important matters.

Coming back from Italy not yet cooled from the February Revolution, I stumbled on the 15th of May, and then lived through the agony of the June days and the state of siege. It was then that I obtained a deeper insight into the *tigre-singe* of Voltaire – and I lost even the desire to become acquainted with the mighty ones of this republic.

On one occasion a possibility almost arose of common work which would have brought me into contact with many persons, but that did not come off either. Count Ksawery Branicki[1] gave 70,000 francs to found a magazine to deal principally with foreign politics and other nations, and especially with the Polish question. The usefulness and appropriateness of such a magazine were obvious. French papers deal little and badly with what is happening outside France; during the republic, they thought it sufficient to encourage all the heathen nations now and then with the phrase *solidarité des peuples*, and the promise that as soon as they had time to turn round at home they would build a world-wide republic based upon universal brotherhood. With the means at the disposal of the new magazine, which was to be called *La Tribune des Peuples*, it might have been made the international *Moniteur* of movement and progress. Its success was the more certain because there is no international periodical at all; there are sometimes excellent articles in *The Times* and the *Journal des Débats* on special subjects, but they are occasional and disconnected. The *Augsburg Gazette* would really be the most international organ if its *black-and-yellow* proclivities were not so glaringly conspicuous.

But it seems that all the good projects of the year 1848 were doomed to be born in their seventh month and to die before cutting their first tooth. The magazine turned out poor and feeble – and died at the slaughter of the innocent papers after the 14th of June, 1849.

When everything was ready and standing by, a house was taken and fitted up with big tables covered with cloth and little sloping desks; a lean French *littérateur*[2] was engaged to watch over international mistakes in spelling; a committee to edit it was set up of former Polish nuncios and senators, and Mickiewicz was appointed head to this with Chojecki[3] as his assistant; – all that was left to arrange was a triumphal opening ceremony, and what date could be more suitable for that than the anniversary of February the 24th,[4] and what form could it more decently take than a supper?

The supper was to take place at Chojecki's. When I arrived I found a good many guests already there, and among them scarcely a single Frenchman; to make up for this other nationalities, from the Sicilians to the Croats, were well represented. I was really interested in one person only – Adam Mickiewicz; I had never seen him before. He was standing by the fireplace with his elbow on the marble mantelpiece. Anyone who had seen his portrait in the French edition of his works, taken, I believe, from the medallion executed by David d'Angers,[5] could have recognised him at once in spite of the great change wrought by the years. Many thoughts and sufferings had passed over his face, which was rather Lithuanian than Polish. The whole impression made by this figure, by his head, his luxuriant grey hair and weary eyes, was suggestive of unhappiness endured, of acquaintance with spiritual pain, and of the exaltation of sorrow – he was the moulded likeness of the fate of Poland. The same impression was made on me later by the face of Worcell,[6] though the features of the latter, while even more expressive of suffering, were more animated and gracious than those of Mickiewicz. It seemed as though Mickiewicz were held back, preoccupied, distracted by something: that 'something' was the strange mysticism into which he retreated further and further.

I went up to him and he began questioning me about Russia: his information was fragmentary; he knew little of the literary movement after Pushkin, having stopped short at the time when he left Russia.[7] In spite of his basic idea of a fraternal league of all the Slavonic peoples – a conception he was one of the first to develop – he retained some hostility to Russia. And indeed it could hardly be otherwise after all the atrocities perpetrated by the Tsar and his satraps; besides, we were speaking at a time when the terrorism of Nicholas was at its very worst.

The first thing that surprised me disagreeably was the attitude to him of the Poles, his followers: they appoached him as monks approach an abbot, with self-abasement and reverent awe; some of them kissed him on the shoulder. He must have been accustomed to these expressions of submissive affection, for he accepted them with great *laisser aller*. To be recognised by people of the same way of thinking, to have influence on them, to see their affection, is desired by everyone who is devoted, body and soul, to his convictions and lives by them; but external signs of sympathy and respect I should not like to accept – they destroy equality and consequently freedom. Moreover, in that respect we can never catch up with bishops, heads of departments, and colonels of regiments.

Chojecki told me that at the supper he was going to propose a toast 'to the memory of the 24th of February, 1848', that Mickiewicz would respond with a speech in which he would expound his views and the spirit of the new magazine; he wished me as a Russian to reply to Mickiewicz. Not being accustomed to public speaking, especially without preparation, I declined his invitation, but promised to propose the health of Mickiewicz and to add a few words describing how I had first drunk his health in Moscow at a public dinner given to Granovsky in the year 1843.[8] Khomyakov had raised his glass with the words, 'To the great Slavonic poet who is absent!' The name (which we dared not pronounce) was not needed; everyone raised his glass and, standing in silence, drank to the health of the exile. Chojecki was satisfied. Having thus arranged our *extempore* speeches, we sat down to the table. At the end of the supper, Chojecki proposed his toast. Mickiewicz got up and began speaking. His speech was elaborate and clever, and extremely adroit – that is to say Barbès[9] and Louis-Napoleon could both have applauded it sincerely; it made me wince. As he developed his thought I began to feel painfully distressed and, that not the slightest doubt might be left, waited for one word, one name: it was not slow to appear![10]

Mickiewicz worked up to the theme that democracy was now preparing to enter a new, open camp, at the head of which stood France; that it would once more rush to the liberation of all oppressed peoples under the same eagles, under the same standards, at the sight of which all tsars and powers had turned pale; and that it would once more be led forward by a member of that dynasty which had been crowned by the people, and, as it seemed, ordained by Providence itself to guide revolution by the well-ordered path of authority and victory.[11]

When he had finished a general silence followed, except for two or three exclamations of approval from his adherents. Chojecki was very well aware of Mickiewicz's blunder and, wishing to efface the effect of the speech as quickly as possible, came up with a bottle, filled my glass and whispered to me:

'Well?'

'I am not going to say a word after that speech.'

'Please do say something.'

'Nothing will induce me.'

The silence continued; some people kept their eyes fixed on their plates, others scrutinised their glasses, others fell into private conversation with their neighbours. Mickiewicz changed colour; he wanted to say something more, but a loud *'Je demande la parole'* put an end to the painful situation. Everyone turned to the man who had risen to his feet. A rather short man of about seventy, grey-haired, with a fine vigorous exterior, stood with a glass in his trembling hand; anger and indignation were apparent in his large, black eyes and excited face. It was Ramon de la Sagra.[12]

'To the 24th of February,' he said: 'that was the toast proposed by our host. Yes, to the 24th of February, and to the downfall of every despotism whatever its name is, king or emperor, Bourbon or Bonaparte. I cannot share the views of our friend Mickiewicz – he can look at things like a poet, and from his own point of view he is right; but I don't want his words to pass without protest in such a gathering'; and so he went on and on, with all the fire of a Spaniard and the authority of an old man.

When he had finished, twenty glasses, mine among them, were held out to clink with his.

Mickiewicz tried to retrieve his position, and said a few words of explanation, but they were unsuccessful. De la Sagra did not give way. Everyone got up from the table and Mickiewicz went away.

There could scarcely have been a worse omen for the new journal; it succeeded in existing after a fashion till the 13th of June, and its disappearance was as little noticed as its existence. There could be no unity in the editing of it. Mickiewicz had rolled up half his imperial banner *usé par la gloire*. The others did not dare to unfurl theirs; hampered both by him and by the committee many of the contributors abandoned the journal at the end of the month; I never sent them a single line. If the police of Napoleon had been more intelligent the *Tribune des Peuples* would never have been prohibited for a few lines on the 13th of June. With Mickiewicz's name and devotion to Napoleon, with its revolutionary mysticism and its dream of a democracy in arms, with the Bonapartes at its head, the journal might have become a veritable treasure for the President, the clean organ of an unclean cause.

Catholicism, so alien to the Slavonic genius, has a destructive effect upon it. When the Bohemians no longer had the strength to resist Catholicism, they were crushed; in the Poles Catholicism has developed that mystical exaltation which supports them perpetually in their world of phantoms. If they are not under the direct influence of the Jesuits, then instead of liberty they either invent some idol for themselves, or come under the influence of some visionary. Messianism, that mania of Wronski's, that delirium of Towjanski's, had turned the brains of hundreds of Poles, Mickiewicz himself[13] among them. The worship of Napoleon stands in the foreground of this insanity. Napoleon had done nothing for them; he had no love for Poland, but he liked the Poles who shed their blood for him with the titanic, poetic courage displayed in their famous cavalry attack of Sommo Sierra.[14] In 1812 Napoleon said to Narbonne:[15] 'I want a camp in Poland, not a forum. I will not permit either Warsaw or Moscow to open a club for demagogues' – and of this man the Poles made a military incarnation of God, setting him on a level with Vishnu and Christ.

Late one winter evening in 1848 I was walking with one of the Polish followers of Mickiewicz along the Place Vendôme. When we reached the column the Pole took off his cap.'Is it possible? . . .' I thought, hardly daring to believe in such stupidity, and meekly asked what was his reason for taking off his cap. The Pole pointed to the bronze emperor. How can we expect men to refrain from domineering or oppressing others when it wins so much devotion!

Mickiewicz's private life was dark; there was something unfortunate about it, something gloomy, some 'visitation of God'. His wife was for a long time out of her mind. Towjanski recited incantations over her, and is said to have done her good; this made a great impression on Mickiewicz, but traces of her illness remained . . . things went badly with them. The last years of the great poet, who outlived himself, were spent in gloom. He died in Turkey while taking part in an absurd attempt to organise a Cossack legion, which the Turkish Government would not permit to be called Polish. Before his death he wrote a Latin ode to the honour and glory of Louis-Napoleon.

After this unsuccessful attempt to take part in the magazine I withdrew even more into a small circle of friends, enlarged by the arrival of new émigrés. Formerly I had sometimes visited a club, and I had participated in three or four banquets, that is I had eaten cold mutton and drunk sour wine, while I listened to Pierre Leroux or Father Cabet and joined in the 'Marseillaise'. Now I was sick of that, too. With profound sorrow I watched and recorded the success of the forces of dissolution and the decline of the republic, of France, of Europe. From Russia came no gleam of light in the distance, no good news, no friendly greeting: people had given up writing to me; personal, intimate, family relations were suspended. Russia lay speechless, as though dead, covered with bruises, like an unfortunate peasant-woman at the feet of her master, beaten by his heavy fists. She was then entering upon those fearful five years from which she is at last emerging now that Nicholas[16] is buried.

Those five years were for me, too, the worst time of my life; I have not now such riches to lose or such beliefs to be destroyed. . . .

. . . The cholera raged in Paris; the heavy air, the sunless heat produced a languor; the sight of the frightened, unhappy population and the rows of hearses which started racing each other as they drew near the cemeteries — all this corresponded with what was happening.

The victims of the pestilence fell near by, at one's side. My mother drove to St Cloud with a friend, a lady of five-and-twenty. When they were coming back in the evening, the lady felt rather unwell; my mother persuaded her to stay the night with us. At seven o'clock the next morning they came to tell me that she had cholera. I went in to see her, and was aghast. Not one feature was unchanged; she was still handsome; but all the muscles of her face were drawn and contracted and dark shadows lay under her eyes. With great difficulty I succeeded in finding Rayer[17] at the Institute, and brought him home with me. After glancing at the sick woman, Rayer whispered to me:

'You can see for yourself what is to be done here.' He prescribed something and went away.

The sick woman called me and asked:

'What did the doctor say? He did tell you something, didn't he?'

'To send for your medicine.'

She took my hand, and her hand amazed me even more than her face: it had grown thin and angular as though she had been through a month of serious illness since she had fallen sick: she fixed upon me a look that was full of suffering and horror and said:

'Tell me, for God's sake, what he said ... is it that I am dying? ... You are not afraid of me, are you?' she added.

I felt fearfully sorry for her at that moment; that frightful consciousness not only of death, but of the infectiousness of the disease that was rapidly sapping her life, must have been intensely painful. Towards the morning she died.

Ivan Turgenev was about to leave Paris; the lease of his flat was up, and he came to me for a night. After dinner he complained of the suffocating heat; I told him that I had had a bath in the morning; in the evening he too went for a bath. When he came back he felt unwell, drank some soda-water with some wine and sugar in it, and went to bed. In the night he woke me.

'I am a lost man,' he said; 'it's cholera.'

He really was suffering from sickness and spasms; fortunately he escaped with ten days' illness.

After burying her friend my mother had moved to the Ville d'Avray. When Turgenev was taken ill I sent Natalie and the children there and remained alone with him; when he was a great deal better I moved there too.

On the morning of June the 12th Sazonov came to see me there. He was in the greatest exaltation: he talked of the popular outbreak that was impending, of the certainty of its being successful, of the glory awaiting those who took part in it, and urgently pressed me to join in reaping the laurels. I told him that he knew my opinion of the present state of affairs – that it seemed to me stupid, without believing in it, to co-operate with people with whom one had hardly anything in common.

To this the enthusiastic agitator remarked that of course it was quieter and safer to stay at home and write sceptical articles while others were in the market-place championing the liberty of the world, the solidarity of peoples, and much else that was good.

A very vile emotion, but one that has led and will lead many men into great errors, and even crimes, impelled me to say:

'But what makes you imagine I am not going?'

'I concluded that from your words.'

'No: I said it was stupid, but I didn't say that I never do anything stupid.'

'That is just what I wanted! That's what I like you for! Well, it's no use losing time; let us go to Paris. This evening the Germans and other refugees are meeting at nine o'clock; let us go to them first.'

'Where are they meeting?' I asked him in the train.

'In the Café Lamblin, in the Palais Royal.'

This was my first surprise.

'In the Café Lamblin?'

'That is where the "reds" usually meet.'

'That's just why I think that today they ought to have met somewhere else.'

'But they are all used to going there.'

'I suppose the beer is very good!'

In the café various *habitués* of the revolution were sitting with dignity at a dozen little tables, looking darkly and consequentially about them from under wide-brimmed felt hats and caps with tiny peaks. These were the perpetual suitors of the revolutionary Penelope, those inescapable actors who take part in every popular demonstration and form its *tableau*, its background, and who are as menacing from afar as the paper dragons with which the Chinese wished to intimidate the English.

In the troubled times of social storms and reconstructions in which states forsake their usual grooves for a long time, a new generation of people grows up who may be called the choristers of the revolution; grown on shifting, volcanic soil, nurtured in an atmosphere of alarm when work of every kind is suspended, they become inured from their earliest years to an environment of political ferment, and like the theatrical side of it, its brilliant, pompous *mise en scène*. Just as to Nicholas marching drill was the most important part of the soldier's business, to them all those banquets, demonstrations, protests, gatherings, toasts, banners, are the most important part of the revolution.

Among them there are good, valiant people, sincerely devoted and ready to face a bullet; but for the most part they are very limited and extraordinarily pedantic. Immobile conservatives in everything revolutionary, they stop short at some programme and do not advance.

Dealing all their lives with a small number of political ideas, they only know their rhetorical side, so to speak, their sacerdotal vestments, that is the commonplaces which successively cut the same figure, *à tour de rôle*, like the ducks in the well-known children's toy – in newspaper articles, in speeches at banquets and in parliamentary devices.

In addition to naïve people and revolutionary doctrinaires, the unappreciated artists, unsuccessful literary men, students who did not complete their studies, briefless lawyers, actors without talent, persons of great vanity but small capability, with huge pretensions but no perseverance or power of work, all naturally drift into this *milieu*. The external authority which guides and pastures the human herd in a lump in ordinary times is weakened in times of revolution; left to themselves people do not know what to do. The younger generation is struck by the ease, the apparent ease, with which celebrities float to the top in times of revolution, and rushes into futile agitation; this inures the young people to violent excitements and destroys the habit of work. Life in the clubs and cafés is attractive, full of movement, flattering to vanity and free from restraint. One must not be left behind, there is no need to work: what is not done today may be done tomorrow, or may even not be done at all.

⋯

... The spectacle of the Café Lamblin was still new to me; at that time I was not familiar with the back premises of the revolution. It is true that I had been about in Rome and in the Cafe delle Belle Arti and in the square; I had been in the Circolo Romano and in the Circolo Popolare; but the movement in Rome had not then that character of political garishness which particularly developed after the failures of 1848. Ciceruacchio and his friends had a *naïveté* of their own, their southern gesticulations which strike one as commonplace and their Italian phrases which seem to us to be rant; but they were in a period of youthful enthusiasm, they had not yet come to themselves after three centuries of sleep. *Il popolano* Ciceruacchio was not in the least a political agitator by trade; he would have liked nothing better than to retire once more in peace to his little house in Strada Ripetta and to carry on his trade in wood and timber within his family-circle like a *pater-familias* and free *civis romanus*.

The men surrounding him were free from that brand of vulgar, babbling pseudo-revolutionism, of that *taré* character which is so dismally common in France.

I need hardly say that in speaking of the café agitators and revolutionary *lazzaroni* I was not thinking of those mighty workers for the emancipation of humanity, those martyrs for the love of their fellow-creatures and fiery evangelists of independence whose words could not be suppressed by prison, exile, proscription, or poverty — of the drivers, the motive powers of events, by whose blood, tears, and words a new historical order is established. I was talking about the incrusted border covered with barren weeds, for which agitation itself is goal and reward, who like the process of national revolution for its own sake, as Chichikov's Petrushka[18] liked the process of reading, or as Nicholas liked military drill.

. . .

In the Café Lamblin, where the desperate *citoyens* were sitting over their *petits verres* and big glasses, I learned that they had no plan, that the movement had no real centre of momentum and no programme. Inspiration was to descend upon them as the Holy Ghost once descended upon the heads of the apostles. There was only one point on which all were agreed — *to come to the meeting-place unarmed*. After two hours of empty chatter we went off to the office of the *True Republic*, agreeing to meet at eight o'clock next morning at the Boulevard Bonne Nouvelle, facing the Château d'Eau.

The editor was not at home: he had gone to the 'Montagnards'[19] for instructions. About twenty people, for the most part Poles and Germans, were in the big, grimy, poorly lit and still more poorly furnished room which served the editorial board as an assembly hall and a committee room. Sazonov took a sheet of paper and began writing something; when he had written it he read it out to us: it was a protest in the name of the *émigrés* of all nationalities against the occupation of Rome, and a declaration of their readiness to take part in the movement. Those who wished to immortalise their names by associating them with the glorious morrow he invited to sign it. Almost all wished to immortalise their names, and signed. The editor came in, tired and dejected, trying to suggest to everyone that he knew a great deal but was bound to keep silent; I was convinced that he knew nothing at all.

'*Citoyens,*' said Thorez,[20] '*la Montagne est en permanence.*'

Well, who could doubt its success – *en permanence*! Sazonov gave the editor the protest of the democracy of Europe. The editor read it through and said:

'That's splendid, splendid! France thanks you, *citoyens*; but why the signatures? There are so few that if we are unsuccessful our enemies will vent all their anger upon you.'

Sazonov insisted that the signatures should remain; many agreed with him.

'I won't take the responsibility for it,' the editor objected; 'excuse me, I know better than you the people we have to deal with.'

With that he tore off the signatures and delivered the names of a dozen candidates for immortality to a holocaust in the candle, and the text he sent to the printer.

It was daybreak when we left the office; groups of ragged boys and wretched, poorly dressed women were standing, sitting, and lying on the pavement near the various newspaper offices, waiting for the piles of newspapers – some to fold them, and others to run with them all over Paris. We walked out on to the boulevard: there was absolute stillness; now and then one came upon a patrol of National Guards, and police-sergeants strolled about looking slyly at us.

'How free from care the city sleeps,' said my comrade, 'with no foreboding of the storm that will wake it up tomorrow!'

'Here are those who keep vigil for us all,' I said to him, pointing upwards – that is, to a lighted window of the *Maison d'Or*.

'And very appropriately, too. Let us go in and have some absinthe; my stomach is a bit upset.'

'And I feel empty; it wouldn't be amiss to have some supper too. How they eat in the Capitole I don't know, but in the Conciergerie the food is abominable.'

From the bones left after our meal of cold turkey no one could have guessed either that cholera was raging in Paris, or that in two hours' time we were going to change the destinies of Europe. We ate at the *Maison d'Or* as Napoleon slept before Austerlitz.

Between eight and nine o'clock, when we reached the Boulevard Bonne Nouvelle, numerous groups of people were already standing there, evidently impatient to know what they were to do; their faces showed perplexity, but at the same time something in the peculiar look of the groups manifested great exasperation. Had those people found real leaders the day would not have ended in a farce.

There was a minute when it seemed to me that something was really going to happen. A gentleman rode on horseback rather slowly down the boulevard. He was recognised as one of the ministers (Lacroix), who probably was having a ride so early not for the sake of fresh air alone. He was surrounded by a shouting crowd, who pulled him off his horse, tore his coat and then let him go — that is, another group rescued him and escorted him away. The crowd grew; by ten o'clock there may have been twenty-five thousand people. No one we spoke to, no one we questioned, knew anything. Chersosi, a *carbonaro* of old days assured us that the *banlieue* was coming to the Arc de Triomphe with a shout of 'Vive la République!'

'Above all,' the elders of the democracy repeated again, 'be un-armed, or you will spoil the character of the affair — the sovereign people must show the National Assembly its will peacefully and solemnly in order to give the enemy no occasion for calumny.'

At last columns were formed; we foreigners made up an honorary phalanx immediately behind the leaders, among whom were E. Arago[21] in the uniform of a colonel, Bastide,[22] a former minister, and other celebrities of 1848. We moved down the boule-vard, voicing various cries and singing the 'Marseillaise'. One who has not heard the 'Marseillaise', sung by thousands of voices in that state of nervous excitement and irresolution which is inevit-able before certain conflict, can hardly realise the overwhelming effect of the revolutionary hymn.

At that minute there was really something grand about the demonstration. As we slowly moved down the boulevards all the windows were thrown open; ladies and children crowded at them and came out on to the balconies; the gloomy, alarmed faces of their husbands, the fathers and proprietors, looked out from be-hind them, not observing that in the fourth storeys and attics other heads, those of poor seamstresses and working girls, were thrust out — they waved handkerchiefs, nodded and greeted us. From time to time, as we passed by the houses of well-known people, various shouts were uttered.

In this way we reached the point where the Rue de la Paix joins

the boulevards; it was closed by a squad of the Vincennes Chasseurs, and when our column came up to it the chasseurs suddenly
moved apart like the scenery in a theatre, and Changarnier,[23]
mounted upon a small horse, galloped up at the head of a squadron
of dragoons. With no summons to the crowd to disperse, with no
beat of drum or other formalities prescribed by law, he threw
the foremost ranks into confusion, cut them off from the others
and, deploying the dragoons in two directions ordered them to
clear the street in quick time. The dragoons in a frenzy fell to
riding down people, striking them with the flat of their swords
and using the edge at the slightest resistance. I hardly had time to
take in what was happening when I found myself nose to nose
with a horse which was almost snorting in my face, and a dragoon
swearing likewise in my face and threatening to give me one with
the flat if I did not move aside. I retreated to the right, and in an
instant was carried away by the crowd and squeezed against the
railings of the Rue Basse des Remparts. Of our rank the only one
left beside me was Müller-Strübing.[24] Meanwhile the dragoons
were pressing back the foremost ranks with their horses, and
people who had no room to get away were thrust back upon us.
Arago leaped down into the Rue Basse des Remparts, slipped and
dislocated his leg; Strübing and I jumped down after him. We
looked at each other in a frenzy of indignation; Strübing turned
round and shouted loudly: '*Aux armes! Aux armes!*' A man in a
workman's blouse caught him by the collar, shoved him out of
the way and said:

'Have you gone mad? Look there!'

Thickly bristling bayonets were moving down the street – the
Chaussée d'Antin it must have been.

'Get away before they hear you and cut off all escape. All is lost,
all!' he added, clenching his fist; he hummed a tune as though
there was nothing the matter, and walked rapidly away. We made
our way to the Place de la Concorde. In the Champs-Élysées there
was not a single squad from the *banlieue*; why, Chersosi must
have known that there was not. It had been a diplomatic lie to save
the situation, and it would perhaps have been the destruction of
anyone who had believed it.

The shamelessness of attacking unarmed people aroused great
resentment. If anything really had been prepared, had there been
leaders, nothing would have been easier than for fighting to have

begun in earnest. Instead of showing itself in its full strength the *Montagne*, on hearing how ludicrously the sovereign people had been dispersed by horses, hid itself behind a cloud. Ledru-Rollin[25] carried on negotiations with Guinard.[26] Guinard, the artillery commander of the National Guard, wanted to join the movement, wanted to give men, agreed to give cannon, but would not on any consideration give ammunition – he seems to have wished to act by the moral influence of the guns; Forestier[27] was doing the same with his legion. Whether this helped them much we saw by the Versailles trial.[28] Everyone wanted to do something, but no one dared; the most foresight was shown by some young men who hoped for a new order – they bespoke themselves prefects' uniforms, which they declined to take after the failure of the movement, and the tailor was obliged to hang them up for sale.

When the hurriedly rigged-up government was installed at the *Arts et Métiers* the workmen, after walking about the streets with inquiring faces and finding neither advice nor leadership, went home, convinced once more of the bankruptcy of the *Montagnard* fathers of the country: perhaps they gulped down their tears like the man who said to us, 'All is lost!' – or perhaps laughed in their sleeves at the way the *Montagne* had been tousled.

But the dilatoriness of Ledru-Rollin, the pedantry of Guinard – these were the external causes of the failure, and were just as *à propos* as are decisive characters and fortunate circumstances when they are needed. The internal cause was the poverty of the republican idea in which the movement originated. Ideas that have outlived their day may hobble about the world for years – may even, like Christ, appear after death once or twice to their devotees; but it is hard for them ever again to lead and dominate life. Such ideas never gain complete possession of a man, or gain possession only of incomplete people. If the *Montagne* had been victorious on the 13th of June, what would it have done? There was nothing new they could call their own. It would have been a photograph in black and white of the grim, glowing Rembrandt or Salvator Rosa picture of 1793 without the Jacobins, without the war, without even the naïve guillotine. . . .

After the 13th of June and the attempted rising at Lyons,[29] arrests began. The mayor came to us with the police at Ville d'Avray to look for Karl Blind[30] and Arnold Ruge; some of our acquaintances were seized. The Conciergerie was full to overflowing.

In one small room there were as many as sixty men; in the middle stood a large slop-bucket, which was emptied once in the twenty-four hours – and all this in civilised Paris, with the cholera raging. Having not the least desire to spend some two months among those comforts, fed on rotten beans and putrid meat, I got a passport from a Moldo-Wallachian and went to Geneva.*

Transport in France in those days was still run by Lafitte and Caillard. The *diligences* were put on the railway line, then taken off – at Châlons, I remember – and put on again somewhere else. A lean, sunburnt gentleman with a clipped moustache and a rather unpleasant appearance got into the carriage with me, and looked at me suspiciously; he had a small travelling-bag, and a sword wrapped in oil-cloth. He was obviously a police-sergeant in disguise. He scanned me carefully from head to foot, then planted himself in the corner and did not utter a single word. At the first station he called up the guard and told him that he had left behind an excellent map, and would be grateful for a scrap of paper and an envelope. The guard said they only had three minutes before the bell would ring; the sergeant jumped out, and returned to inspect me more suspiciously than ever. For four hours the silence continued: even my permission to smoke he asked for without speaking; I answered in the same way with my head and my eyes, and took out a cigar. When it began to grow dark he asked me,

'Are you going to Geneva?'

'No, to Lyons,' I answered.

'Ah!' With that the conversation ended.

A little while later the door opened and the guard with difficulty thrust in a bald-headed individual, in a roomy, pea-green overcoat and a waistcoat of various colours, with a thick stick, a bag, an umbrella and a huge belly. When this typical figure of the virtuous uncle had installed himself between the sergeant and me, I asked him, before he had time to recover his breath:

'*Monsieur, vous n'avez pas d'objection?*'

* How well founded my apprehensions were was shown by a police search of my mother's house at Ville d'Avray two days after my departure. They seized all the papers, even the correspondence of her maid with my cook. I thought it inopportune to publish my account of the 13th of June at the time.

Coughing, mopping his face, and tying a silk handkerchief round his head, he answered:

'Not in the least, by all means; my son who is in Algiers is always smoking, *il fume toujours*'; and with this good opening he began chatting and telling us stories. Half an hour later, he was already asking me where I had come from and where I was going. Hearing that I came from Wallachia, he added with characteristic French politeness, 'Ah, *c'est un beau pays*,' though he did not know for certain whether it was in Turkey or in Hungary.

My neighbour answered his questions very laconically.

'*Monsieur est militaire?*'

'*Oui, monsieur.*'

'*Monsieur a été en Algérie?*'

'*Oui, monsieur.*'

'My eldest son, too, he is there now. In Oran,[31] I suppose?'

'*Non, monsieur.*'

'And in your country are there *diligences?*'

'Between Jassy and Bucharest,' I answered with inimitable assurance. 'Only with us *diligences* are drawn by oxen.'

This greatly astonished my neighbour, and I am sure he would have taken his oath that I was a Wallachian; after this fortunate piece of detail, even the sergeant was softened and became more conversational.

At Lyons I took my portmanteau and went at once to another *diligence* booking-office, climbed up to the *impériale* and five minutes later was galloping along the road to Geneva. At the last big town before the frontier a *commissaire* of police was sitting with a clerk in the square before the police-station; gendarmes were standing about, and a preliminary examination of passports was held. The description in my passport did not quite fit me and so, getting down from the *impériale*, I said to the gendarme:

'*Mon brave*, where could we quickly drink a glass of wine together? Show me; the heat is insufferable.'

'Why, there's my own sister's café not two steps away.'

'But what about my passport?'

'Give it here. I'll hand it over to my comrade; he will bring it to us.'

A minute later the gendarme and I had emptied a bottle of Beaune in his own sister's café, and five minutes later his friend brought the passport. I presented him with a glass, he put his hand to his hat and we returned to the *diligence* friends. The first time all went well. We reached the frontier; there was a river, and over the river a bridge, and on the other side of the bridge the Piedmontese custom-house. French gendarmes were running about in all directions on the bank, looking for Ledru-Rollin, who had crossed the frontier long before, or at least for Félix Pyat,[32] who would nevertheless cross it later, and like me with a Wallachian passport.

The guard told us that here our papers would be finally looked at, that this would take rather a long time – about half an hour – and so he advised us to have something to eat at the posting inn. We went in, and had no sooner sat down than another Lyons *diligence* galloped up; the passengers came in, and foremost among them was my sergeant. The devil! I had told him that I was going to Lyons. We bowed frigidly; he, too, seemed surprised; however, he did not say a word.

A gendarme came in and distributed passports; the *diligences* were already on the other side of the river. 'Kindly cross the bridge on foot, gentlemen.'

Now there will be trouble, I thought. We went out ... and there we were on the bridge – no trouble; and now we were over the bridge – still no trouble.

'Ha–ha–ha!' the sergeant laughed nervously. 'So we've got across! Ough! it's like a load off one's back.'

'What?' said I, 'are you ...'

'Why, you too, it seems?'

'Upon my word,' I answered, laughing heartily. 'I am straight from Bucharest; came all the way as if by ox-cart.'

'You were lucky!' the guard said to me, holding up his finger. 'You must be more careful next time. Why did you give two francs as a tip to the boy who brought you to the office? It's a good thing he is *one of us* too; he said to me at once, "He must be a red; he didn't stop a minute at Lyons, and he was so pleased to get a seat that he gave me two francs." "You hold your tongue, it's not your business," I said to him, "or some beast of a policeman will hear, and perhaps he'll stop him." '

Next day we arrived at Geneva, the old haven of refuge for the persecuted. 'At the time of the king's death a hundred and fifty families,' says Michelet in his history of the sixteenth century, 'escaped to Geneva; a little later, another fourteen hundred. The refugees from France and the refugees from Italy founded the real Geneva, that wonderful sanctuary among three nations; with no support, afraid of the very Swiss, it maintained itself by its moral force alone.'

Switzerland at this time was the meeting-place in which the survivors from European political movements gathered together from all parts. Representatives of all the unsuccessful revolutions were shifting about between Geneva and Basle, crowds of militia-men were crossing the Rhine, others were descending the St Gothard or coming from beyond the Jura. The cowardly Federal Government did not yet dare to expel them openly; the cantons still held fast to their ancient, sacred right of sanctuary.

All the people whose names were on everybody's lips, whom I loved at a distance and was now hurrying to meet, were passing through Geneva as though on parade at a review, stopping there to rest and going on again. . . .

*

CHAPTER IV

WESTERN-EUROPEAN ARABESQUES

Notebook the Second

IL PIANTO

AFTER the June days I saw that the revolution was vanquished, but I still believed in the vanquished, in the fallen; I believed in the wonder-working power of the relics, in their moral strength. At Geneva I began to understand more and more clearly that the revolution not only had been vanquished, but had been bound to be vanquished.

My head was dizzy with my discoveries, an abyss was opening before my eyes and I felt that the ground was giving way under my feet.

It was not the reaction that vanquished the revolution. The reaction showed itself everywhere densely stupid, cowardly, in its dotage; everywhere it retreated ignominiously round the corner before the shock of the popular tide, furtively biding its time in Paris, and at Naples, Vienna, and Berlin. The revolution fell, like Agrippina, under the blows of its own children, and, what was worse than anything, without their being conscious of it; there was more heroism, more youthful self-sacrifice, than good judgment; and the pure, noble victims fell, not knowing for what. The fate of the survivors was almost more grievous. Absorbed in wrangling among themselves, in personal disputes, in melancholy self-deception and consumed by unbridled vanity, they kept dwelling on their unexpected days of triumph, and were unwilling to take off their faded laurels or wedding garments, though it was not the bride who had deceived them.

. . .

My heart almost broke at these painful truths; I had to live through a difficult page of my education.

... I was sitting mournfully one day in my mother's dining-room at gloomy, disagreeable Zürich; this was at the end of

December 1849. I was going next day to Paris. It was a cold, snowy day; two or three logs, smoking and crackling, were unwillingly burning on the hearth. Everyone was busy packing; I was sitting quite alone. My life at Geneva floated before my mind's eye; everything ahead looked dark; I was afraid of something, and it was so unbearable that if I could have, I would have fallen on my knees and wept and prayed; but I could not and instead of a prayer I wrote my *curse* – my *Epilogue to 1849*.[1]

'Disillusionment, fatigue, *Blasiertheit!*' The democratic critics said of those lines I vomited up. Disillusionment, yes! Fatigue, yes! ... Disillusionment is a vulgar, hackneyed word, a veil under which lie hidden the sloth of the heart, egoism posing as love, the noisy emptiness of vanity with pretensions to everything and strength for nothing. All these exalted, unrecognised characters, weazened with envy and wretched from pretentiousness, have long wearied us in life and in novels. All that is perfectly true; but is there not something real, peculiarly characteristic of our times, at the bottom of these frightful spiritual sufferings which degenerate into absurd parodies and vulgar masquerades?

The poet who found words and voice for this malady was too proud to pose and to suffer for the sake of applause; on the contrary, he often uttered his bitter thought with so much humour that his kind-hearted readers almost died of laughing. Byron's disillusionment was more than caprice, more than a personal mood; Byron was shattered because life deceived him. And life deceived him not because his demands were unreal, but because England and Byron were of two different ages, of two different educations, and met just at the epoch when the fog was dispersing.

This rupture existed in the past, too, but in our age it has come to consciousness; in our age the impossibility of the intervention of any beliefs is becoming more and more manifest. After the break-up of Rome came Christianity; after Christianity, the belief in civilisation, in humanity. Liberalism is the *final religion*, though its church is not of the other world but of this. Its theology is political theory; it stands upon the earth and has no mystical conciliations, for it must have conciliation in fact. Triumphant and then defeated liberalism has revealed the rift in all its nakedness; the painful consciousness of this is expressed in the irony of modern man, in the scepticism with which he sweeps away the fragments of his shattered idols.

Irony gives expression to the vexation aroused by the fact that logical truth is not the same as the truth of history, that as well as dialectical development it has its own development through chance and passion, that as well as reason it has its romance.

Disillusionment* in our sense of the word was not known before the Revolution; the eighteenth century was one of the most religious periods of history. I am no longer speaking of the great martyr Saint-Just or of the apostle Jean-Jacques; but was not Pope Voltaire, blessing Franklin's grandson in the name of God and Freedom, a fanatic of his religion of humanity?

Scepticism was proclaimed together with the republic of the 22nd of September, 1792.

The Jacobins and revolutionaries in general belonged to a minority, separated from the life of the people by their culture: they constituted a sort of secular clergy ready to shepherd their human flocks. They represented the *highest* thought of their time, its *highest* but not its *general consciousness*, not the *thought of all*.

This new clergy had no means of coercion, either physical or fancied: from the moment that authority fell from their hands, they had only one weapon – conviction; but for conviction to be *right* is not enough; their whole mistake lay in supposing so; something more was necessary – *mental equality*.

So long as the desperate conflict lasted, to the strains of the hymn of the Huguenots and the hymn of the 'Marseillaise', so long as the faggots flamed and blood flowed, this inequality was not noticed; but at last the oppressive edifice of feudal monarchy crumbled, and slowly the walls were shattered, the locks struck off . . . one more blow struck, one more wall breached, the brave men advance, gates are opened and the crowd rushes in . . but it is not the crowd that was expected. Who are these men; to what age do they belong? These are not Spartans, not the great *populus Romanus. Davus sum, non Œdipus!*[2] An irresistible wave of filth flooded everything. The inner horror of the Jacobins was expressed in the Terror of 1793 and 1794: they saw their fearful mistake,

---

* On the whole 'our' scepticism was not known in the last century; England and Diderot alone are the exceptions. In England scepticism has been at home for long ages, and Byron follows naturally on Shakespeare, Hobbes, and Hume.

and tried to correct it with the guillotine; but, however many heads they cut off, they still had to bow their own before the might of the rising stratum of society. Everything gave way before it; it overpowered the Revolution and the Reaction, it submerged the old forms and filled them up with itself because it constituted the one effective majority of its day. Sieyès was more right than he thought when he said that the *petite bourgeoisie was every-thing*.

The *petits bourgeois* were not produced by the Revolution; they were ready with their traditions and their customs, which were alien, in a different mode, to the revolutionary idea. They had been held down by the aristocracy and kept in the background; set free, they walked over the corpses of their liberators and established their own régime. The minority were either crushed or dissolved in the *petite bourgeoisie*.

A few men of each generation remained, in spite of events, as the tenacious preservers of the idea; these Levites, or perhaps ascetics, are unjustly punished for their monopoly of an exclusive culture, for the mental superiority of the well-fed castes, the leisured castes that had time to work not only with their muscles.

We were angered, moved to fury, by the absurdity, by the in-justice of this fact. As though someone (not ourselves) had promised that everything in the world should be just and elegant and should go like clockwork. We have marvelled enough at the abstract wisdom of nature and of historical development; it is time to perceive that in nature as in history there is a great deal that is fortuitous, stupid, unsuccessful and confused. Reason, fully developed thought, comes last. Everything begins with the dull-ness of the newborn child; potentiality and aspiration are innate in him, but before he reaches development and consciousness he is exposed to a series of external and internal influences, deflections and checks. One has water on the brain; another falls and flattens it; both remain idiots. A third does not fall nor die of scarlet fever – and becomes a poet, a military leader, a bandit or a judge. On the whole we know best, in nature, in history and in life, the advances and successes: we are only now beginning to feel that all the cards are not so well prearranged as we had thought, because we are ourselves a failure, a losing card.

It mortifies us to realise that the idea is impotent, that truth has no binding power over the world of actuality. A new sort of Manichaeism takes possession of us, and we are ready, *par dépit*, to believe in rational (that is, purposive) evil, as we believed in rational good – that is the last tribute we pay to idealism.

The anguish will pass with time; its tragic and passionate character will calm down: it scarcely exists in the New World of the United States. This young people, enterprising and more practical than intelligent, is so busy building its own dwelling-place that it knows nothing at all of our agonies. Moreover, there are not two cultures there. The persons who constitute the classes in the society of that country are constantly changing, they rise and fall with the bank balance of each. The sturdy breed of English colonists is multiplying fearfully; if it gets the upper hand people will not be more fortunate for it, but they will be better contented. This contentment will be duller, poorer, more arid than that which hovered in the ideals of romantic Europe; but with it there will be neither Tsars nor centralisation, and perhaps there will be no hunger either. Anyone, who can put off from himself the old Adam of Europe and be born again a new Jonathan, had better take the first steamer to some place in Wisconsin or Kansas; there he will certainly be better off than in decaying Europe.

Those who *cannot* will stay to live out their lives, as patterns of the beautiful dream dreamt by humanity. They have lived too much by fantasy and ideals to fit into the age of American good sense.

There is no great misfortune in this: we are not many, and we shall soon be extinct.

But how is it men grow up so out of harmony with their environment? ...

Imagine a hothouse-reared youth, the one, perhaps, who has described himself in *The Dream*;[3] imagine him face to face with the most boring, with the most tedious society, face to face with the monstrous Minotaur of English life, clumsily welded together of two beasts – the one decrepit, the other knee-deep in a miry bog, weighed down like a Caryatid whose muscles, under a constant strain, cannot spare one drop of blood for the brain. If he could have adapted himself to this life he would, instead of dying in Greece at thirty, now have been Lord Palmerston or Lord John Russell. But since he could not it is no wonder that, with his own Childe Harold, he says to his ship:

'Nor care what land thou bearest me to,
But not again to mine.'

But what awaited him in the distance? Spain cut up by Napoleon, Greece sunk back into barbarism, the general resurrection
after 1814 of all the stinking Lazaruses; there was no getting away
from them at Ravenna or at Diodati. Byron could not be satisfied
like a German with theories *sub specie aeternitatis*, nor like a
Frenchman with political chatter; he was broken, but broken like
a menacing Titan, flinging his scorn in men's faces and not
troubling to gild the pill.

The rupture of which Byron, as a poet and a genius was conscious forty years ago, now, after a succession of new experiences,
after the filthy transition from 1830 to 1848, and the abominable
one from 1848 to the present, shocks many of us. And we, like
Byron, do not know what to do with ourselves, where to lay
our heads.

The realist Goethe, like the romantic Schiller, knew nothing of
this rending of the spirit. The one was too religious, the other too
philosophical. Both could find peace in abstract spheres. When the
'spirit of negation' appears as such a jester as Mephistopheles, then
the swift *disharmony* is not yet a fearful one; his mocking and for
ever contradictory nature is still blended in the higher harmony,
and in its own time will ring out with everything – *sie ist gerettet*.
Lucifer in *Cain* is very different; he is the rueful angel of darkness
and on his brow shines with dim lustre the star of bitter thought;
he is full of an inner disintegration which can never be put
together again. He does not make a jest of denial, he does not seek
to amuse with the impudence of his unbelief, he does not allure by
sensuality, he does not procure artless girls, wine or diamonds; but
he quietly prompts to murder, draws towards himself, towards
crime – but that incomprehensible power with which at certain
moments a man is enticed by still, moonlit water, that promises
nothing in its comfortless, cold, shimmering embraces, nothing
but death.

Neither Cain nor Manfred, neither Don Juan nor Byron, makes
any inference, draws any conclusion, any 'moral'. Perhaps from
the point of view of dramatic art this is a defect, but it gives a
stamp of sincerity and indicates the depth of the gulf. Byron's
epilogue, his last word, if you like, is *The Darkness*; here is the

finish of a life that began with *The Dream*. Complete the picture for yourselves.

Two enemies, hideously disfigured by hunger, are dead, they are devoured by some crab-like animals ... their ship is rotting away – a tarred rope swings in the darkness of dim waters; there is fearful cold, the beasts are dying out, history has died already and space is being cleared for new life: our epoch will be reckoned as belonging to the fourth geological formation – that is, if the new world gets as far as being able to count up to four.

Our historical vocation, our work, consists in this: that by our disillusionment, by our sufferings, we reach resignation and humility in face of the truth, and spare following generations from these afflictions. By means of us humanity is regaining sobriety; we are its headache next morning, we are its birth-pangs; but we must not forget that the child or mother, or perhaps both, may die by the way, and then – well, then history, like the Mormon it is, will start a new pregnancy.... *E sempre bene*, gentlemen!

We know how Nature disposes of individuals: later, sooner, with no victims or on heaps of corpses, she cares not; she goes her way, or goes any way that chances. Tens of thousands of years she spends building a coral reef, every spring abandoning to death the ranks that have run ahead too far. The polyps die without suspecting that they have served the *progress* of the reef.

We, too, shall serve something. To enter into the future as an element in it does not yet mean that the future will fulfil our ideals. Rome did not carry out Plato's idea of a republic nor the Greek idea in general. The Middle Ages were not the development of Rome. Modern Western thought will pass into history and be incorporated in it, will have its influence and its place, just as our body will pass into the composition of grass, of sheep, of cutlets, and of men. We do not like that kind of immortality, but what is to be done about it?

Now I am accustomed to these thoughts; they no longer frighten me. But at the end of 1849 I was stunned by them; and in spite of the fact that every event, every meeting, every contact, every person vied with each other to tear away the last green leaves, I still frantically and obstinately sought a *way out*.

That is why I now prize so highly the courageous thought of Byron. He saw that there is *no way out*, and proudly said so.

I was unhappy and perplexed when these thoughts began to

haunt me; I tried by every means to run away from them . . . like a lost traveller, like a beggar, I knocked at every door, stopped people I met and asked the way, but every meeting and every event led to the same result — to *meekness* before the *truth*, to self-sacrificing acceptance of it.

Three years ago I sat by Natalie's sick-bed and saw death drawing her pitilessly, step by step, to the grave; that life was my whole fortune. Darkness spread around me; I was a savage in my dull despair, but did not try to comfort myself with hopes, did not betray my grief for one moment by the stultifying thought of a meeting beyond the grave.

So it is less likely that I should be false to myself over the impersonal problems of life.

*

CHAPTER V

MONEY AND THE POLICE

IN the December of 1849 I learnt that the authorisation for the mortgage of my estate sent from Paris and witnessed at the Embassy had been destroyed, and that after that a distraint had been laid on my mother's fortune. There was no time to be lost and [...] I at once left Geneva and went to my mother's.

It would be stupid and hypocritical to affect to despise property in our time of financial disorder. Money is independence, power, a weapon; and no one flings away a weapon in time of war, though it may have come from the enemy and even be rusty. The slavery of poverty is frightful; I have studied it in all its aspects, living for years with men who have escaped from political shipwrecks in the clothes they stood up in. I thought it right and necessary, therefore, to take every measure to extract what I could from the bear's paws of the Russian Government.

Even so I was not far from losing everything. When I left Russia I had had no definite plan; I only wanted to stay abroad as long as possible. The Revolution of 1848 arrived and drew me into its vortex before I had done anything to secure my property. Worthy persons have blamed me for throwing myself headlong into political movements and leaving the future of my family to the will of the gods. Perhaps it was not altogether prudent; but if, when I was living in Rome in 1848, I had sat at home considering ways and means of saving my property while an awakened Italy was seething before my windows, then I should probably not have remained in foreign countries, but have gone to Petersburg, entered the service once more, might have become a vice-governor, have sat at the head prosecutor's table, and should have addressed my secretary with insulting familiarity and my minister as 'Your Exalted Excellency'.

I had no such self-restraint and good sense, and I am infinitely thankful for it now. My heart and my memory would be the poorer if I had missed those shining moments of faith and enthusiasm! What would have compensated me for the loss of them? Indeed, why speak of me? What would have compensated her whose broken life was nothing afterwards but suffering that ended in the grave? How bitterly would my conscience have reproached me if, from over-prudence, I had robbed her of almost the last minutes of untroubled happiness! And after all I did do the important thing: I did save almost all our property except the Kostroma estate.

After the June days my situation became more dangerous. I made the acquaintance of Rothschild, and proposed that he should change for me two Moscow Savings Bank bonds. Business then was not flourishing, of course, and the exchange was very bad; his terms were not good, but I accepted them at once, and had the satisfaction of seeing a faint smile of compassion on Rothschild's lips – he took me for one of the innumerable *princes russes* who had run into debt in Paris, and so fell to calling me *Monsieur le Comte*.

On the first bonds the money was paid promptly; but on the later ones for a much larger sum, although payment was made, Rothschild's agent informed him that a distraint had been laid on my capital – luckily I had withdrawn it all.

In this way I found myself in Paris with a large sum of money in very troubled times, without experience or knowledge what to do with it. Yet everything was settled fairly well. As a rule, the less impetuosity, alarm, and uneasiness there is in financial matters, the better they succeed. Grasping money-grubbers and financial cowards are as often ruined as spendthrifts.

By Rothschild's advice I bought myself some American shares, a few French ones and a small house in the Rue Amsterdam which was let to the Havre Hôtel.

One of my first revolutionary steps, which cut me off from Russia, plunged me into the respectable class of conservative idlers, brought me acquaintance with bankers and notaries, taught me to keep an eye on the Stock Exchange – in short, turned me into a Western European *rentier*. The rift between the modern man and the environment in which he lives brings a fearful confusion into private behaviour. We are in the very middle of two currents which are getting in each other's way; we are flung and shall continue to be flung first in one direction and then in the other, until one current or the other finally wins and the stream, still restless and turbulent but now flowing in one direction, makes things easier for the swimmer by carrying him along with it.

Happy the man who knows how to manoeuvre so that, adapting and balancing himself among the waves, he still swims on his own course!

On the purchase of the house I had the opportunity of looking more closely into the business and *bourgeois* world of France. The bureaucratic pedantry over completing a purchase is not inferior to ours in Russia. The old notary read me several documents, the statute concerning the reading of them *main levée*, then the actual statute itself – all this making up a complete folio volume. In our final negotiations concerning the price and the legal expenses, the owner of the house said that he would make a concession and take upon himself the very considerable expenses of the legal conveyance, if I would immediately pay the whole sum to him personally. I did not understand him, since from the very first I had openly stated that I was buying it for ready money. The notary explained to me that the money must remain in his hands for at least three months, during which a notice of sale would be published and all creditors who had any claims on the house would be called upon to state their case. The house was mortgaged for 70,000, but there might be further mortgages in other hands. In three months' time, after inquiries had been made, the *purge hypothécaire* would be handed to the purchaser and the former owner would receive the purchase money.

The owner declared that he had no other debts. The notary confirmed this.

'Your honour and your hand on it,' I said to him: 'you have no other debts which would concern the house?'

'I willingly give you my word of honour.'

'In that case I agree, and shall come here tomorrow with Rothschild's cheque.'

When I went next day to Rothschild's his secretary flung up his hands in horror:

'They are cheating you! This is impossible: we will stop the sale if you like. It's something unheard of, to buy from a stranger on such terms.'

'Would you like me to send someone with you to look into the business?' Baron James himself suggested.

I did not care to play the part of an ignorant boy, so I said that I had given him my word, and took a cheque for the whole sum. When I reached the notary's I found there, besides the witnesses, the creditor who had come to receive his 70,000 francs. The deed of purchase was read over, we signed it, the notary congratulated me on being a Parisian house-owner — all that was left was to hand over the cheque. . . .

'How vexing!' said the house-owner, taking it from my hands; 'I forgot to ask you to draw it in two cheques. How can I pay out the 70,000 separately now?'

'Nothing is easier: go to Rothschild's, they'll give it you in two cheques; or, simpler still, go to the bank.'

'I'll go if you like,' said the creditor.

The house-owner frowned and answered that that was his business, and he would go.

The creditor frowned. The notary good-naturedly suggested that they should go together.

Hardly able to refrain from laughter I said to them:

'Here's your receipt; give me back the cheque, I will go and change it.'

'You will infinitely oblige us,' they said with a sigh of relief; and I went.

Four months later the *purge hypothécaire* was sent me, and I gained about 10,000 francs by my rash trustfulness.

After the 13th of June, 1849, Rébillaud, the Prefect of police, laid information against me; it was probably in consequence of his report, that some unusual measures were taken by the Petersburg Government against my estate. It was these, as I have said, that made me go with my mother to Paris.

We set off through Neuchâtel and Besançon. Our journey began with my forgetting my greatcoat in the posting-station yard at Berne; since I had on a warm overcoat and warm galoshes I did not go back for it. All went well till we reached the mountains, but in the mountains we were met by snow up to the knees, eight degrees of frost, and the cursed Swiss *bise*. The *diligence* could not go on and the passengers were transferred by twos and threes into small sledges. I do not remember that I have ever suffered so much from cold as I did on that night. My feet were simply in agony, and I dug them into the straw; then the driver gave me a collar of some sort, but that was not much help. At the third stage I bought a shawl from a peasant woman for fifteen francs, and wrapped myself in it; but by that time we were already on the descent, and with every mile it became warmer.

This road is magnificently fine on the French side; the vast amphitheatre of immense mountains, so varied in outline, accompanies one as far as Besançon itself; here and there on the crags the ruins of fortified feudal castles are visible. In this landscape there is something mighty and harsh, solid and grim; with his eyes upon it, there grew up and was formed a peasant boy, the descendant of old country stock, Pierre-Joseph Proudhon. And indeed one may say of him, though in a different sense, what was said by the poet of the Florentines:

'*E tiene ancor del monte e del macigno.*'[1]

Rothschild agreed to take my mother's bond, but would not cash it in advance, referring to Gasser's letter. The Board of Trustees did in fact refuse payment. Then Rothschild instructed Gasser to request an interview with Nesselrode[2] and to inquire of him what was wrong. Nesselrode replied that, though there was no doubt about the bonds and Rothschild's claim was valid, the Tsar had ordered the money to be stopped, for secret, political reasons.

I remember the surprise in Rothschild's office on the reception of this reply. The eye involuntarily sought at the bottom of the document for the mark of Alaric or the seal of Genghis Khan. Rothschild had not expected such a trick even from so celebrated a master of despotic affairs as Nicholas.

'For me,' I said to him, 'it is hardly surprising that Nicholas should wish to purloin my mother's money in order to punish me, or hope to catch me with it as a bait; but I could not have imagined that your name would carry so little weight in Russia. The bonds are yours and not my mother's; when she signed them she transferred them to the bearer (*au porteur*), but ever since you endorsed them that *porteur** has been you; and you have received the insolent answer: "The money is yours, but master orders me not to pay." '

My speech was successful. Rothschild grew angry, and walked about the room saying:

'No, I shan't allow myself to be trifled with; I shall bring an action against the bank; I shall demand a categorical reply from the Minister of Finance!'

* This endorsement is done for security in sending cheques, in order that a cheque may not be sent unendorsed, by means of which anybody would be able to receive the money.

'Well,' thought I, 'Vronchenko won't understand this at all. A "confidential" reply would still have been all right, but not a "categorical" one!'

'Here you have a sample of how familiarly and *sans gêne* the autocracy, upon which the reaction is building such hopes, disposes of property. The communism of the Cossack is almost more dangerous than that of Louis Blanc.'

'I shall think it over,' said Rothschild; 'we can't leave it like this.'

Three days or so after this conversation, I met Rothschild on the boulevard.

'By the way,' he said, stopping me, 'I was speaking of your business yesterday to Kiselëv.* You must excuse me, but I ought to tell you that he expressed a very unfavourable opinion of you, and does not seem willing to do anything for you.'

'Do you often see him?'

'Sometimes, at evening parties.'

'Be so good as to tell him that you have seen me today, and that I have the worst possible opinion of him, but that even so I don't think it would be at all just to rob his mother on that account.'

Rothschild laughed; I think that from that time he began to surmise that I was not a *prince russe*, and now he took to addressing me as Baron; he elevated me thus, I imagine, to make me worthy of conversing with him.

Next day he sent for me; I went at once. He handed me an unsigned letter to Gasser, and added:

'Here is the draft of our letter; sit down, read it carefully and tell me whether you are satisfied with it. If you want to add or change anything, we shall do it at once. Allow me to go on with my work.'

At first I looked about me. Every minute a small door opened and one Bourse agent after another came in, uttering a number in a loud voice; Rothschild, going on reading, muttered without raising his eyes: 'Yes – no – good – perhaps – enough –' and the number walked out. There were various gentlemen in the room, rank-and-file capitalists, members of the National Assembly, two or three exhausted tourists with youthful moustaches and elderly

* This was not P. D. Kiselëv, who was in Paris later, the well-known Minister of Crown Property, a very decent man; but another one; N. D. Kiselëv, afterwards transferred to Rome.

cheeks, those everlasting figures who drink – wine – at watering-
places and are presented at courts, the feeble, lymphatic suckers
that drain the sap from aristocratic families, and shove their way
from the gaming table to the Bourse. They were all talking
together in undertones. The Jewish autocrat sat calmly at his
table, looking through papers and writing something on them,
probably millions, or at least hundreds of thousands.

'Well,' he said, turning to me, 'are you satisfied?'

'Perfectly,' I answered.

The letter was excellent, curt and emphatic as it should be when
one power is addressing another. He wrote to Gasser telling him to
request an immediate audience with Nesselrode and the Minister
of Finance, he was to tell them that Rothschild was not interested
to know to whom the bonds had belonged; that he has bought
them and demands payment, or a clear legal declaration why payment had been stopped; that in case of refusal he would submit the
affair to the judgment of the legal authorities; and he advised
careful reflection on the consequences of a refusal, which was
particularly strange at a time when the Russian Government was
negotiating through him for the conclusion of a new loan. Rothschild wound up by saying that in case of further delays he would
have to give the matter publicity through the press, in order to
warn other capitalists. He recommended Gasser to show the letter
to Nesselrode.

'I'm very glad ... but ...' he said, holding a pen in his hand
and looking me straight in the face with a somewhat ingenuous
air ... 'but, my dear baron, do you really think that I shall sign
this letter which, *au bout du compte,* might put me on bad terms
with Russia – and that for a commission of one half of one per
cent?'

I was silent.

'In the first place,' he continued, 'Gasser will have disburse-
ments – nothing is done for nothing in your country – and of
course they must be at your expense; and in addition to that – how
much do you propose?'

'I think,' I said, 'it is for you to propose and for me to agree.'

'Well, five: what do you say? That's not much.'

'Let me think about it. . . .'

I simply wanted to calculate.

'As long as you like. Besides,' he added with an expression of

Mephistophelean irony, 'you can manage this business for nothing. Your mother's rights are incontestable. She is a subject of Württemberg: apply to Stuttgart – the Minister for Foreign Affairs is bound to support her and exert himself to procure payment. For my part, to tell you the truth, I shall be very glad to get this unpleasant affair off my shoulders.'

We were interrupted. I left the office impressed by all the old-fashioned simplicity in his look and his question. If he had asked for ten or fifteen per cent I should have agreed then and there. His help was essential to me, and he knew this so well that he even put himself out for a Russified subject of Württemberg; but, allowing myself to be guided as of old by the Russian rules of political economy, which ordain that, for whatever distance an *izvozchik* asks for twenty kopecks, one should still try to get him to take fifteen, I told Schomburg, on no sufficient basis, that I proposed that a commission of one per cent might be added. Schomburg promised to tell him and asked me to come back in half an hour.

When half an hour later I was mounting the staircase of the Winter Palace of Finance in the Rue Laffitte, the rival of Nicholas was coming down it.

'Schomburg has told me,' said His Majesty, smiling graciously, and majestically holding out his own august hand, 'that the letter has been signed and sent off. You will see how they will come round. I'll teach them to trifle with me.'

'Only not for half of one per cent,' I thought, and I felt inclined to drop on my knees and to offer an oath of allegiance together with my gratitude, but I confined myself to saying: 'If you feel perfectly certain of it, allow me to open an account, if only for half of the whole sum.'

'With pleasure,' answered His Majesty the Emperor, and went his way into the Rue Laffitte.

I made my obeisance to His Majesty and, since it was so close, went into the *Maison d'Or*.

Within a month or six weeks Nicholas Romanov, that Petersburg merchant of the first guild, who had been so stingy about paying up, now terrified of competition and of publication in the newspapers, did at the Imperial command of Rothschild pay over the illegally detained money, together with the interest and the interest on the interest, justifying himself by his ignorance of the laws, which in his social position he certainly could not be expected to know.

From that time forth I was on the best of terms with Rothschild. He liked in me the field of battle on which he had beaten Nicholas; I was for him something like Marengo or Austerlitz, and he several times recited the details of the action in my presence, smiling faintly, but magnanimously sparing his vanquished opponent.

While this action of mine was going on – and it occupied about six months – I was staying at the Hôtel Mirabeau, in the Rue de la Paix. One morning in April I was told that a gentleman was waiting for me in the hall and wished to see me without fail. I went in there. A cringing figure that looked like an old government clerk was standing in the hall.

'The *Commissaire* of Police of the *Tuileries arrondissement*: So-and-so.'

'Pleased to see you.'

'Allow me to read you a decree of the Ministry of Home Affairs, communicated to me by the Prefect of police, and relating to you.'

'Pray do so; here is a chair.'

'We, the Prefect of police:* – In accordance with paragraph seven of the law of the 13th and 21st of November and 3rd of December of 1849, giving the Ministry of Home Affairs the power to expel (*expulser*) from France any foreigner whose presence in France may be subversive of order and dangerous to public tranquillity, and in view of the ministerial circular of the 3rd of January, 1850,

'Do command as follows:

'The here-mentioned' (*le N——é*, that is, *nommé*, but this does not mean 'aforesaid' because nothing has been said about me before; it is merely an illiterate attempt to designate a man as rudely as possible) 'Herzen, Alexandre, aged 40' (they added two years), 'a Russian subject, living in such a place, is to leave Paris at once after this intimation, and to quit the boundaries of France within the shortest possible time.

'It is forbidden for him to return in future on pain of the penalties laid down by the eighth paragraph of the same law (imprisonment from one to six months and a money fine).

---

* I translate it word for word.

'All necessary measures will be taken to secure the execution of these orders.

'Done (*Fait*) in Paris, April 16th, 1850.
'Prefect of police,
'A. Carlier.

'Confirmed by the general secretary of the *Préfecture*.
'Clément Reyre.'

On the margin:

'Read and approved April 19th, 1850,
'Minister of Home Affairs,
'G. Baroche.

'In the year eighteen hundred and fifty, April the twenty-fourth.

'We, Émile Boullay, *Commissaire* of Police of the city of Paris and in particular of the *Tuileries arrondissement*, in execution of the orders of M. *le Préfet de Police* of April 23rd:

'Have notified the Sieur Alexandre Herzen, telling him in words as written herewith.' Here follows the whole text over again. It is just as children tell the story of the White Bull, prefacing it every time they tell it with the same phrase: 'Shall I tell you the tale of the White Bull?'

Then: 'We have invited *le dit Herzen* to present himself in the course of the next twenty-four hours at the Prefecture for the obtaining of a passport and the assignment of the frontier by which he will quit France.

'And that *le dit Sieur Herzen n'en prétende cause d'ignorance* (what jargon!) *nous lui avons laissé cette copie tant du dit arrêté en tête de cette présente de notre procès-verbal de notification.*'

Oh, my Vyatka colleagues in the secretariat of Tyufyayev; oh, Ardashov, who would write a dozen sheets at one sitting, Veprëv, Shtin, and my drunken head clerk! Would not their hearts rejoice to know that in Paris, after Voltaire, Beaumarchais, George Sand, and Hugo, documents are written like this?

And, indeed, not only they would be delighted, but also my father's village foreman, Vasily Yepifanov, who from profound considerations of politeness would write to his master: 'Your commandment by this present preceding post received, and by the same I have the honour to report . . .'

Ought there to be left one stone upon another of this stupid, vulgar temple *des us et coutumes*, only fitting for a blind, doting old goddess like Themis?

The reading of this document did not produce the result expected; a Parisian thinks that exile from Paris is as bad as the expulsion of Adam from Paradise, and without Eve into the bargain. To me, on the contrary, it was a matter of indifference, and I had already begun to be sick of Parisian life.

'When am I to present myself at the Prefecture?' I asked, assuming a polite air in spite of the wrath which was tearing me to pieces.

'I advise ten o'clock tomorrow morning.'

'With pleasure.'

'How early the spring is beginning this year!' observed the *commissaire* of the city of Paris, and in particular of the *Tuileries arrondissement*.

'Extraordinarily.'

'This is an old-fashioned hotel. Mirabeau used to dine here; that is why it bears his name. Have you really been well satisfied with it?'

'Very well satisfied. Only fancy what it must be to leave it so abruptly!'

'It's certainly unpleasant. ... The hostess is an intelligent, beautiful woman – Mlle Cousin; she was a great friend of the celebrated Le Normand.'[3]

'Imagine that! What a pity I did not know it! Perhaps she has inherited her art of fortune-telling and might have predicted my *billet doux* from Carlier.'

'Ha, ha! ... It is my duty, you know. Allow me to wish you good-day.'

'To be sure, anything may happen. I have the honour to wish you goodbye.'

Next day I presented myself in the Rue Jérusalem, more celebrated than Le Normand herself. First, I was received by some sort of a youthful spy, with a little beard, a little moustache, and all the manners of an abortive journalist and an unsuccessful democrat. His face and the look in his eyes bore the stamp of that refined corruption of soul, that envious hunger for enjoyment, power, and acquisition, which I have so well learned to read on Western European faces, and which is completely absent from those of the English. It cannot have been long since he had taken up his appointment; he still took pleasure in it, and therefore spoke somewhat condescendingly. He informed me that I must leave within three days, and except for particularly important reasons it was impossible to defer the date. His impudent face, his accent and his gestures were such that without entering into further discussion with him I bowed and then asked, first putting on my hat, when I could see the Prefect.

'The Prefect only receives persons who have asked him for an audience in writing.'

'Allow me to write to him at once.'

He rang the bell, and an old *huissier* with a chain on his breast walked in; saying to him with an air of importance, 'Pen and paper for this gentleman,' the youth nodded at me.

The *huissier* led me into another room. There I wrote to Carlier that I wished to see him in order to explain to him why I had to defer my departure.

On the evening of the same day I received from the Prefecture the laconic answer: 'M. *le Préfet* is ready to receive So-and-So tomorrow at two o'clock.'

The same repulsive youth met me next day: he had his own room, from which I concluded that he was something in the nature of the head of a department. Having begun his career so early and with such success, he will go far, if God grants him a long life.

This time he led me into a big office. There a tall, stout, rosy-cheeked gentleman was sitting in a big easy-chair at a huge table. He was one of those persons who are always hot, with white flesh, fat but flabby, plump, carefully tended hands, a necktie reduced to a minimum, colourless eyes and the jovial expression which is usually found in men who are completely immersed in love for their own well-being, and who can have recourse, coldly and without great effort, to extraordinary infamies.

'You wished to see the Prefect,' he said to me; 'but he asks you to excuse him; he has been obliged to go out on very important business. If I can do anything in any way for your pleasure I ask nothing better. Here is an easy-chair: will you sit down?'

All this he brought out smoothly, very politely, screwing up his eyes a little and smiling with the little cushions of flesh which adorned his cheek-bones. 'Well, this fellow has been in the service for a long time,' I thought.

'You surely know what I've come about.' He made that gentle movement of the head which everyone makes on beginning to swim, and did not answer.

'I have received an order to leave within three days. Since I know that your minister has the right of expulsion without giving a reason or holding an inquiry, I am not going to inquire why I am being expelled, nor to defend myself; but I have, besides my own house . . .'

'Where is your house?'

'14, Rue Amsterdam . . . very important business in Paris, and it is difficult for me to abandon it at once.'

'Allow me to ask, what is your business? Is it to do with the house or . . . ?'

'My business is with Rothschild. I have to receive 400,000 francs.'

'What?'

'A little over a 100,000 silver roubles.'

'That's a considerable sum!'

'C'est une somme ronde.'

'How much time do you need for completing your business?' he asked, looking at me more blandly, as people look at pheasants stuffed with truffles in the shop windows.

'From a month to six weeks.'

'That is a terribly long time.'

'My action is being settled in Russia. I should not wonder if it is thanks to that that I am leaving France.'

'How so?'

'A week ago Rothschild told me that Kiselëv spoke ill of me. Probably the Petersburg Government wishes to hush up the business; I dare say the ambassador has asked for my expulsion as a favour.'

'D'abord,' observed the offended patriot of the Prefecture, assuming an air of dignity and profound conviction, 'France will not permit any other Government to interfere in her domestic affairs. I am surprised that such an idea could enter your head. Besides, what can be more natural than that the Government, which is doing its utmost to restore order to the suffering people, should exercise its right to remove from the country, in which there is so much inflammable material, foreigners who abuse the hospitality she grants them?'

I determined to get at him by money. This was as sure a method as the use of texts from the Gospel in discussion with a Catholic, and so I answered with a smile:

'For the hospitality of Paris I have paid a 100,000 francs, and so I considered I had almost settled my account.'

This was even more successful than my *somme ronde*. He was embarrassed, and saying after a brief pause, 'What can we do? It is our duty,' he took my *dossier* from the table. This was the second volume of the novel, the first part of which I had once seen in the hands of Dubelt. Stroking the pages, as though they were good horses, with his plump hand:

'Now look,' he observed, 'your connections, your association with ill-disposed journals' (almost word for word what Sakhtyn-sky had said to me in 1840), 'and finally the considerable *subventions* which you have given to the most pernicious enter-prises, have compelled us to resort to a very unpleasant but necessary step. That step can be no surprise to you. Even in your own country you brought political persecution upon yourself. Like causes lead to like results.'

'I am certain,' I said, 'that the Emperor Nicholas himself has no suspicion of this solidarity; you cannot really approve of his administration.'

'*Un bon citoyen* respects the laws of his country, whatever they may be. . . .'*

'Probably on the celebrated principle that it is in any case better there should be bad weather than no weather at all.'

'But to prove to you that the Russian Government has no hand in it, I promise to try to get the Prefect to grant a postponement for one month. You will surely not think it strange if we make inquiries of Rothschild concerning your business; it is not so much a question of doubting. . . .'

'Do by all means make inquiries. We are at war, and if it had been of any use for me to have resorted to stratagem in order to remain, do you supose I should not have employed it?'

But this nice *alter ego* of the Prefect, this man of the world, would not be outdone.

'People who talk like you never say what is untrue,' he replied. A month later my business was still not completed. We were

* Later on Professor Chicherin preached a doctrine somewhat similar at Moscow University.[4]

visited by an old doctor, Palmier, whose agreeable duty it was to make a weekly examination of an interesting class of Parisian women at the Prefecture. Since he gave such a number of certificates of health to the fair sex, I thought he would not refuse to write me out a certificate of sickness. Palmier was acquainted, of course, with everyone in the Prefecture: he promised me to give X. personally the history of my indisposition. To my extreme surprise Palmier came back without a satisfactory answer. This trait is worth noting because there is in it a fraternal similarity between the Russian and French bureaucracies. X. had given no answer but had shuffled, being offended at my not having come in person to inform him that I was ill, in bed, and unable to get up. There was no help for it: I went next day to the Prefecture, glowing with health.

X. asked me most sympathetically about my illness. As I had not had the curiosity to read what the doctor had written, I had to invent an illness. Luckily I remembered Sazonov who, with his great corpulence and insatiable appetite, complained of aneurism. I told X. that I had heart disease and travelling might be very bad for me.

X. was sorry to hear it, and advised me to take care of myself; then he went into the next room, and returned a minute later, saying:

'You may stay for another month. The Prefect has charged me to tell you at the same time that he hopes and desires that your health may be restored during that period; if this should not be so, he would greatly regret it, for he cannot postpone your departure a third time.'

I understood this, and made ready to leave Paris about the 20th of June.

• • •

The last two months I spent in Paris were insufferable. I was literally *gardé à vue*; my letters arrived shamelessly unsealed and a day late; wherever I went I was followed at a distance by a loathsome individual, who at the corners passed me on with a wink to another.

It must not be forgotten that this was the time of the most frenzied activity of the police. The stupid conservatives and revolutionaries of the Algiers-Lamartine persuasion helped the rogues and knaves surrounding Napoleon, and Napoleon himself, to prepare a network of espionage and surveillance, in order that, by spreading it over the whole of France, they might at any given minute reach out by telegraph from the Ministry of Home Affairs and the Élysée and catch all the active forces in the country and strangle them. Napoleon cleverly used the weapon entrusted to him against these men themselves. The 2nd of December meant the elevation of the police to the rank of a state authority.

There has never anywhere, even in Austria or in Russia, been such a political police as existed in France after the time of the Convention. There are many causes for this, apart from the peculiar *national* bent for a police. Except in England, where the police have nothing in common with Continental espionage, the police are everywhere surrounded by hostile elements and consequently thrown on their own resources. In France, on the contrary, the police is the most popular institution. Whatever government seizes power, its police is *ready*; part of the population will help it with a zest and a fanaticism which have to be restrained and not intensified, and will help it, too, with all the frightful means at the disposal of private persons which are impossible for the police. Where can a man hide from his shopkeeper, his house-porter, his tailor, his washerwoman, his butcher, his sister's husband or his brother's wife, especially in Paris, where people do not live in separate houses as they do in London, but in something like coral reefs or hives with a common staircase, a common courtyard and a common porter?

. . .

CHAPTER VI

TOWARDS SWISS CITIZENSHIP

(1850–1851)

A year after our arrival at Nice from Paris I wrote: '*In vain
I rejoiced at my quiet seclusion, in vain I drew the penta-
gram on my doors: I have not found a quiet haven nor the
peace I desired. Pentagrams are a protection from unclean
spirits: no polygon delivers from unclean men, except only
the square floor of a prison cell.*

'*A tedious, wearisome, and extraordinarily empty
period, the exhausting journey between the post-stage of
1848 and the post-stage of 1852, – there is nothing new
except that each personal misfortune comes nearer to
breaking the heart and some wheel of life crumbles to bits.*'
('Letters from France and Italy', 1 June 1851.)

INDEED, going over that time makes my heart ache as it does at
the memory of funerals, operations, or agonising illnesses. Without yet touching here upon my inner life, which was more and
more overcast by dark storm-clouds, public events and the news
in the papers were enough to make one want to escape to somewhere in the steppe. France was dropping with the speed of a
falling star towards the 2nd of December [1851]. Germany lay
at the feet of Nicholas, to which Hungary, luckless and betrayed,
had dragged her. The *condottieri* of the police met at their oecumenical councils, and secretly took counsel together about common measures of international espionage. The revolutionaries
continued their vain agitation. The men at the head of the movement had been disappointed in their hopes and lost their heads.
Kossuth[1] returned from America somewhat less nationalistic,
Mazzini together with Ledru-Rollin and Ruge was founding in
London the Central European Committee ... while the reaction
was growing more and more ferocious.

After our meeting at Geneva, and later at Lausanne, I saw
Mazzini in Paris in 1850; he was secretly in France, staying in

some aristocratic household, and sent one of his intimate associates to fetch me. He then told me of his project of an international *junta* in London, and asked whether I would like to take part in it *as a Russian*; I declined to talk about it. A year[2] later Orsini[3] came to me at Nice and handed me the programme, various manifestoes of the European Central Committee and a letter from Mazzini renewing his proposal. I could not even think of joining the Committee; what element of Russian life could I have represented at that time, completely cut off as I was from everything Russian? But this was not the only reason why the European Committee did not attract me. It seemed to me that its basis lacked depth of thought and unity, that there had not even been a need for its foundation, and that its form was plainly unsound.

The side of the movement which the Committee represented, that is, the restoration of the oppressed nationalities, was not strong enough in 1851 to be openly represented by a *junta*. The existence of such a Committee demonstrated only the tolerance of the English legislature; and partly too that the Home Office did not believe in its power; otherwise it would have squashed it, either by an Aliens Bill or by a motion for the suspension of *habeas corpus*.

The European Committee, though it scared all the governments, did nothing, without perceiving it. Even the most earnest people are terribly easily led away by formal procedure, and persuade themselves that they are doing something by having periodical meetings, masses of papers, minutes, conferences, voting, accepting resolutions, printing manifestoes, *professions de foi*, and so on. A revolutionary bureaucracy dissolves things into words and forms just as our official bureaucracy does. In England there is a profusion of various associations which hold solemn meetings attended by dukes and lords, clergymen and secretaries. Treasurers collect funds, literary men write articles, and all of them together do absolutely nothing. These meetings, for the most part philanthropic and religious, on the one hand serve as an entertainment and on the other soothe the Christian conscience of people who are devoted to worldly interests. But a revolutionary senate *en permanence* in London could not play such a meek and mild rôle. It was a public conspiracy, a conspiracy with open doors: that is, an impossible one.

A conspiracy must be secret. The period of secret societies is over only in England and America. Everywhere where there is a minority that has forestalled the understanding of the masses, and desires to realise an idea that they have grasped, secret societies will be formed, if there is no freedom of speech or right of free assembly. I speak of this quite objectively; after my youthful attempts, which ended in my exile in 1835, I have never had a hand *in any secret society*, but not at all because I consider the spending of energy on individual efforts more worth while. I have taken no part because I have not happened to come upon a society which would have tallied with my own aspirations and in which I could have achieved anything. If I had come across Pestel's or Ryleyev's union,[4] I should have flung myself into it head first.

Another error or another misfortune of the Committee lay in its lack of unity. This focusing together of heterogeneous aspirations could only in an effective unity have developed the power of its component parts. If each member of the Committee had contributed nothing but his own exclusive nationality, this would still have been no hindrance; they would have had a unity in their hatred for the chief enemy they had in common, the Holy Alliance. But their views which agreed on two negative principles, opposition to the authority of a monarch and to socialism, differed on everything else. To act in unison they must have made compromises, and compromises of that kind a e narmful to the unilateral power of each party, tying up, for the sake of general euphony, just the strings which sound most sharply of all, and so leaving the combined harmony trite, blurred and wavering.

After reading the papers which Orsini had brought I wrote the following letter to Mazzini:

'Dear Mazzini. – I have a sincere respect for you and so I am not afraid to tell you my opinion frankly. In any case you will give me a patient and indulgent hearing.

'You are almost the only one of the chief political leaders of recent times whose name has remained surrounded by sympathy and respect. One may differ from you in opinion or in method, but cannot fail to respect you. Your past and the Rome of 1848 and 1849 oblige you to bear with pride your great bereavement until events call back the champion who anticipated them. That is why it is painful to me to see your name coupled with the names of incapable men who have ruined the cause, with names which remind us only of the calamities they have brought down upon us.

'What kind of organisation can this be? It is nothing but a medley.

'These men are of no use to you nor to history; all that one can do for them is to forgive them their trangressions. You want to cover them with your name, you want to share with them your influence and your past; they will share with you their unpopularity and their past.

'What is there new in the manifestoes or in the *Proscrit*? Where are the signs of the menacing lessons that should have been learnt from the 24th of February? This is the continuation of the old liberalism and not the beginning of a new freedom: it is an epilogue and not a prologue. Why is there not in London the organisation you desire? Because it cannot be formed on the basis of undefined aspirations, but only on a profound concept held in common: and where is that?

'The first publication made under such conditions as the manifesto you have sent ought to have been full of sincerity; well, but who can read without a smile the signature of Arnold Ruge on a manifesto which speaks in the name of Divine Providence? Since 1838 Ruge has been preaching philosophic atheism; for him (if his brain is constructed logically) the idea of Providence ought to represent in embryo all reactions. This is a compromise, a bit of diplomacy, of policy, a weapon in the hands of our enemies. Moreover, all this is unnecessary. The theological part of the manifesto is a pure luxury; it adds neither to its understanding nor to its popularity. The people have a positive religion and Church. Deism is the religion of the rationalists, the representative system applied to faith, a religion surrounded by atheistic institutions.

'For my part I advocate a complete rupture with incomplete revolutionaries. One scents reaction a hundred yards from them. Having burdened their shoulders with thousands of blunders they go on justifying them to this day – the surest proof that they will repeat them.

In the *Nouveau Monde* there is the same *vacuum horrendum*, the same sad chewing of the cud, at once green and dry, which still is not digested.

'Please do not imagine that I am saying this in order to get out of doing anything. No, I am not sitting with my arms folded. I have too much blood in my veins and energy in my nature to be satisfied with the part of a passive spectator. From my thirteenth year I have served one idea and under one banner – of war against every oppressive authority, against every form of bondage, in the name of absolute personal freedom. I should like to continue my little guerrilla war – like a true Cossack ... *auf eigene Faust*, as the Germans say, alongside the great revolutionary army – not joining its regular ranks until they are completely transformed.

'While I am waiting for this, I write. Perhaps this waiting will last for a long time. It is not in my power to alter the capricious development of men; but to speak, to appeal, to exhort is in my power; and I am doing this with all my heart and with all my mind.

'Forgive me, dear Mazzini, both the candour and the length of my letter, and do not cease to love me a little and to reckon me a man devoted to your cause – but also devoted to his own convictions.'

'Nice, 13 September 1850.'

To this letter Mazzini answered with a few friendly lines in which, without touching on the essential point, he spoke of the necessity of uniting all forces in a single field of action, deplored people's differences of opinion, and so on.

In the same autumn in which Mazzini and the European Committee remembered me, the anti-European Committee of Nicholas Pavlovich finally remembered me too.[5]

One morning our maid, with a somewhat anxious expression, told me that the Russian consul was downstairs and was asking whether I could receive him. I looked upon my relations with the Russian Government as so completely at an end that I was surprised myself at this honour, and could not imagine what he wanted of me.

A second-rate official figure, looking like a German office-clerk, walked in.

'I have a communication to make to you.'

'Although,' I replied, 'I have no idea of its nature, I am almost certain that it will be unpleasant. I beg you to be seated.'

The consul flushed and was rather disconcerted; then he sat down on the sofa, took a document out of his pocket, unfolded it and, after reading, 'Adjutant-General Count Orlov has notified Count Nesselrode that His Im ...' rose to his feet again.

Here, fortunately, I remembered that a secretary at our Embassy in Paris had risen from his chair on announcing to Sazonov the Tsar's order that he should return to Russia, and Sazonov suspecting nothing had also stood up, though the secretary had done this from a deep sense of duty which required that a loyal subject should be on his feet with his head a little bowed when listening to his sovereign's will. Therefore, the more stiffly erect the consul stood, the more comfortably I buried myself in my arm-chair and, wishing him to observe the fact, said with a nod:

'Pray go on; I am listening.'

'. . . perial Majesty,' he went on, resuming his seat, 'has been graciously pleased to order that So-and-so shall return to Russia at once of which he is to be informed, accepting from him no reasons for delaying his departure and granting him no postponement under any circumstances.'

He stopped. I continued to say nothing.

'What am I to answer?' he asked, folding up the paper.

'That I am not going.'

'How do you mean "not going"?'

'What I say: I'm simply not going.'

'Have you considered that such a step . . . ?'

'I have considered.'

'But how can this be? . . . Kindly tell me what I am to write. For what reason . . . ?'

'You have been commanded not to accept any reasons.'

'What am I to say, then? Why, this is disobedience to the will of His Imperial Majesty!'

'Then say so.'

'This is impossible. I should never venture to write that . . .' and he blushed deeper than ever. 'Really, you had better change your mind while it is all still within four walls.' (The consul evidently thought the Third Division was a monastery.)

. . .

'Surely,' I said to him, 'when you were coming here you could not for one second have imagined that I should go? Forget that you are a consul and consider the position yourself. My estate has been sequestrated and my mother's fortune detained, and all this without my being asked whether I wished to return. Can I go back after this without taking leave of my senses?'

He hesitated, blushing continually, and at last hit on an idea that was wise, adroit, and above all new.

'I cannot,' he said, 'enter into . . . I understand the difficulty of your situation; on the other hand, the clemency! . . .'

I looked at him: he blushed again. '. . . besides, why burn your bridges like this? Write to me that you are very ill; I'll send it on to the Count.'

'That's too stale; besides, why should I tell a lie unnecessarily?'

'Well, then, will you be good enough to give me your answer in writing?'

'Certainly. Will you leave me a copy of the notice you read to me?'

'That is not usual.'

'A pity. I am making a collection of them.'

Simple as my written answer was, the consul was alarmed by it. He seemed to think he might be transferred on account of it to Beirut or Tripoli, or I do not know where; he positively declined to venture either to accept or to forward it. In spite of my assurances that no responsibility could fall on him, he refused, and begged me to write another letter.

'That's impossible,' I answered. 'I am not doing this for a joke, and I am not going to write nonsensical reasons: here is the letter for you, and you can do what you like with it.'

'Excuse me,' said the mildest-mannered consul since the days of Junius Brutus and Calpurnius Bestia: 'write the letter not to me but to Count Orlov, and I'll simply forward it to his office.'

'That's an easy matter; I've only to put *M. le Comte* instead of *M. le Consul*. I agree to that.'

As I was copying my letter it struck me that there was no reason for me to write to Orlov in French. If it were in Russian some *cantonist*[6] in his office or in the office of the Third Division might read it; it might be sent to the Senate, and a young head secretary might show it to his clerks: why deprive them of this pleasure? So I translated the letter, and here it is:

'Dear Sir, Count Alexey Fëdorovich!

'The Imperial Consul at Nice has notified me of His Imperial Majesty's will concerning my return to Russia. With every inclination to do so, I find myself unable to carry it out without making my position clear.

'Before any summons to return, more than a year ago, an injunction was placed on my estate; business papers of mine which were in private hands were confiscated and finally some money, a sum of 10,000 francs sent to me from Moscow, was seized. Such severe and extraordinary measures against me prove that I am not merely accused of some crime but, before any inquiry or any trial has been held, am found guilty and punished by the deprivation of part of my property.

'I cannot hope that my mere return might save me from the melancholy consequences of a political trial. It is easy for me to explain every one of my actions, but in cases of this kind it is opinions and theories that are the judges. It is upon them that sentences are based. Can I, should I, expose myself and all my family to such a trial? ... Your Excellency will appreciate the simplicity and candour of my answer, and will bring to the consideration of His Imperial Majesty the reasons that compel me to remain in foreign parts in spite of my deep and genuine desire to return to my country.'

'Nice, 23 September 1850.'

I really do not know whether it was possible to answer more simply and discreetly; but the habit of slavish silence is so deeply rooted among us that the consul at Nice thought even this letter monstrously audacious, and probably Orlov himself thought the same.

· · ·

Having rid myself of the emperor and the consul, I wanted to find my way out of the category of people without a passport.

The future was dark and dismal. ... I might die, and the thought that that same blushing consul would arrive to dispose of my house and to seize my papers, compelled me to think of obtaining the rights of citizenship somewhere. I need hardly say that I chose Switzerland, in spite of the fact that it was just about this time that the Swiss police played a trick on me.

A year after the birth of my second son we noticed with horror that he was completely deaf. Various consultations and experiments soon proved that it was impossible to cure him. But then the question arose whether we ought to leave him to become dumb, as is always done. The schools I had seen in Moscow I had found far from satisfactory. Conversation on one's fingers and by signs is not conversation; talking must be done with the mouth and lips. I knew by what I had read that attempts had been made in Germany and Switzerland to teach deaf mutes to speak as we speak, and to listen by watching the lips. In Berlin I saw for the first time a lecture given orally to deaf mutes and heard them recite verses. This was a huge step forward from the method of the Abbé de l'Épée.

In Zürich this teaching was carried to great perfection. My mother, who was passionately fond of Kolya, decided to settle with him for a few years in Zürich in order to send him to the school.

The child was gifted with unusual abilities: the everlasting stillness about him, by concentrating his lively, impulsive nature, assisted his development splendidly, and at the same time encouraged an exceptional power of plastic observation. His eyes glowed with intelligence and interest; at five years old he could tease everyone who came to see us by intentionally caricaturing them, and with such a comic touch that it was impossible not to laugh.

In six months he made great progress at the school. His voice was *voilée*; he scarcely marked the stress-accent, but already spoke German respectably and understood everything that was said to him slowly; nothing could have gone better. On my way through Zürich I thanked the director and council of the school and paid them various civilities, and they did the same to me.

But after I had gone away the elders of the town of Zürich learnt that I was not a Russian Count at all but a Russian *émigré* and, moreover, friendly with the radical party, which they could not endure; and even with socialists, too, whom they hated; and, what was worse than all that put together, that I was a man without religion and openly admitted the fact. This last they learned from an awful little book, *Vom andern Ufer*, which, as though to mock them, had been issued under their very noses by the best firm of publishers at Zürich. On learning this their conscience troubled them at the thought that they were giving an education to the son of a man who believed neither in Luther nor in Loyola, and they set to work to find means of getting rid of him. Since Providence was interested in the question, it at once showed them the way. The town police suddenly demanded the child's passport; I answered from Paris, supposing that it was a simple formality,

that Kolya really was my son and that he appeared on my passport, but that I could not obtain a separate one for him from the
Russian Embassy, because I was not on the best of terms with
them. The police were not satisfied and threatened to turn the
child out of the school and out of the town. I told the story of
this in Paris and one of my acquaintances published a paragraph
about it in *Le National*. Shamed by the publicity the police said
that they did not demand the expulsion of the child, but only the
payment of an insignificant sum of caution-money as a guarantee
that the child was himself and not somebody else. What guarantee
is there in a few hundred francs? On the other hand, if my mother
and I had not had the money, would the child have been turned
out? (I asked them about that through *Le National*.) And this
could happen in the nineteenth century in free Switzerland! After
what had taken place I disliked the idea of leaving the child in this
den of asses.

But what was to be done? The best teacher in the institution, a
young man who devoted himself enthusiastically to the training
of deaf mutes, a man of a thorough university education, luckily
did not share the views of the police Sanhedrin, and was a great
admirer of the very book which had so stirred the wrath of the
pious police-constables of the canton of Zürich. We suggested to
him that he should leave the school and enter my mother's household as tutor in order to go with her to Italy. Of course he consented. The authorities of the school were furious, but could do
nothing. My mother went with Kolya and with this young man,
Spielmann, to Nice. Before leaving she sent for her deposit; it was
not given to her, on the pretext that Kolya was still in Switzerland. I wrote from Nice. The Zürich police demanded to be
informed whether Kolya had the legal right to live in Piedmont.

This was too much, and I wrote the following letter to the
president of the canton of Zürich:

'M. le Président!

'In 1849, I placed my son, aged five years, in the Zürich Institute for the Deaf and Dumb. A few months later the Zürich police
demanded his passport from my mother. Since with us passports
are not required from newborn babies or from children going to
school, my son had not a separate one but was entered upon mine.
This explanation did not satisfy the Zürich police. They demanded

a deposit. My mother, fearing that the child who had brought down upon himself such dangerous suspicions on the part of the Zürich police would be expelled, paid it.

'In August 1850 my mother, wishing to leave Switzerland, asked for the deposit, but the Zürich police did not return it; they wished to ascertain first that the child had actually left the canton. On reaching Nice my mother asked Messieurs Avigdor and Schulthess to receive the money, enclosing a certificate that we, and above all my suspect six-year-old son, were at Nice and not at Zürich. The Zürich police, loth to pay back the deposit, then demanded another certificate, in which the police here were to attest, "that my son is officially permitted to live in Piedmont" (*que l'enfant est officiellement toléré*). M. Schulthess communicated this to M. Avigdor.

'Seeing this odd inquisitiveness of the Zürich police I refused M. Avigdor's proposal to send the new certificate, which he very kindly offered to take himself. I did not want to afford the Zürich police this satisfaction because, for all the importance of its position, it yet has no right to consider itself an international police, and still less because its demand is offensive not only to me but also to Piedmont.

'The Sardinian Government, M. *le Président*, is a free and civilised one; how is it possible that it should not permit (*ne tolerât pas*) a sick child of six years old to live in Piedmont? I really do not know how I am to take this demand of the Zürich police, whether as a strange joke or as the result of a partiality for deposits in general.

'Presenting this affair for your scrutiny, M. *le Président*, I shall ask you as a special favour, in case of a fresh refusal, to explain this occurrence to me, for it is too curious and interesting for me to consider that I have the right to conceal it from the knowledge of the public.

'I have written once more to M. Schulthess about receiving the money, and I can confidently assure you that neither my mother nor myself nor the suspect child have the smallest desire to return to Zürich after these unpleasant attentions from the police. There is not the faintest risk of it.

'Nice, 9 September 1850.'

I need hardly say that after this the police of the town of

Zürich, in spite of their oecumenical pretensions, paid back the deposit....

... Except for Swiss naturalisation, I would not have accepted citizenship in any European country, not even in England; I was repelled by the idea of voluntarily becoming anybody's subject. I did not want to change a bad master for a good one, but to escape from serfdom into being a free husbandman. This was possible only in two countries: America and Switzerland.

America I greatly respect. I believe that she is destined to a great future; I know that she is now twice as near to Europe as she was; but American life is antipathetic to me. It is very likely that from her angular, coarse, chilly elements another way of life will be formed. America has not yet settled down: she is an unfinished edifice. In her labourers and craftsmen in their workaday clothes are dragging about beams and stones, sawing, hewing, hammering nails. Why should an outsider warm her damp house?

Moreover America, as Garibaldi said, is the 'land for forgetting one's own'; let those who have no faith in their fatherland go there: they ought to get away from their own graveyards. With me it was quite the contrary: the more I lost all hope of a Latin-German Europe, the more my faith in Russia revived; but to think of returning under Nicholas would have been madness.

So there was nothing left but to enter into an alliance with the free men of the Helvetian Confederation.

As early as 1849 Fazy[7] had promised to naturalise me at Geneva, but kept putting it off; perhaps he simply did not want to increase by me the number of socialists in his canton. I got sick of this. I was passing through a black period: the last walls were tottering and might crumble about my head; calamity is never far off.... Karl Vogt[8] offered to correspond about my naturalisation with J. Schaller, who was at that time president of Freiburg canton and leader of the radical party there.

...

Schaller promised Vogt to take steps about my naturalisation, that is, to find a commune which would consent to receive me, and then to support the case in the Great Council. For naturalisation in Switzerland it is essential that some town or village commune should previously agree to accept the new citizen, a regulation quite in keeping with the self-government of each canton and each little district. The village of Châtel near Morat (Murten) agreed to receive my family into the number of its peasant families in return for a small money contribution to the village community. This village is not far from the Lake Murten, near which Charles the Bold was defeated and killed, whose unhappy death and name were so adroitly used by the Austrian censorship (and afterwards the Petersburg one) to replace the name of William Tell in Rossini's opera.[9]

When the case came before the Great Council, two Jesuitical deputies raised their voices against me, but did nothing. One of them said that it ought to be ascertained why I was in exile, and how I had incurred the anger of Nicholas. 'Why, but that's a recommendation in itself!' somebody answered, and they all laughed. Another, from far-sighted prudence, asked for fresh guarantees that in case of my death the education and maintenance of my children should not fall on the poor commune. This son in Jesus too was satisfied by Schaller's answer. My rights of citizenship were accepted by a vast majority, and I was transformed from a Russian aulic councillor to a taxable peasant of the village of Châtel near Murten, *originaire de Châtel près Morat*, as the Freiburg clerk wrote on my passport.

Naturalisation, however, is no hindrance at all to a career at home. I have two illustrious examples before my eyes: Louis Bonaparte became a citizen of Thurgau, and Alexander Nikolayevich a burgher of Darmstadt;[10] both became emperors after their naturalisation. I am not going so far as that.

On receiving the news of the ratification of my rights, it was almost indispensable for me to go and thank my new fellow-citizens and to make their acquaintance. Moreover, just at that time I had an intense desire to be alone, to look into myself, to revise the past, to try to discern something in the fog of the future, and I was glad of this impulse from outside myself.

On the eve of my departure from Nice I received a summons from the head of the police *di la sicurezza pubblica*. He informed me that I was ordered by the Minister of the Interior to leave the domains of Sardinia immediately. This strange step on the part of the tame and evasive Sardinian Government surprised me far more than my banishment from Paris in 1850; besides, there was no sort of occasion for it.

I am told that I was indebted for it to the zeal of two or three faithful Russian subjects living in Nice, and among them it is pleasant for me to name the Minister of Justice, Panin; it was more than he could tolerate that a man who had brought upon himself the Imperial wrath of Nicholas Pavlovich was not only living in peace, and even the same town as himself, but was also writing articles, though aware that His Majesty the Tsar did not look upon this with favour. On arriving at Turin, Justice, I am told, asked the minister Azeglio, as a friend, to banish me. Azeglio's heart probably had some intuition that when I was learning Italian in the Krutitsky Barracks I had read his *La Disfida di Barletta* – a novel neither 'classical nor old-fashioned',[11] though nevertheless tedious; and so he did nothing, or perhaps he hesitated to send me out because such friendly attentions should have been preceded by sending in a Russian ambassador, and Nicholas was still sulking over the rebellious ideas of Charles Albert.[12]

In return for this the Nice *Intendant* and the ministers in Turin took advantage of the suggestion at the first opportunity. Some days before I was turned out, there was a 'popular agitation' in Nice, in which the boatmen and shopkeepers, carried away by the eloquence of the banker Avigdor, protested, and rather insolently too, against the suppression of the free port, talking of the independence of the County of Nice, and its inalienable rights. The imposition of a light customs-duty on the whole kingdom diminished their privileges, without consideration of the 'independence of the County of Nice', and its rights 'traced on the tables of history'.

Avigdor, that O'Connell of the Paillon (that is the name of the dry river that runs through Nice), was cast into prison, patrols paraded the streets at night, and so did the people, and both sang songs, the same songs too; and that was all. Need I say that neither I nor any other foreigner took any part in this domestic quarrel over tariffs and customs-duties? Nevertheless, the *Intendant* pitched upon several of the refugees as ringleaders, me among them. The ministry, wishing to set an example of salutary severity, ordered me to be expelled with the rest.

I went to the *Intendant* (a Jesuit) and, observing to him that it was a superfluous luxury to expel a man who was going of himself and had his passport already viséd in his pocket, asked him what was wrong. He declared that he was as surprised as I was, and that the measure had been taken by the Ministry of the

Interior without even any preliminary reference to himself. At the same time he was so extremely polite that no doubt remained in my mind that he was responsible for the whole nasty business. I wrote an account of my conversation with him to the well-known deputy in opposition, Lorenzo Valerio, and went off to Paris.

Valerio made a savage attack upon the minister in his interpellation, and demanded the reason for my deportation. The minister was disconcerted, denied any influence of Russian diplomacy, threw everything upon the reports of the *Intendant*, and meekly concluded by saying that if the ministry had acted too hastily and imprudently it would with pleasure alter its decision.

The opposition applauded; consequently, *de facto*, the prohibition was withdrawn, but though I wrote to the minister he made no answer. I read Valerio's speech and the answer to it in the newspapers, and resolved simply to go to Turin without more ado on my way back from Freiburg. In order that I might not be refused a visa I went without one; on the frontier between Switzerland and Piedmont passports are not examined with the savage zeal of French gendarmes. At Turin I went to the Minister of the Interior: I was received instead by his deputy, who administered the senior police, Count Ponsa de la Martino, a man well known in those parts, clever, crafty, and devoted to the Catholic party.

His reception surprised me. He said to me everything I had meant to say to him; something similar had happened to me in one of my interviews with Dubelt, but Count Ponsa far outdid that.

He was a very elderly, thin, sickly-looking man of repellent appearance, with malicious, sly-looking features, rough, grey hair, and a somewhat clerical aspect. Before I had managed to say a dozen words about the reason why I was asking for an interview with the minister, he interrupted me with the words:

'Why, upon my word, what doubt can there be about this? ... Go to Nice, go to Genoa, stay here – only without the slightest *rancune* ... we shall be very glad ... it was all the doing of the *Intendant* ... you see, we are still learning our business, we are not accustomed to legality, to constitutional order.[13] If you had done anything contrary to the laws, there is a law-court for that; then you would have no cause to complain of injustice, would you?'

'I entirely agree with you.'

'Instead of that *they take* steps which cause irritation ... and excite an uproar – and without any necessity!'

After this speech against *himself*, he promptly snatched up a piece of paper with the ministerial imprint and wrote: *Si permette al Sig. A. H. di ritornare a Nizza e di restarvi quanto tempo credera conveniente. Per il ministro: S. Martino. 12 luglio, 1851.*

'Here, take this just in case, though you may rest assured that you will never need it. I am very glad, very glad indeed, that we have settled this business with you.'

As this was equivalent, in the vulgar tongue, to 'Go, and God bless you', I left my Ponsa, smiling at the thought of the face of the *Intendant* at Nice; but God did not favour me with the sight of it: he had been removed.

But to return to Freiburg and its canton: when, like all mortals who have been at Freiburg, we had listened to the celebrated organ and driven over the celebrated bridge, we set off for Châtel, accompanied by a good-natured old man, the treasurer of Freiburg canton. At Murten the Prefect of police, a vigorous man and a radical, asked us to stay with him, telling us that the elder had charged him to warn him of our arrival, because he and the other householders would be very much disappointed if I arrived without their knowing, when they were all in the fields at work. After walking about Morat or Murten for a couple of hours we set off, and the Prefect with us.

Near the head-man's house several elderly peasants were waiting for us, with the elder himself in front of them, a tall, venerable, grey-headed old man, rather bent but muscular. He stepped forward, took off his hat, held out to me his broad, powerful hand, and saying, '*Lieber Mitbürger ...*' delivered a speech of welcome in such Swiss-German that I did not understand a word. It was possible to make a rough guess at what he might say to me and therefore, reflecting that, if I concealed that I did not understand him, he would conceal that he did not understand me, I boldly replied to his speech:

'Dear Citizen Elder, and dear fellow-citizens of Châtel! I am come to thank you for giving a refuge to me and my children in your commune, and putting a term to my homeless wandering. I, dear citizens, did not leave my native land to seek another; I love the Russian people with my whole heart, but I left Russia because I could not be a dumb, inactive witness of her oppression. I left it after exile, pursued by the ferocious despotism of Nicholas. His arm, which has reached me everywhere where there is a king or a sovereign, is not long enough to reach me in your commune! I come tranquilly to put myself under your shelter and protection, as into a haven where I can always find peace. You, citizens of Châtel, you a handful of men, you accepting me into your midst, have been able to arrest the lifted hand of the Russian Emperor armed with a million bayonets. You are stronger than he! But you are strong only through the free republican institutions that have been yours for ages! With pride I enter into your alliance, and long live the Helvetian Republic!'

'*Dem neuen Bürger hoch! Es lebe der neue Bürger!*' answered the old man, and warmly pressed my hand; I was somewhat moved myself!

The elder invited us into his house.

We went in and sat down on benches at a long table on which there was bread and cheese. Two peasants dragged in a bottle of terrific size, bigger than those classic bottles which sweat away for whole winters in our old-fashioned houses in some corner by the stove, filled with home-made liqueurs and cordials. This bottle was covered with basket-work and full of white wine. The elder told us that this was the local wine, but that it was very old; that he remembered the bottle for thirty years, and that this wine was drunk only on very unusual occasions. All the peasants sat down with us to the table except two, who were busy with the cathedral-like bottle. They poured wine from it into a large jug, and the elder poured it from the jug into the glasses; there was a glass before every peasant, but to me he brought a grand crystal goblet, observing as he did so to the treasurer and the Prefect:

'You must excuse me this time; today we shall give the cup of honour to our fellow-citizen: you are old friends.'

While the elder was filling the glasses, I noticed that one of the company, dressed not quite like a peasant, was very restless, mopping his face, blushing and looking unwell; when the elder proposed my health he jumped up with a kind of desperate courage, turned towards me and began a speech.

'That,' the elder whispered in my ear with a significant glance, 'is the citizen teacher in our school.' I stood up.

The teacher spoke not Swiss but German, and not simply but on the model of eminent orators and writers: he referred both to

William Tell and to Charles the Bold (what would the Austrian and Russian stage censorship have done about this? – perhaps they would have called them William the Bold and Charles Tell), and at the same time did not forget the comparison, not so much new as expressive, of bondage with a gilded cage from which the bird will still strive to be free. Nicholas Pavlovich caught it from him properly; he ranked him with very disreputable persons from Roman history. I almost interrupted him at that point to say, 'Don't insult the dead' but, as though from a presentiment that Nicholas would soon be among them, I held my peace.

The peasants listened to him, craning their wrinkled, sunburnt necks and putting up their hands to their ears like sunshades; the treasurer had a little nap, and to conceal the fact was the first to praise the orator.

Meanwhile the elder was not sitting idle, but zealously pouring wine and proposing toasts like the most practised master of ceremonies:

'To the Confederation!' 'To Freiburg and its radical government!' 'To President Schaller!'

'To my kindly fellow-citizens of Châtel!' I proposed at last, feeling that the wine, though its taste was not strong, was far from weak in its effects. They stood up. . . . The elder said:

'No, no, *lieber Mitbürger*, a full glass, as we drank a full glass to you.' My venerable friends were becoming expansive, the wine was warming them up. . . .

'Bring your children,' said one.

'Yes, yes,' others chimed in; 'let them see how we live: we are simple people, they will learn no harm from us, and we shall have a look at them.'

'Certainly!' I answered, 'certainly!'

Here the elder began apologising for the poorness of their reception, saying that it was all the treasurer's fault, that he ought to have let them know a couple of days before, that then it would have been very different; they might even have got hold of a band, and most important, they would have received and seen me off with a volley of rifle-fire. I very nearly said to him, *à la* Louis-Philippe: 'But what has happened after all? Only one peasant more at Châtel.'

We parted great friends. I was rather surprised that I had seen not one woman, neither an old one nor a girl, nor even one young man. It was a working day, however. It is noteworthy, too, that to a festivity so unusual for them the pastor had not been invited.

In this I considered they had done me a great service. The pastor would certainly have spoilt it all; he would have delivered a stupid sermon, and with his decorous piety would have been like a fly in a glass of wine, which must be removed without fail in order to drink with pleasure.

Eventually we took our places once more in the treasurer's little carriage, or rather chaise; we gave the Prefect a lift to Morat, and set off for Freiburg. The sky was covered with storm-clouds; I felt sleepy and giddy. I made an effort not to go to sleep: surely it cannot be their wine? I wondered with some contempt for myself. . . . The treasurer smiled slyly, and then began dozing himself; drops of rain began falling, I covered myself with my overcoat, was just about falling asleep . . . then woke up at the contact of cold water. . . . The rain was pouring down in bucketsful, black storm-clouds seemed striking fire from craggy peaks, far-away peals of thunder rolled about the mountains. The treasurer was standing in the hall laughing loudly and talking with the host of the Zöringer Hof.

'Well,' the host asked me, 'it seems our simple peasant wine is very different from the French, eh?'

'Why, can we have arrived?' I asked, emerging drenched from the chaise.

'There's nothing so strange in that,' observed the treasurer; 'what is strange is that you have slept through a storm such as we have not had for a long time. Did you really hear nothing?'

'Nothing!'

I found out afterwards that the simple Swiss wines, which do not taste at all strong, acquire great strength with age and act particularly powerfully on those unaccustomed to them. The treasurer had purposely avoided telling me this; besides, even if he had told me I should not have refused the peasants' good-natured hospitality and their toasts, still less could I have ceremoniously moistened my lips for the show of the thing. That I did the right thing is proved by the fact that when a year later, on my way from Berne to Geneva, I met the Prefect of Morat at the station, he said to me:

'Do you know how you earned particular popularity among our Châtel peasants?'

'No.'

'To this day they tell with proud self-satisfaction how their new fellow-citizen, after drinking their wine, slept through a storm and drove in a downpour of rain from Morat to Freiburg, knowing nothing about it.'

So that is how I became a free citizen of the Swiss Confederation and got drunk on Châtel wine.*

*

* I cannot forbear adding that I happened to be revising this very page at Freiburg, and in the same Zöringer Hof. And the host was still the same, looking like a regular innkeeper. and the dining-room in which I had sat with Sazonov in 1851 was the same, and the room in which a year later I wrote my will, making Karl Vogt my executor, and this page which brings back to me so many details.

Fifteen years !

Involuntarily, unaccountably, one is seized with fear. . . .

14 October 1866.

A FAMILY DRAMA[1]

CHAPTER I

THE YEAR 1848

1

'To understand so much' (Natalie wrote to Ogarëv at the end of 1846) 'and not to have the strength to deal with it, not to have the fortitude to drink the bitter and sweet alike, but to stay at the former – it is pitiful! And all this I understand as well as possible, and yet I cannot earn for myself enjoyment or even indulgence. I understand what is good outside myself, I give it its due, but in my soul only what is sombre is reflected and torments me. Give me your hand and say with me that nothing satisfies you, that there is much that discontents you; and then teach me to rejoice, to be gay, to enjoy myself, I have everything for this, if only I could develop the faculty.'

These lines and the fragment of her diary relating to the same time, and quoted elsewhere,[2] were written under the influence of misunderstandings in Moscow. The dark side of her nature had got the better of her once more; the estrangement of the Granov-skys frightened Natalie; it seemed to her that our whole coterie was falling to pieces and that we were being left alone with Ogarëv. A young woman[3] hardly more than a child, whom she loved like a younger sister, had drifted further from us than any. To break out of this circle at all costs became at that time a passionate *idée fixe* with Natalie.

We went away.

At first the novelty of Paris, then an awakening Italy and a revolutionary France, occupied our whole souls. Personal doubts were vanquished by history. So we lived on until the days of June.

Even before those frightful, bloody days, the 15th of May [1848] had swung a scythe across the second shoots of our hopes. . . .

'Three full months had not passed since the 24th of February, men had not had time to wear out shoes in which they had built the barricades, but now a tired France was asking for pacification.'[4]

Not a drop of blood was spilt on that day; it was the dull clap of thunder from a clear sky that portends a fearful storm. On that day I seemed to look with clairvoyance into the soul of the *bourgeois*, into the soul of the working man – and was horrified. I saw the savage lust for blood on both sides – concentrated hatred on the side of the working man, and ferocious, carnivorous self-preservation on the side of the *petit bourgeois*. Two such camps could not stand side by side, jostling each other every day just as though they were working neighbouring strips of land – at home, in the street, in the workshop, in the market-place. A fearful, bloody conflict, foreboding nothing good was at hand. Nobody saw this except the conservatives who had called it up; my nearest friends spoke with a smile of my irritable pessimism. It was easier for them to snatch up a musket and to go to die on a barricade than to look events boldly in the face; on the whole they wanted not to understand things but to triumph over their opponents; they wanted to have their own way.

I drifted further and further away from everyone. There was the menace of emptiness in that – but suddenly the beating of drums, the clatter of the *levée* in the streets in the early morning, announced the beginning of the catastrophe.

Those June days, and the days that followed them, were awful; they drew a line across my life. I shall repeat a few lines that I wrote a month later.

'Women weep to relieve their hearts: we do not know how to weep. Instead of tears I want to write – not in order to describe and explain the bloody events, but simply to talk about them, to give rein to speech, to tears, to thoughts, to spleen. Where is the place here for description, for the collection of information, for deliberation? One's ears are still ringing with the sound of shots, the tramp of charging cavalry, the heavy, muffled sound of the wheels of gun carriages along the dead streets; individual details flash upon the memory – a wounded man on a stretcher holds his side with his hand, and drops of blood trickle down his arm; the omnibuses filled with corpses, the prisoners with bound hands, the cannon on the Place de la Bastille, the camp at the Porte St Denis in the Champs-Élysées, and the gloomy call at night: *"Sentinelle, prenez garde à vous!"* How can one describe these things when the brain is too much inflamed and the blood too bitter?

'To stay in one's room with one's arms folded, unable to go out beyond the gate, and to hear near and far all round one musketry, cannonading, shouts, drums beating, and to know that blood is flowing nearby, men are being hacked and stabbed and dying at your side – that is enough to kill one, to drive one mad. I did not die, but it aged me; I am recovering after the days of June as if after a serious illness.

'And yet they began impressively. On the 23rd, at four o'clock, before dinner, I was walking along the banks of the Seine towards the Hôtel de Ville; the shops were being shut, columns of the National Guard with ominous faces were marching in different directions, the sky was covered with clouds, a light rain was falling. I stopped on the Pont Neuf: a vivid flash of lightning gleamed out of a storm-cloud, claps of thunder followed one after another and in the midst of all this the slow, measured sound of the tocsin rang out from the belfry of St Sulpice, with which the deceived proletariat were once more summoning their brothers to arms. The cathedral and all the buildings along the bank of the river were unexpectedly lit up by a few rays of sunshine which issued brightly from behind a storm-cloud, the drums sounded from all sides, the guns were hauled along from the direction of the Place du Carrousel.

'I heard the thunder and the tocsin, and could not gaze enough at the panorama of Paris, as though I had been taking leave of it; I passionately loved Paris at that moment; this was my last tribute to the great city – after the days of June it repelled me.

'On the other side of the river in all the streets and alleys barricades were being built. I can see now those dark figures dragging stones; women and children were helping them. A young Polytechnic student climbed on to one barricade that apparently was finished, hoisted a banner and began singing the "Marseillaise" in a gentle, mournful voice; all those who were working joined in, and the chorus of the grand song resounding from the stones of the barricades laid hold of one's heart . . . and the tocsin still rang out. Meanwhile, the artillery thudded over the bridge, and from the bridge General Bédau scanned the *enemy*'s position through a telescope. . . .'

At that time it was still possible to prevent it all, it was still possible then to save the republic and the freedom of all Europe, it was still possible then to make peace. The dull-witted, clumsy government could not do this, the Assembly would not, the reactionaries sought revenge, blood, atonement for the 24th of February, and the *enactments* of the 'National' gave them the agents to do it with.

On the evening of the 26th of June, after the victory of the *National* over Paris, we heard regular salvos at short intervals. . . . We glanced at one another and all our faces were green. . . .

'They are shooting people,' we said with one voice, and turned away from one another. I pressed my forehead against the window-pane. Such moments provoke ten years of hatred, a lifetime of thirst for vengeance: *woe to him who forgives at such moments!*

After the fighting, which lasted for four days and four nights, there followed the calm and stillness of a state of siege; the streets were still cordoned off and only very rarely did one meet a carriage; the haughty National Guards, with brutal and stupid malignity in their faces, guarded their shops, threatening with bayonets and the butts of their fire-arms. Exultant crowds of drunken *gardes mobiles* paraded the boulevards singing *Mourir pour la Patrie*; lads of sixteen or seventeen boasted of their brothers' blood baked and impasted on their hands; women of the citizenry ran out from behind the counter and threw flowers to them, to greet the conquerors. Cavaignac took round with him in a carriage a monster who had killed dozens of Frenchmen. The *bourgeoisie* was triumphant. Yet the houses of the Fauborg St Antoine were still smoking, the walls battered by cannon-balls were in ruins, the interior of rooms thrown open displayed wounds in the stones, broken furniture was smouldering, bits of shattered looking-glass glittered . . . and where were the owners, the inhabitants? No one even thought of them. . . . In places sand was being scattered, but blood still leaked out. People were not allowed to approach the Panthéon, which had been damaged by cannon-balls; there were tents pitched in the boulevards, horses were nibbling the carefully tended trees of the Champs-Élysées; in the Place de la Concorde hay, cuirassiers' breastplates, and saddles were everywhere; in the Tuileries gardens soldiers were making soup by the fence. Paris had not seen this even in 1814.

A few more days passed – and Paris began to assume its customary aspect. Crowds of *flâneurs* appeared again on the boulevards, smartly-dressed ladies drove about in carriages and cabriolets *to look* at the ruins of houses and the traces of desperate fighting. ... It was only the frequent patrols and gangs of prisoners that reminded one of the frightful days, only then did the past begin to grow clear. Byron has a description of a battlefield at night; its blood-stained details are hidden by the darkness; at dawn, when the battle has long been over, the traces of it are to be seen – a sword-blade and some bloodstained clothing. It was just such a dawn that rose now in the soul and illuminated the fearful desolation. Half our hopes, half our beliefs had been slain, ideas of negation and despair haunted the brain and took root in it. One could never have supposed that, after passing through so many trials, after being schooled by the scepticism of our times, we had so much left in our souls to be destroyed.

Natalie wrote about this time to Moscow: 'I look at the children and weep; I am frightened: I no longer dare to wish that they may live, for perhaps for them too there is a fate as awful in store.'

In these words is the echo of all that she had been through: in them one seems to see the omnibuses crammed with corpses, and prisoners with bound hands, escorted with oaths, and the poor deaf-and-dumb boy shot a few steps from our gate because he did not hear: '*Passez au large!*'

And how could it make any other impression on the heart of a woman who had unhappily so profound an understanding of everything sorrowful? ... Even joyous natures grew sombre and full of gall, the heart ached with an angry pain, and a predestined shame made life an encumbrance.

It was not a chimerical grief for ideals, not a revival of the tears and romantic religion of her girlhood, which surged up once more over Natalie's whole soul, but a real sorrow, too heavy a burden for a woman's shoulders. Natalie's living interest in public affairs did not grow colder: on the contrary, it turned into a living agony. It had been the shattering of a sister, the tears of a mother, on the sorrowful field of the battle that had just taken place. She was in reality what Rachel falsely played at being with her 'Marseillaise'.

Weary of fruitless discussions, I seized my pen and, with an inward fury, slew my former hopes and expectations. The energy that was breaking and tormenting me spent itself in these pages of resentment and exorcism, in which even now, when I read them over, I am conscious of the fevered blood and the indignation that passed beyond all bounds ... in this way I gave vent to my feelings.

She had no such outlet. In the morning there were the children, in the evening our irritable, spiteful discussions, the post-mortem arguments of prosectors with bad doctors.

She was suffering – and I, in place of healing, handed her the bitter cup of scepticism and irony. If I had tended her sick soul with half the care I gave afterwards to her sick body, I should never have let this rankling sorrow send out roots in all directions. I helped to feed and strengthen them without finding out whether she could bear them or could cope with them.

Our life itself was oddly arranged. We rarely had quiet evenings of intimate talk and peaceful rest. We did not yet know how to shut our doors to outsiders. Towards the end of the year those who had been driven out from all countries, homeless wanderers, began to appear from everywhere. In their ennui and loneliness they sought a friendly roof and a warm welcome.

Here is what Natalie wrote about this:

21 November 1848. – 'I am sick of magic-lantern shadows. I do not know whom I see and why I see them, I only know that I see too many people – they are all good people; I fancy I might take pleasure in their company *sometimes*, but, as it is, it is too often; life is so like the dripping water in springtime: drip, drip, drip.

'All the morning I am looking after Sasha, after Tata, and that goes on all day. I cannot concentrate for one minute. I am so distracted that it sometimes makes me frightened and ill; the evening comes, the children are put to bed – well, one might think I should get some rest. . . . No, good people begin strolling in, and that they are good people makes it even harder to bear; otherwise I should be quite alone, and as it is I am not alone, and I do not feel their presence, it's as though there were smoke drifting about the room, one's eyes smart, and it is hard to breathe. Then they go away and nothing is left. . . . Tomorrow comes – always the same thing; another tomorrow – still the same. To no one else would I say this; others would take it for complaining, would think I am dissatisfied with life. You understand me, you *know* that I would not change with anyone on earth. This is a moment's exasperation, weariness . . . a breath of fresh air, and I rise up again as strong as ever.

'If I am to say all that passes in my mind, I am sometimes frightened when I look at the children . . . what boldness, what temerity, to *make* a new creature live, and to have nothing, nothing to make its life happy – this is frightful, and sometimes I think I am a criminal; it would be easier to take life away than to give it, if this were done with full consciousness. I have never yet met anyone of whom I could say, "If only my child were like him," that is, if only his life were such as his. ... My view of things grows simpler and simpler. Soon after Sasha's birth, I wanted him to be a great man, and later on to be this or that; finally I want him to ...'

Here the letter was interrupted when Tata's typhoid fever became acute; but on the 15th of December these words were added: 'Well, what I meant to say then was that now I don't want to make anything out of the children; so long as their lives are gay and good – all the rest is trifling....'

24 January 1849. 'How I should sometimes like to run about like a mouse as others do, and to be interested in that trotting to and fro, instead of being so idle, so idle in the midst of this bustle, in the midst of these essentials, while to occupy myself as I should like is out of the question; how agonising it is to feel oneself always in such disharmony with one's surroundings – I am not speaking of our most intimate circle, but if only one could confine oneself to it: one cannot. One longs to get away, far away – but it was all right to get away when we were in Italy. But now: what good would it be? At thirty to have the same yearnings, the same thirst, and the same dissatisfaction – yes, I said that aloud: and Tata came up at that word and hugged me so warmly. Dissatisfaction? – I am too happy, *la vie déborde* ... But

' "Why is it on the world
  One longs to gaze,
  Why to fly over it
  Does the soul crave?"[5]

'It is only to you that I talk like this; you will understand me, because you are just as weak as I am – but with others, who are stronger or weaker, I should not like to talk like this, I should not care for them to *hear* me talking. For them I shall find something else. Then I am frightened by my indifference; so few things, so few people interest me. ... Nature – only not in the kitchen; history – only not *in camera*; and then my own family, then two or three others – that's all. And yet how kind they all are – they take so much interest in my health and Kolya's deafness.'

27 *January*. 'After all I really have not the strength to go on watching the death-agonies: they last too long, and life is so short; I am possessed by egoism, because one does no good by self-sacrifice, except perhaps to prove the truth of the proverb, "Two in distress make sorrow less". But enough of dying. I should like to live: I should like to run away to America. ... What we believed in, what we took for accomplishment, was only a prediction, and a very premature one. How hard, how cheerless it is! I want to cry like a child. What is personal happiness? Common happiness is all round one like the air, but that air is filled only with a pestilent breathing, the antecedent of death.'

1 *February*. 'Natasha,[6] if you knew, my dear, how dark, how cheerless it is outside the threshold of the personal and private! Oh, if one could shut oneself inside it, and forget, forget everything except that intimate circle. ... How insufferable is the ferment, the result of which won't come for some centuries! My being is too feeble to rise above that ferment and look so far into the future – it shrinks and is annihilated.'

This letter ends with the words: 'I should like to have so little strength as not to feel my own existence; when I do feel it, I feel all the disharmony of everything that exists. ...'

2. TOKENS

The reaction was triumphant; through the pale blue republic could be seen the features of pretenders; the National Guard went hunting after working men's blouses, the Prefect of police sent stalking parties through the groves and catacombs, searching for those who had gone into hiding. Less martial men eavesdropped and informed.

Until the autumn we were surrounded by our own friends, and gave vent to our grief and anger in our own language; the Tuchkovs[7] lived in the same house and Maria Fëdorovna Korsh[8] was with us, Annenkov[9] and Turgenev used to come every day; but they were all looking into the distance and our little circle was breaking up. After its bath of blood Paris had no more hold on them; they were all prepared to go away, from no special necessity, but probably thinking to escape from an inward burden, from the days of June which had become ingrained in them, and which they carried about with them.

Why did not I go too? Much would have been saved, and I should not have had to offer so many human sacrifices and so much of myself as an immolation to a cruel and merciless god.

The day of our parting with the Tuchkovs and Maria Fëdorovna gave a sort of special raven's croak in my life, but I let that cry of warning pass unheeded, like hundreds of others.

Every man who has had much experience remembers days, hours, a succession of scarcely noticed moments, at which a break begins, at which the wind blows from a different quarter; these signs or warnings do not come by chance at all: they are consequences, from which in turn come the first embodiments of what is ready to burst into life, manifestations of what is secretly fermenting and exists already. We do not notice these physical tokens, but laugh at them as we do over a spilt salt-cellar, or an extinguished candle, because we consider ourselves incomparably more independent than we really are, and proudly desire to govern our lives ourselves.

On the eve of our friends' departure they gathered at our house together with three or four other intimate friends. The travellers were to be at the railway-station at seven o'clock in the morning; it was not worth the trouble of going to bed and we all preferred to spend the last hours together. At first everything went animatedly with that nervous excitement which is always there at partings; but little by little a dark cloud began to overshadow all of us. . . . Conversation flagged, everyone began to feel dispirited, the wine in our glasses went flat and the forced jokes did not cheer us. Someone, seeing the dawn, pulled back the curtain and threw a pale bluish light on all our faces, as though we had been at the Roman orgy of Couture.[10]

Everyone was sad, and I was so grieved that I could hardly breathe. My wife was sitting on a small sofa, and Tuchkov's younger daughter, 'Consuelo di sua alma', as Natalie used to call her, was kneeling before her with her face hidden in Natalie's bosom.[11] She was passionately fond of my wife, and was unwillingly leaving her to go and live in the wilds of the country; her sister stood sadly near at hand. Consuelo was whispering something through her tears, while two steps away Maria Fëdorovna was sitting in sombre silence; she had long before become accustomed to submit to destiny, she knew life and in her eyes one could read simply 'goodbye', while through the young girl's tears there still gleamed 'till we meet again'.

Then we went to see them off. It was piercingly cold in the high, empty, stone terminus, doors slammed violently and draughts blew from all directions. We sat down on a seat in the corner. Tuchkov went to look after the luggage. Suddenly the door opened, and two drunken old men came noisily into the room. Their clothes were filthy, their faces distorted and they had an air of savage debauchery. They came in, swearing. One tried to strike the other, who dodged and, swinging his fist with all his might, hit him in the face. The old drunkard was sent flying. His head cracked against the stone floor with a sharp, jarring sound; he cried out and raised his head, while the blood poured in streams over his grey hair and the stone floor. The police and passengers fell with fury upon the other old man.

Though we had been exasperated and agitated ever since the previous evening, and our nerves were over-strained, we maintained our self-control, but the fearful echo that resounded through the huge hall when the skull struck the floor produced an effect upon all of us that was something like hysteria. Our household and all our circle were at all times sane and free from neurotic and hysterical exhibitions, but this was more than we could bear. I felt myself shuddering all over, and my wife was almost fainting. And then the bell – time was up! And we were left suddenly outside the barrier – alone.

Nothing could be more churlish and offensive for parting friends than the police arrangements at railway-stations in France; they rob those left behind of the last two or three minutes; the friends are still there, the engine has not yet whistled, the train has not moved off, but between you is a fence, a barrier, and the arm of a policeman – and you want to see them settled, to see the train move off, then to follow with your eyes its departure, its dust, the smoke, the dot that it becomes, to follow with your eyes when there is no longer anything to be seen. . . .

In silence we drove home. My wife was weeping quietly all the way: she was sorry for her Consuelo; at times, wrapping herself in her shawl, she asked me, 'Do you remember that sound? it is in my ears still.'

At home I persuaded her to lie down, while I sat down to read the newspapers; I read, read the *premiers-Paris* and the *feuilletons* and the miscellanies, and glanced at the time – it was not yet twelve; what a day! I went to see Annenkov; he, too, was going away in a few days. We started for a walk together, and the streets were even more dreary than reading; my anguish was as though I was overcome by the gnawings of conscience. 'Come to dinner with me,' I said, and we went home. My wife was really ill; the evening was disjointed and stupid.

'And so it is settled,' I asked Annenkov as we parted, 'that you go at the end of the week?'

'Yes.'

'You'll be wretched in Russia.'

'I can't help that, I must go. I shall not stay in Petersburg, I shall go into the country. Why, even here it's not so wonderfully nice now. You may still regret staying.'

At that time I could still have returned; I had not yet burnt my boats. Rébillaud and Carlier had not yet written their secret police reports, but in my mind the matter was settled. Yet Annenkov's words jarred unpleasantly on my exposed nerves; I thought for a moment and answered:

'No, there is no choice for me, I must stay; and if I regret anything, it is rather that I did not take the musket when a working man offered it to me behind a barricade on the Place Maubert.'

Many times in moments of weakness and despair, when the cup of bitterness was too full, when my whole life seemed to me nothing but one prolonged blunder, when I doubted of myself, of 'the last thing, all that is left', those words came into my head; 'Why did I not take the gun from the workman and stay at the barricade?' Struck down by a sudden bullet, I should have borne two or three beliefs with me to the grave.

And again the time dragged on ... day after day ... grey and wearisome. ... People flitted in, made friends for a day, passed by, vanished, were lost. Towards the winter exiles from other lands, survivors from other shipwrecks, began to arrive; full of hope and self-confidence, they took the reaction that had sprung up all over Europe for a fleeting wind, a minor reverse; they expected that their turn would come, next day, next week.

I felt that they were wrong, but I was glad of their mistake. I tried to be inconsistent, struggled with myself and lived in a state of alarming irritability. That time has remained in my memory like a day of suffocating charcoal-fumes. . . . In my misery I turned hither and thither, seeking distraction – in books, in noise, in seclusion at home, in company, but always there was something lacking, laughter did not make me merry, wine only made me heavily drunk, music cut me to the heart and lively talk almost always ended in sombre silence.

Everything within was outraged, everything was turned upside down, all was chaos and glaring contradictions; again there was a break-up, again there was nothing. The principles of one's moral existence, settled long ago, were once more turned into questions, facts had rudely risen up on all sides and refuted them. Doubt had set its heavy foot on our last, prized convictions, it was now giving a thorough shaking, not to the church vestry nor the robes of learned doctors, but to the banners of the revolution. . . . From general ideas doubt was making its way into our life. There is a deep gulf between theoretical scepticism and doubt that is transferred to conduct; thought is bold, the tongue is reckless, it readily utters words which the heart fears; hopes and beliefs still smoulder in the breast while the mind, outrunning them, shakes its head. The heart lags behind because it loves; and, while the mind is sentencing and punishing, it is still taking its leave.

Perhaps in youth, when all is ferment and commotion, when there is so much of the future, when the loss of some beliefs only clears a space for others; perhaps in old age, when all grows indifferent from weariness; these breaks grow easier – but *nel mezzo del cammin di nostra vita*, we pay dearly for them.

What does it all mean in the end – is it a jest? Everything sacred that we loved, to which we aspired, for which we made sacrifices, has been betrayed by life, betrayed by history, betrayed for its own ends – it needs madmen as a leaven, and cares not what becomes of them when they recover; they have served its turn – let them live out their lives in a hospital for the disabled! The shame, the heartbreak of it! And here at one's side simple-hearted friends shrug their shoulders, wonder at one's pusillanimity, at one's impatience, look forward to the morrow and, for ever fussing, for ever busy with the same thing, understanding nothing, stop for nothing, go on for ever and are never a step forwarder. . . . They judge you, comfort you, chide you – what tedium! what chastisement!

'Men of faith, men of love', as they call themselves in contra-

distinction to us men 'of doubt and negation', do not know what
it is to tear out by the roots the cherished hopes of a lifetime, they
know nothing of the *sickness of truth*, they have surrendered no
treasure with that loud wail of which the poet speaks:

> '*Ich riss sie blutend aus dem wunden Herzen*
> *Und weinte laut und gab sie hin.*'[12]

Happy are the lunatics who have no lucid intervals; they know
nothing of the inner conflict, they suffer from external causes,
from evil men and evil chances, but within all is whole, conscience
is at rest, they are content. And so the worm of despair that gnaws
at others seems to them caprice, the self-indulgence of a satiated
mind, frivolous irony. They see that the wounded man laughs at
his wooden leg, and conclude that the operation meant nothing to
him; it does not enter their heads to wonder why he is prematurely
old and how the amputed leg aches at a change in the weather or
a puff of wind.

My logical confession, the history of the malady through which
my affronted mind made its way, remains in the series of articles
that make up *From the Other Shore*. In them I pursued the last
idols I had left. With irony I avenged myself on them for the pain
and the betrayal. I did not jeer at my neighbours, but at myself
and, carried away again, was already dreaming of being free, but
there too I came to grief. Losing faith in words and banners, in
canonised humanity and the one church of salvation, that is,
Western civilisation, I still believed in a few persons, I believed in
myself.

Seeing that all was tumbling into ruins, I wished to save myself,
to begin a new life, to get away with two or three others, to run,
to hide from ... superfluous people. And haughtily I headed my
last article: *Omnia mea mecum porto!*

Life, undisciplined, scorched, half blighted in the slough of
events, in the vortex of public interests, held aloof, was reduced
anew to the period of youthful lyricism, but without youth, without faith. With this *faro da me*,[13] my boat was bound to be
wrecked on submerged rocks, and wrecked it was. I survived, it
is true, but lost everything. ...

3. TYPHOID

In the winter of 1848 my little daughter, Tata, was ill. She had been ailing for a long time, then had a slight attack of fever, and it seemed to have passed off. Rayer, the well-known doctor, advised that she should be taken for a drive, in spite of the wintry weather. It was a beautiful day, but not warm. When she was brought home, she was unusually pale; she asked for something to eat and fell asleep beside us on the sofa before some broth was brought; several hours passed, and she went on sleeping. Adolf Vogt, a medical student, a brother of the naturalist, happened to be with us.

'Look at the child,' he said; 'why, that's not natural sleep.' The deathly, somewhat bluish pallor of the face alarmed me; I put my hand on the child's forehead: it was stone-cold. I rushed to Rayer myself, luckily found him at home and brought him back with me. The child had not woken up. Rayer lifted her up, shook her violently and made me call her loudly by her name. . . . She opened her eyes, said a couple of words and dropped back once more into the same heavy, deathlike sleep; her breathing was hardly perceptible. She remained in that condition, with slight changes, for several days, without food and almost without drink. Her lips turned black, her nails became dark blue and spots came out on the body – it was typhoid. Rayer did scarcely anything, waited, watched the illness and did not give us too much hope.

The appearance of the child was frightening; I expected the end from hour to hour. Pale and silent, my wife sat day and night by the little bed, her eyes covered with that pearly lustre which betrays fatigue, suffering, exhausted strength and nerves unnaturally strained. Once, between one and two o'clock in the morning, it seemed to me that Tata was not breathing. I looked at her, concealing my horror. Natalie guessed it.

'My head is going round,' she said, 'give me some water.'

When I handed her the glass, she was unconscious. Ivan Turgenev, who had come to share our dark hours, ran to the chemist's for spirits of ammonia. I stood motionless between two unconscious bodies, gazed at them, and did nothing. Our maid rubbed Natalie's hands and moistened her temples; in a few minutes she came to herself.

'Well?' she asked.

'I think Tata opened her eyes,' said our good, kind Louisa.

I looked: she seemed to be waking; I spoke her name in a whisper; she opened her eyes and smiled with her black, dry, cracked lips. From that minute she began to recover.

There are poisons which destroy a man more cruelly, more agonisingly than children's illnesses. I know them, too, but there is nothing worse than the slow poison that works by exhaustion, that saps the strength in the stillness and insults one with the fearful rôle of an idle spectator.

The man who has once carried a child in his arms and felt it growing cold and heavy, turning to stone; who has heard the last moan with which the frail creature implores mercy, rescue, begs to stay on earth; who has seen on his table the pretty little coffin covered with pink satin, and the little white frock with lace on it, which contrasts so with the little yellow face, will think at every childish illness: 'Why may there not be another little coffin on that table?'

Misfortune is the worst school! Of course a man who has undergone much has more endurance, but that is only because his soul has been kneaded and weakened. A man is worn out and grows more cowardly from what he has endured. He loses that confidence in the morrow without which nothing can be done: he becomes more indifferent because he grows used to frightening thoughts; eventually he is afraid of misfortunes, that is, he is afraid of feeling all over again the succession of racking pains, the successive sinkings of the heart, of which the memory is never dispersed with the clouds.

The moan of a sick child causes such inward horror in me, chills me so, that I have to make great efforts to overcome this purely nervous effect of memory.

It was on the morning after that same night that I went out for the first time for a walk; out of doors it was cold, the pavements were lightly powdered with hoar frost, but in spite of the cold and the early hour the boulevards were crowded with people; shouting street-boys were selling *bulletins* – over five million voices laid France bound at the feet of Louis Napoleon. The bereaved servants' hall had found its master at last!

... It was just at this time of bitter trial and strain that a person came into our circle who brought with him another train of misfortunes, who brought more ruin into our private life than the black days of June into our public. This person approached us rapidly; he forced himself upon us without giving us time for consideration. ... At ordinary times I am quick to become acquainted and slow to become intimate, but this, I repeat, was not an ordinary time.

All our nerves were laid bare and smarting; trivial meetings, insignificant reminders of the past, made one quiver in every fibre. I remember, for instance, that three days after the cannonade I was wandering along the Faubourg St Antoine; everything still bore traces still fresh of fierce fighting: ruined walls, barricades still standing, pale, frightened women looking for something, and children rummaging in the *débris*. ...

I sat down before a little café and looked with an aching heart at the frightful picture. A quarter of an hour passed, and someone softly laid a hand on my shoulder: it was Dowiat, a young enthusiast who had preached in Germany, *à la* Ruge, a Neo-catholicism of a kind peculiar to itself, and in 1847 had gone to America.

He was pale, his features were disordered, his long hair was dishevelled and he was dressed as though for a journey.

'My God!' he said, 'how we meet again!'

'When did you come?'

'Today. Having heard in New York of the Revolution of February and of all that was being done in Europe, I hurriedly sold everything I could, got my money together, and rushed on to a steamer with a light heart full of hope. Yesterday at Le Havre I learned about the latest events, but I had not enough imagination to conceive of *this*.'

We both looked at it once more, and the eyes of both of us were full of tears.

'Not a day, not one single day in this accursed city!' said the agitated Dowiat, and he actually resembled a young prophesying Levite.

'Hence! Away! Farewell, I go to Germany!'

He did go, and found his way into a Prussian prison, where he spent six years.

I still remember the performance of *Catiline* which the unflinching Dumas put on the stage at that time in his Theâtre Historique. The forts were packed with prisoners; the surplus was being sent in droves to the Château d'If to be deported; their relations wandered like shadows from one police office to another, imploring to be told who was killed, who had survived, and who had been shot; but Alexandre Dumas was already producing the days of June in the Roman *laticlavia*[14] on the stage. I went to see it. At first it was nothing much; Ledru-Rollin as Catiline and Cicero as Lamartine – classical sentences with rhetorical padding. The rising is put down. Lamartine passes across the stage with his 'Vixerunt'; the scene changes. A square is covered with corpses, in the distance is a red glow; the dying, in the agonies of death, are lying among the dead; the dead are covered with bloody rags. . . . I could hardly breathe. So recently outside the walls of that playhouse, in the streets leading up to it, we had seen the very same thing, and the corpses were not of cardboard, and the streaming blood was not coloured water, but trickled from living bodies.

I rushed away in a fit of hysteria, cursing the wildly applauding *petit bourgeois*.

In such days of convulsion, when a man cannot stay in restaurant or theatre, at home or in his study, but goes out in a fever, with brain inflamed, inwardly stunned, deeply offended, and ready to offend the first man he meets – at such times every word of sympathy, every tear over the same grief, every curse that springs from the same hatred, has a fearful power.

Sore places quickly draw people together when the original hurts were alike.

In the early days of my youth I was struck by a French novel, which I have not met since: it was called *Arminius*. Possibly it has no great merits, but at the time it influenced me greatly, and it haunted my mind for years afterwards. I remember the main outlines of it to this day.

We all know from the history of the first centuries of the meeting and collision between two different worlds: one, the ancient, classical world of culture, corrupt and effete; the other savage as a wild beast of the forest, but full of untried powers and chaotic impulses – that is, we all know the official, published side of this collision, but not that side which was completed by *minutiae*, in the stillness of home life. We know the events in a lump, but not the fortunes of the persons who were directly dependent on them, though it was in those events that, with no perceptible noise, lives were broken and perished in the collisions. Tears took the place of blood, scattered families of depopulated towns and forgotten tombs of battlefields. The author of *Arminius* (I have forgotten his name) tried to reproduce these two worlds – one coming from the forest into history, the other going from history to the tomb – as they met at the domestic hearth.

Universal history, when thinned out into tales, comes nearer to us, become more nearly measurable by our standards, more alive. I was so taken with *Arminius* that I began writing about 1833, a series of historical scenes in the same manner, and in 1834 the *ober-politsmeyster*, Tsynsky,[15] made a critical analysis of them. But I need hardly say that when I wrote them it never entered my head that I should be caught in a similar collision, and that my own hearth would be desolated, crushed at the meeting of two wheels of world-history.

Whatever may be said there, similar sides exist in our relationships to Europeans. Our civilisation is skin-deep, our depravity is crude, our coarse hair bristles up under the powder, and the sunburn shows through the ceruse; we have the cunning of savages, the lewdness of animals, the suppleness of slaves, among us money and skinflints are appearing everywhere; but we are far behind the hereditary, intangible subtleties of West European corruption. Among us intellectual development serves as a purification and a guarantee. Exceptions are rare. Culture among us, until recent times, constituted a barrier which much that was base and vicious could not cross.

In Western Europe this is not so. And that is how it is that we readily yield to a man who touches upon our holy things, who understands our secret thoughts, who boldly utters what we are wont to pass over in silence or to speak of in whispers to the ear of a friend. We do not take into account that half the sayings which set our hearts beating and our bosoms heaving have become for Europe truisms or mere phrases; we forget how many other corrupted passions, the artificial senile passions, are entangled in the soul of a modern man belonging to that effete civilisation. From his earliest years he is trying to get in first, worn out with importuning, sick with envy, *amour-propre*, insatiable self-indulgence and petty egoism, to which every relationship, every feeling is subordinated. . . . He wants to play a part, to pose on the stage, he wants at all costs to retain his place, to satisfy his passions. We sons of the steppes, when we receive one blow and a second, often not seeing whence they come and stunned by them, take a long time to come to, and then fling ourselves about like wounded bears, smashing up the trees to kill this man. Seven years have passed since then; a true son of our opponent points the finger of scorn at us. Much hatred yet will accrue and much blood yet be spilt through this difference in the two stages of growth and education.

There was a time when I sternly and passionately judged the man who shattered my life. There was a time when I genuinely desired to kill this man. Seven years have passed since then; a true son of our age, I have outworn all desire for vengeance, and have cooled my passionate outlook by long, uninterrupted analysis. In those seven years I have learnt both my own limit and the limits of many others, and instead of a knife I have a scalpel in my hand, and instead of curses and abuse I set to work to tell a narrative from psychological pathology.

CHAPTER II

A DAILY VISITOR

ON coming home one evening, a few days before the 23rd of June 1848,[1] I found in my room an unknown person who came to meet me with a melancholy and embarrassed air.

'Why, is it you?' I said eventually, laughing and holding out my hands to him. 'Is it possible? I should never have known you ...'

It was Herwegh, shaven, shorn, with no moustache or beard.

His luck had very quickly turned. Two months before, surrounded by admirers, he and his wife had driven in a comfortable *dormeuse* from Paris to the Baden campaign,[2] to proclaim the German republic. Now he had come back fleeing from the field of battle, pursued by showers of caricatures, ridiculed by his enemies, blamed by his friends. In a trice everything had changed, everything had fallen to pieces and, to finish it all off, ruin could be glimpsed through the cracks in the scenery.

When I left Russia Ogarëv gave me a letter of introduction to Herwegh, whom he had known in the days of his greatest glory. Ogarëv, always penetrating in questions of thought and of art, had no judgment of men. To him all who were not boring or vulgar were excellent people, and particularly all artists. I found Herwegh an intimate friend of Bakunin's and Sazonov's, and I was soon familiar with him rather than intimate. In the autumn of 1847 I went away to Italy; on my return to Paris I did not find him there. I read of his misfortunes in the newspapers. Almost on the eve of the days of June he arrived in Paris and, since it was from me that he first met with a friendly reception after his blunder in Baden, he took to coming to see us more and more often.

There was a good deal at first to prevent my becoming intimate with the man. He had not that simple, open nature, that complete *abandon* which is so in keeping with everything strong and gifted, and which in Russians almost invariably accompanies talent. He was reserved and wily, afraid of other people, and liked to enjoy himself on the sly; he had a sort of unmanly tenderness, a pitiful dependence on trifles and on the conveniences of life, and a limitless *rücksichtslos* egoism which extended to *naïveté* and cynicism. For all this I attributed to him only half the blame.

Fate had put beside him a woman who by her sophisticated love and her exaggerated solicitude fanned his egoistical propensities and encouraged his weaknesses by making them seem something fine in his eyes. Before his marriage he had been poor: she had brought him wealth, surrounded him with luxury, had become his nurse, his housekeeper, his attendant, a permanent necessity of the lowest order. Humbled to the dust in a sort of perpetual adoration, *Huldigung*, before the poet 'come to fill the place of Goethe and Heine', she took the edge off his talent at the same time, and stifled it in the feather-bed of *petit bourgeois* sybaritism. It used to annoy me that he accepted so eagerly his position of *kept* husband, and I admit that it was not without satisfaction that I saw the financial ruin for which they were inevitably making, and I looked pretty coolly at the weeping Emma when she had to give up her 'gilt-edged' lodging, as we used to call it, and to sell off, one by one, and at half price, her 'amours and cupids'[3] – not serfs, fortunately, but bronze figures.

I shall pause here to say a few words about their previous life and about their marriage itself, which bears a wonderfully well defined imprint of modern German sentiment.

Germans, and still more German women, have cerebral passions galore – that is contrived, spectral, forced literary passions; this is a sort of *Ueberspanntheit* ... a bookish proneness to enthusiasm, a cold-blooded, sham exaltation, always ready to be amazed beyond measure or moved without sufficient reason. It is not pretence, but a false truth, a psychological incontinence, aesthetical hysterics which cost nothing but procure many tears, much joy and grief, many distractions and sensations, *Wonne*. Even an intelligent woman like Bettina von Arnim[4] was unable all her life to get rid of this German malady. The forms it takes may change, the subjects with which it deals may be different, but the psychological treatment of the material, so to speak, is the same. Everything comes down to different variations, to different *nuances* of sensual pantheism, that is, of a religiously sexual and theoretically enamoured attitude to nature and to men, which by no means excludes romantic chastity or theoretical sensuality, either in the secular priestesses of the Cosmos or the monastic brides of Christ, abandoning themselves to erotic ecstasies in their prayers. They

are both burning to be the elect sisters of the real Magdalens.
They do this from curiosity and sympathy for lapses to which
they never bring themselves, and every time they forgive them
their sins, even when they are not asked to. The most enthusiastic
of them go through the whole programme of passions without
descending to practice, and they are tempted by all the sins as it
were by proxy, *per contumaciam*, in the books of others and their
own notebooks.

One of the commonest features of all ecstatic German women is
their idolatrous worship of geniuses and of great men; this religion
proceeds from Weimar, from the days of Wieland, Schiller, and
Goethe. But since men of genius are rare and Heine lived in Paris,
and Humboldt was too old and too much of a realist, they flung
themselves, with a sort of hungry despair, on good musicians and
passable painters. The image of Franz Liszt passed like an electric
spark through the hearts of all the women of Germany, branding
on them his high forehead and long hair combed back
from it.

Eventually, for lack of great men common to all Germany, they
accepted, so to speak, provincial geniuses, who had distinguished
themselves in whatever it might be; all the women fell in love
with him, all the girls *schwärmten für ihn*. They all embroidered
braces and slippers for him, and sent him various souvenirs – in
secret, anonymously.

In the 'forties there was a great intellectual awakening in Germany. It might have been expected that this people, grown grey
over books like Faust, might eventually wish, as he did, to come
out into the market-place and look upon the light of day. We
know now that these were false birth-pangs, that the new Faust
went back from Auerbach's cellar to the *Studierzimmer*. It seemed
otherwise at the time, particularly to the Germans, and therefore
every manifestation of the revolutionary spirit met with ardent
recognition. It was in the full flush of this period that Herwegh's
political songs appeared. I never saw any great talent in them;
no one but his wife could compare Herwegh with Heine, but
Heine's malicious scepticism was out of harmony with the mood
of the day. The Germans in the 'forties did not want Goethes or
Voltaires, but Béranger's songs and the 'Marseillaise', adapted to
transrhenane ways. Herwegh's verses ended sometimes *in crudo*
with a French shout, with the refrain *Vive la République!*, and

that put people in ecstasies in 1842. In 1852 they were forgotten. It is impossible to read them over again.

Herwegh, the poet-laureate of democracy, went from one banquet to another all over Germany, and eventually appeared in Berlin. Everyone rushed to invite him, dinners and parties were given in his honour, everyone wanted to see him, even the King himself experienced such a desire to have a talk with him that the Court doctor, Schenlein, thought it necessary to present Herwegh to the King.

A few paces away from the palace in Berlin there lived a banker. His daughter had for a long time been in love with Herwegh. She had never seen him and had no definite idea of him, but as she read his verses she felt that it was her vocation to make him happy and to plait the roses of domestic bliss into his laurel wreath. When she did actually see him for the first time, at an evening party given by her father, she became finally convinced that this was *he*, and he did in fact become her *he*.

The enterprising, resolute girl opened her attack impetuously. At first the poet, aged twenty-four, shied at the thought of marriage, and especially of marriage with a very plain person, with rather *Junker* manners and a loud voice; the future was opening before him both leaves of her front door: what should a wife do here, or domestic quietude? But the banker's daughter, on her side, was opening *in the real present* sacks of gold ducats, a tour of Italy, Paris, Strasbourg pies, and Clos de Vougeot. The poet was as poor as Irus; he could not go on living at the Follens' for ever – he hesitated and hesitated – and ... accepted the offer, forgetting to say 'thank you' to old Follen,[5] Vogt's grandfather.

Emma herself used to tell me how circumstantially and precisely the poet conducted negotiations about the dowry. He even sent from Zürich sketches of the furniture, the curtains and so on, and required that it should all be despatched *before* the wedding – he insisted upon that.

There was no thought of love; there had to be something to take its place. Emma understood this and made up her mind to secure her power by other means. They spent some time at Zürich; she took her husband to Italy and then settled with him in Paris. There she assigned to her *Schatz* a study with soft sofas, heavy velvet hangings, expensive rugs and bronze statuettes, and arranged for him a whole life of empty idleness. This was new to him and he liked it, but meanwhile his talent grew dim, and he produced nothing; she was angry at this, scolded him for it and at the same time drew him more and more into *bourgeois* epicureanism.*

In her own way she was not at all stupid and had far more strength and energy than he. Her education had been purely German: she had read an immense amount, but not what one should; and had studied a miscellaneous hotch-potch without reaching the summit in anything. One was struck unpleasantly by her lack of womanly grace. From her sharp voice to her angular movements and angular features, from her cold eyes to her eagerness to drag the conversation down to ambiguous topics, everything about her was masculine. Openly, before everyone, she dangled about her husband as elderly men dangle about young girls; she looked into his eyes, drew attention to him with her glance, straightened his necktie, smoothed his hair and praised him with shocking immodesty. Before strangers he was embarrassed by this, but in his own circle he paid no attention to it, as the master, when he is absorbed in his work, does not notice the devotion with which his dog licks his boots and fawns upon him. They sometimes even had scenes over this when visitors had gone away; but next day the adoring Emma would once more begin the same fond pestering, and he would once more put up with it for the sake of the conveniences of life and for the sake of her careful supervision of everything. The following anecdote will show better than anything the extremes to which she went in spoiling her darling. One day after dinner Ivan Turgenev came to see them. He found Herwegh lying on the sofa. Emma was rubbing his foot; she left off when their visitor entered.

'Why have you stopped? Go on!' said the poet wearily.

'Are you ill?' Turgenev asked.

'No, not at all, but it's very pleasant. Well, what's the news?'

---

* Here is an example of the extremes to which her solicitude would go. On one occasion in Italy Herwegh was dissatisfied with the eau de Cologne. At once his wife wrote to Jean Marie Farina to send the very purest eau de Cologne to Rome. Meanwhile they went away from Rome, leaving instructions that letters and parcels should be sent on to Naples; in just the same way they left Naples. . . . Some months later a box containing the eau de Cologne reached them in Paris with an immense bill to pay for its travelling expenses.

They went on talking and Emma went on rubbing his feet.

Confident that everyone admired her husband, she chattered about him incessantly, without observing either that this was very boring, or that she did him no good with anecdotes of his delicate nerves and capricious exigence; to her all this seemed endlessly pleasant and worth being imprinted for ever and ever in men's memory; other people were revolted by it.

'My Georg is a fearful egoist and quite spoilt (*zu verwöhnt*),' she used to say; 'but who has more right to be spoilt than he? All great poets have been eternally capricious children, and they have always been spoilt. The other day he bought me a superb camellia; when he got it home he was so sorry to part with it that he did not even show it me, but hid it in his cupboard and kept it there till it was quite withered – *so kindisch!*'

That is word for word what she told us. By this idolising of him Emma brought her Georg to the edge of an abyss; he fell into it, indeed, and, if he did not perish, he nevertheless covered himself with shame and disgrace.

The noise made by the February Revolution roused Germany. Talk, murmurs, and beating hearts were heard from one end to another of the One German Fatherland divided into thirty-nine parts. In Paris the German working men formed a club, and considered what they were to do. The Provisional Government encouraged them – not to rise, but to leave France; their sleep was somewhat troubled as it was by the French working men. Of course, it might well happen that after Flocon's[6] parting blessing, and Caussidière's[7] strong language about tyrants and despots, these poor fellows would be shot or hanged, or thrown into prison for twenty years; but that was not their affair.

The Baden expedition was resolved upon; but who was to be the deliverer? Who was to lead this new *armée du Rhin*, consisting of a few hundred peaceful working men and foremen? Who, thought Emma, if not the great poet, with his lyre on his back and his sword in his hand, on the 'battle steed' of which he dreamed in his poems? He would sing after battles and conquer after his songs; he would be chosen dictator, he would be in the assembly of kings and dictate to them the will of *his* Germany; in Berlin, *Unter den Linden*, his statue would be set up, and would be visible from the old banker's house; his name would be sung through the ages, and ... in those lyrics the good, self-denying Emma, who had accompanied him as sword-bearer, page, orderly, and watched over him *in der Schwertfahrt*, would perhaps not be forgotten. And she ordered herself, in the Rue Neuve-des-Petits-Champs, a martial riding-habit of the three national colours, black, red, and gold, and bought herself a black velvet *béret* with a cockade of the same three colours.

Through friends Emma indicated the poet to the working men; having no one in view, and remembering Herwegh's poems calling them to rebellion, they chose him as their leader. Emma persuaded him to accept the call.

Upon what basis did this woman thrust the man she so loved into this dangerous position? Where, how, when had he shown that presence of mind, that faculty of being inspired by circumstances which gives a man power over them, that rapidity of reflection, that clearness of vision, and that fervour, in fine, without which no surgeon can perform an operation, no guerrilla chief command his followers? Where was this weakling to find strength to rouse one part of his nerves to redoubled activity and to truss the other up till it was insensible? She herself had both determination and self-control, and that makes it the more unpardonable in her to have forgotten how he started at the slightest sound and turned pale at any unforeseen contingency; how his spirits fell at the least physical pain and how he lost his head before any danger.

What made her lead him into the fearful ordeal, where pretence is useless, where there is no saving yourself by prose or verse, where on one hand the laurel wreath dangles from the tomb, and on the other is flight and the pillory of disgrace? Her expectations were utterly different: she unconsciously described them in conversations and letters afterwards.

The republic in Paris had been proclaimed almost without a battle, the revolution was getting the upper hand in Italy, the news from Berlin and even from Vienna clearly said that those thrones, too, were tottering; it was hard to imagine that the Duke of Baden or the King of Württemberg could stand firm against the current of revolutionary ideas. It might have been expected that at the first shout of freedom the soldiers would fling away their weapons and the people would open their arms to the insurgents; that the poet would proclaim the republic, the republic would proclaim the poet dictator — had not Lamartine been dictator? All that would be left then would be for the dictator-poet to pass in triumphal procession through the whole of Germany, with his black, red, and gold Emma in her *béret*, to be covered with military and civic glory. . . .

In reality things turned out otherwise. The dull-witted soldier of Baden or Swabia knows nothing of poets or republics, but his discipline and his sergeant-major he knows very well and, from his innate servility, loves them and blindly obeys his senior officers. The peasants were taken unawares, the liberators had thrust themselves in without any positive plan and with nothing prepared. Here even brave men like Hecker[8] and Willich[9] could do nothing – they too were defeated, but they did not flee from the field of battle and, luckily for them, had no adoring German woman beside them.

When the firing began Emma saw her pale, frightened Georg, with tears of fear in his eyes, ready to fling away his sword and hide anywhere – and she completed his ruin. She stood in front of him under fire and called on his comrades to save the poet. The government soldiers were winning. Emma, covering her husband's flight, ran the risk of being wounded, killed or taken prisoner, that is, imprisoned for twenty years at Spandau or Rastadt, besides a preliminary flogging.

He hid in a small neighbouring village[10] at the very beginning of the defeat. There he rushed to a peasant, beseeching, adjuring the latter to hide him. The peasant hesitated at first, for fear of the soldiers; but eventually he called him into his yard and, after looking all round, hid the future dictator in an empty barrel and covered him with straw, exposing his house to the risk of pillage and himself to the risk of *Fuchtel*[11] and imprisonment. The soldiers appeared: the peasant did not betray him, but sent word to Emma, who came to fetch him, hid her husband in a cart, changed her clothes, seated herself in the driver's seat and drove him over the frontier.

'What was the name of the man who saved you?' we asked him.

'I forgot to ask him,' Herwegh answered calmly.

His exasperated and embittered comrades now fell upon the luckless singer to tear him to pieces, paying him out at the same time for having grown rich and for his lodgings being covered with 'gilt carving' and for his aristocratic effeminacy and the rest of it. His wife was so far from understanding the *portée* of what she had done that four months later she published, in her husband's defence, a brochure in which she recounted her own exploits, oblivious of the cloud that this must cast over him.

Accusations were soon made against him, no longer only of running away, but of wasting and embezzling the party funds. I think that the money was not appropriated by him, but I am equally sure that it was irregularly squandered, and partly on unnecessary fancies of the marital pair. P. Annenkov was an eyewitness of the purchase of turkeys stuffed with truffles, and of pasties at Chevet's, and of wines and so on being packed in the 'general's' travelling carriage. Money was given by Flocon in accordance with the instructions of the Provisional Government: there are very strange variations in the estimate of this sum; the French talked of 30,000 francs, but Herwegh declared that he had not received even half that, but that the government had paid his railway fare. To this charge the returning insurgents added that, when they reached Strasbourg, hungry, ragged and without a farthing after the defeat, they applied to Herwegh for assistance, and were refused. Emma did not even allow them to see him — while he was living in an expensive hotel and wearing 'yellow morocco slippers'. Why they regarded that as a sign of luxury I do not know, but I heard about those yellow slippers a dozen times.

All this happened as though in a dream. At the beginning of March the liberators *in spe* had been still feasting in Paris; by the middle of May, knocked to pieces, they crossed the French frontier on their way back. Herwegh, coming to his senses in Paris, saw his former garden-path to glory choked up . . . real life had harshly reminded him of his limitations. He understood that his situation — as a poet kept by his wife and a dictator who had run away from the field of battle — was awkward. . . . He had to begin his life afresh, or go to the bottom. I thought (and this was where I made my worst mistake) that he would outgrow the petty side of his character. I thought that I could help him in this more than anyone.

And could I think otherwise when the man said to me every day (and later on wrote it): 'I know the pitiful weakness of my character — your character is serener and stronger than mine — support me, be an elder brother, a father to me. . . . I have no one near or dear to me — I concentrate on you all my affections — by love and friendship you can make anything of me — do not be severe, but kind and indulgent. Let me keep your hand ... and, indeed, I shall not let it go, I shall cling to you. ... In one thing I am not only not inferior to you, but perhaps I am more intense than you: in my unbounded love for anyone who is close to my heart.'

He was not lying, but what he said committed him to nothing. Why, he did not go to the Baden rising, either, meaning to desert his comrades in the hour of conflict: but at the sight of danger he fled. So long as there is no clash, no conflict, so long as neither effort nor sacrifice is demanded, all may go splendidly – for years at a time, for a whole lifetime ... but let nothing happen to you on the way or there will be trouble: crime or shame.

Why did I not know this then?

Towards the end of 1848 Herwegh took to spending nearly every evening with us. He was bored at home. Indeed Emma did bother him fearfully. She had come back from the Baden expedition the same as she went. She had no inner misgivings over what had happened. As before, she was in love, pleased, talkative, as though they had come back from a victory, or at any rate with no wounds in the back. Only one thing worried her – shortage of money and the positive expectation of soon having none at all. The revolution which she had helped so unsuccessfully had not liberated Germany and had not covered the poet's brow with laurels, but it had completely ruined the old banker, her father. She was continually trying to dissipate her husband's sombre thoughts; it never entered her head that perhaps only these melancholy reflections might be his salvation.

The superficial, volatile Emma had no desire for the profound, inward thought which apparently produces nothing but pain. Hers was one of those uncomplicated natures (like simple tunes in two time) which cut every Gordian knot with their *entweder oder*; to left or to right does not matter to them so long as they can extricate themselves somehow and hurry on again ... where to? That they do not know themselves. She would break into the middle of what someone was saying, either with an anecdote or with a pertinent observation, though its pertinent bearing was of the lowest order. Convinced that no one among us was gifted with so much practical sense as she, instead of concealing her business abilities through coquetry, she coquettishly displayed them. It must be said, moreover, that she never showed any serious, practical, good sense. Fussing about, discussing prices and cooks, furniture and materials, is something very different from intelligent application. Everything in her house was at sixes and sevens, because everything was subordinated to her monomania; she was for ever on the *qui vive*, watching her husband's face, and she put his caprices before all the real necessities of life, and even before the health and upbringing of the children.

Herwegh naturally longed to get away from home, and sought peace and harmony with us; he saw in us a sort of ideal family in which he loved and worshipped everything, the children as much as ourselves. He dreamed of going with us somewhere far away and thence calmly watching the fifth act of the dark European tragedy. And for all that, except for having the same or very similar understanding of things in general, there was very little resemblance between us. Herwegh somehow collated everything in the world with himself; he gave himself up to his own ends, sought attention, was timidly and conceitedly diffident, and at the same time was convinced of his own superiority. All this together made him coquettish and capricious, and sometimes designedly melancholy, attentive or inattentive. He was constantly in need of someone who would be guide, friend, confidant and slave all at once (precisely as Emma was), who would put up with coldness and reproaches when his services were not needed, and who would be ready at the first beck to rush headlong back and do what he was ordered with smiling submission.

I, too, have sought love and friendship, have sought sympathy and even applause, and tried to evoke them; but I have never had a trace of this feminine-feline playing at *dépit* and scenes of explanation, this eternal thirst for attention and caresses. Possibly the unconstrained truthfulness, excessive self-confidence, and healthy simplicity of my behaviour – my *laisser aller* – also arose from vanity, and perhaps it is through them that I have called down misfortunes on my head, but so it is. In laughter and in sorrow, in love and in public interests, I surrendered myself sincerely, and could rejoice and grieve with no thought of self. With my strong nerves and muscles I stood independent and self-reliant, and was ready and eager to give a helping hand to another, but I never asked myself for help or support as charity. With such opposite temperaments it could hardly be expected that there should not sometimes be unpleasant jars between Herwegh and me. But in the first place he was much more on his guard with me than with others; in the second, he completely disarmed me by his melancholy admission that he was at fault; he did not try to justify himself, but in the name of friendship asked for indulgence for the weakness of character which he himself recognised and censured. I played the part of a sort of guardian, defended him from others and gave him rebukes which he received submissively. His submissiveness was extremely distasteful to Emma, and she teased him jealously about this.

The year 1849 arrived.

CHAPTER III

DELIRIUM OF THE HEART

LITTLE by little I began in 1849 to notice various changes in Herwegh. His uneven temper became still more uneven; he had attacks of intolerable melancholy and ineffectuality. His wife's father lost his fortune for good and all; the remnants that were saved were needed for other members of the family ... poverty knocked more rudely at the poet's door. .... He could not think of it without shuddering and losing every trace of manliness. Emma exhausted herself, borrowed right and left, obtained goods on credit and sold her things ... all that he might not notice the true state of affairs. She not only denied herself necessaries, but even had no underclothes made for the children that he might dine at the Provençal Brothers and buy himself rubbish. He took money from her, not knowing where it came from, and not wishing to know. I quarrelled with her about this, told her that she was ruining him, and hinted the same to him: he stubbornly refused to understand, she lost her temper, and everything went on as before.

Although he was ludicrously afraid of poverty, yet this was not the cause of his depression. In his lamentations about himself one note was continually recurring, and I eventually grew sick of it; it annoyed me to hear Herwegh's everlasting repetition of complaints of his own feebleness, accompanied with reproaches for my not needing kindness and endearments – that he was fading away and perishing without a helping hand, that he was so lonely and unhappy that he would like to die, that he had a profound respect for Emma, but his sensitive, differently attuned soul shrank from her brusque, jarring touch and 'even from her loud voice'. Then followed passionate protestations of his friendship for me. .... In this feverish, nervous condition I began to discern a feeling that alarmed me for him as much as for myself. It seemed to me that his affection for Natalie was assuming a more passionate character. There was nothing for me to do: I remained silent, and began sorrowfully to foresee that in this way we should soon get into great trouble, and that something in our life would be shattered.

... Everything was shattered.

The everlasting talk of despair, the everlasting entreaty for attention, for a warm word, the dependence on it, and the tears, the tears – all this had a powerful effect on a woman who had only just lost her hardly won serenity, and who was suffering from the profoundly tragic surroundings in which we were living.

'You have a bruised place which is still sensitive,' Natalie said to me; it is something which goes very well with your character; you don't understand the yearning for the tender care of a mother, a friend, a sister, which so torments Herwegh. I understand him because I feel it myself. He is a big child, while you are a grown-up man; he can be distressed or made happy by a trifle. A chilling word is almost death to him, he must be spared. ... And to make up for this with what infinite gratitude he thanks one for the smallest attention, for warmth, for sympathy. ...'

Could it be? ... But no, he would have told me himself before speaking to her ... and I kept his secret inviolate and did not hint at it with a single word, regretting that he had not spoken to me.

A secret may be kept if it is entrusted to no one, but only if it is to *no one*. If he had spoken of his love he could not have been silent with a man with whom he was living in such deep intimacy, and about a secret so closely affecting the latter. So he had not spoken. ... I had forgotten for the time the old novel entitled *Arminius*.

At the end of 1849 I went from Zürich to Paris to take some steps about money of my mother's, which was held up by the Russian Government. I had parted from Herwegh when I left Geneva. On the way I visited him at Berne. I found him reading to Simon[1] of Trèves passages from *Vom andern Ufer*, from the proofs. He rushed at me as though we had not met for months. I was going on the evening of the same day. He did not leave me for a single minute, repeating again and again declarations of the most enthusiastic and passionate affection. Why did he not find the strength then to make a straightforward and open confession to me? ... I was in a gentle mood then, and everything might have gone off humanely.

He saw me to the posting-station, said goodbye and, leaning against the gate by which the post-chaise drives out, stayed there wiping away tears. ... That was almost the last minute at which I still really loved this man. ... I thought all night long, and it was only then that I got as far as a single word, which would not leave my brain: 'Calamity! Calamity! ... What will come of this?'

Soon after this my mother left Paris. I remained at Emma's, but in reality I was completely alone. This solitude was essential for me; I had to be alone, in order to think what to do. A letter from Natalie, in which she spoke of her sympathy for Herwegh, gave me an opportunity and I made up my mind to write to her. My letter was sad but gentle; I asked her to examine her heart calmly and attentively and to be open with herself and with me; I reminded her that we were too closely bound together by all the past and by our whole life to leave anything unspoken.

'I got your letter of the 9th,' Natalie wrote (and this letter has survived: almost all the rest were burnt at the time of the *coup d'état*), 'and I, too, sit and think only: Why is this? And I weep and weep. Perhaps it is all my fault, perhaps I don't deserve to live – but I feel just as I wrote to you on that evening when I was left alone. Clean before you and before the whole world, I have not heard a single reproach in my soul. I have lived in my love for you as in God's world; outside it, it seemed, there was nowhere for me to live! To be cast out of that world – where to? – I must be born again. I am inseparable from it as from nature, come from it and go back into it again. I have never for one instant felt differently. It is a wide, rich world, I know no inner world more rich. Perhaps it is too wide, perhaps it has too much enlarged my being and its needs – in this fullness there have been minutes, and there have been from the very beginning of our life together, in which imperceptibly, somewhere deep down in the very bottom of my soul, something like a fine hair has fretted me, and afterwards everything has become luminous again.'

'This dissatisfaction, this something left unused, and wasted,' Natalie writes in another letter, 'sought sympathy elsewhere, and found it in Herwegh's friendship.'

This was not enough for me, and I wrote to her: 'Don't be diverted from simply looking into the depths of yourself, don't try to find explanations. You won't escape the whirlpool by dialectics, it will sweep you away just the same. There is a new note in your letters which is unfamiliar to me, not the note of sadness, but another. ... Now everything is still in our hands ... let us have the courage to go on to the end. Consider that since we have put into words the secret which troubled the peace of our souls, *Herwegh will sound as a false note in our harmony – or I shall. I am ready to go with Sasha to America*; then we shall see what to do and how ... it will be hard for me, but I shall try to bear it; *here* it will be harder still, and I shall not bear it!'

To this letter she answered with a cry of horror: the thought of separation from me had never occurred to her. 'How can you! How can you! I – and to part from you ... as though that were possible! No, no, I want to come to you, to come to you at once. ... I am going to pack, and in a few days the children and I will be in Paris!'

On the day when they left Zürich she wrote again: 'As though after storm and shipwreck I am coming back to you, to my native land, with complete faith, with complete love. If only the state of your soul were like that in which I am now! I am happier than ever! My love for you is always the same, but I have learnt to know your love better, and all my accounts with life are settled: I expect nothing, I desire nothing. Misunderstandings! – I am grateful to them for they have made so much clear to me, and even they will pass and disperse like storm-clouds.'

Our meeting in Paris was not a joyful one, but it was filled with the feeling of profound, genuine recognition that the roots of the tree had gone too deep for the storm to tear it up, that it was not easy to part us. In the long conversations of that period one thing surprised me, and I examined it several times, and each time was convinced that I was right. Together with the warm sympathy that she still retained for Herwegh, Natalie seemed to breathe more freely, as though she had stepped out of some black magic circle; she was afraid of him, she felt that there were dark forces in his soul, she was frightened by his boundless egoism and she looked to me for a bulwark and defence.

Though he knew nothing of my correspondence with Natalie, Herwegh saw that there was something wrong with my letters. As a matter of fact, apart from everything else, I was very much displeased with him. Emma was struggling, weeping, trying to please him and get him money, and he either did not answer her letters or wrote caustically and kept asking for more and more money.

His letters to me, which I have kept, are more like the letters of an alarmed lover than a friendly correspondence. He tearfully reproaches me for coldness; he beseeches me not to abandon him; he cannot live without me, without my former complete, unclouded sympathy; he curses our misunderstandings and the interference of the 'frantic woman' (that is Emma); he thirsts to begin a new life, a life far away, a life with us – and again he calls me his father, his brother, his twin. I pulled out various stops when I wrote to him in answer to all this: 'Consider whether you are capable of beginning a new life, whether you are able to shake off . . . the corruption, the vitiated civilisation' – and once or twice I reminded him of Aleko,[2] to whom the old Gypsy says: 'Leave us, proud man: you want freedom for yourself alone.' He answered this with reproaches and tears, but he did not speak out. His letters of 1850 and our first conversations at Nice make up a fearful indictment . . . of what? Of deceit, of cunning, of lying? . . . No: and that would not have been anything new – but of that pusillanimous duplicity of which I have so often accused the Western European. Going many times over all the details of our melancholy drama, I have always been brought up short with amazement at how this man never once, never by one word, never by one straightforward impulse of the heart, betrayed himself. How could he, when he felt the impossibility of being frank, try to enter into closer and closer intimacy with me, and touch in his talk on those forbidden places of the soul which can only be touched without sacrilege in perfect and mutual openness?

From the moment when he guessed my misgiving – and not only kept silent, but assured me more and more of his friendship, and at the same time worked more and more powerfully, by means of his despair, on the woman whose heart was shaken – from the moment when he began acting a lie to me by his silence and beseeching her (as I afterwards learned) not to deprive him of my friendship by an unguarded word – from that moment his crime begins.

Crime! . . . Yes . . . and all the calamities that follow come as the simple, inevitable consequence of it; they come unchecked by deaths, unchecked by penitence, because they are not a punishment but a consequence. . . . The consequences come upon a generation through the frightful indestructibility of what has once been done. Punishment may redeem a man and reconcile him to himself and others, and penitence may redeem him, but the consequences of his action go their own fearful way. For an escape from them religion has invented paradise and its ante-chamber, the monastery.

I was expelled from Paris,[3] and almost at the same time Emma was expelled, too. We intended to spend a year or two at Nice – it was then Italy – and Emma went there also. Within a short time, that is, towards winter, my mother was to come to Nice, and Herwegh with her.

Why did I go with Natalie to just that town? This question has occurred to me and to others, but it is really a petty one. Apart from the fact that wherever I had gone Herwegh could have come likewise, would it have been possible to achieve anything but mortification by geographical and other superficial measures?

Within a fortnight or three weeks of his arrival Herwegh assumed the aspect of a Werther in the last stages of despair, and so obviously that a Russian doctor, who was passing through Nice, was convinced that he was becoming insane. His wife used to appear with tear-stained eyes – he was treating her shockingly. She would come to weep for hours together in Natalie's room, and both were convinced that today or tomorrow he would fling himself into the sea or shoot himself. Natalie's pale cheeks and agitated air, and her alarming preoccupation, even with regard to the children, showed me clearly what was passing within her.

Not a word had been said yet, but already through the external calm there was the gleam of something sinister coming nearer and nearer – like two glittering points continually disappearing and appearing again at the edge of the forest that show that a wild beast is near. Everything was moving rapidly towards a *dénouement*. It was delayed by the birth of Olga.[4]

CHAPTER IV

ANOTHER YEAR: 1851

BEFORE the New Year Natalie brought for me to see a water-colour she had ordered from the painter Guyot.[1]

It was a sketch of our terrace, part of the house and the yard, with the children playing in the yard and Tata's goat lying there; Natalie herself was in the background, on the terrace. I thought that the sketch was meant for me, but Natalie said that she meant to give it to Herwegh for a New Year's present. . . . I was vexed.

'Do you like it?' asked Natalie.

'I like it so much,' I said, 'that if Herwegh will allow it I shall order a copy to be made for myself.'

From my pallor and my voice Natalie understood that these words were both a challenge and evidence of a violent storm within me. She looked at me, and there were tears in her eyes.

'Take it for yourself,' she said.

'Nothing would induce me to. What an absurd idea!'

We said no more.

We celebrated the New Year, 1851, at my mother's. I was in a state of acute irritability; I sat beside Vogt and, filling his wine-glass and mine over and over again, I poured out a stream of tart and caustic remarks. Vogt was rolling with laughter. Herwegh looked sadly from under his brows: at last he understood. After we drank to the New Year he raised his glass and said that the only thing he desired was 'that the coming year might be no worse than the last'; that he desired this with all his heart, but did not hope for it. On the contrary, he felt that 'all, all was collapsing in ruins'.

I said nothing.

Next morning I took up my old tale, *Who Is At Fault?*[2] and read over Lubenka's diary and the last chapters. Could this be a prophecy of my fate, just as Onegin's duel was a presage of Pushkin's? . . . But an inner voice said to me: 'What sort of Krutsi-fersky are you? yes, and he is a queer sort of Beltov! Where is his noble sincerity, where is my tearful self-sacrifice?' And in the midst of my conviction of Natalie's momentary inclination I was even more convinced that I should measure up to him, that he would not supplant me in her heart.

What happened was what I expected: Natalie herself invited an explanation. After the scene over the sketch, and the New Year's party at my mother's, it was impossible that it should be put off.

The conversation was painful. *We were neither of us at that high pitch at which we had been the year before.* She was confused, afraid of my going away, afraid of his going away, wanted to go away to Russia for a year herself and was afraid to go. I saw her wavering and saw that with his egoism he would destroy her ... and she would not find sufficient strength. I was beginning to hate him for his silence.

'Once more,' I repeated, 'I put my fate in your hands. Once more I beseech you to weigh everything, to consider everything. I am still ready to accept any decision; I am ready to wait a day, a week, only *the decision must be final.* I feel,' I said, 'that I have reached the limit of my strength; I can still behave well – but I feel, too, that I shan't be able to for long.'

'You shan't go away, you shan't go away,' she said, bursting into tears. 'I could not survive that.' On her tongue such words were not to be taken lightly. '*He* must go.'

'Natalie, don't be in a hurry – don't be in a hurry to take the final decision, because it is final ... think, as much as you like, but give me a final answer. These ebbs and flows are more than I can stand ... they are making me stupid. I am growing petty, I shall go off my head. .... Ask anything you like of me, but it must be once for all....'

At that point my mother and Kolya drove up to invite us to Mentone; when we went out to take our seats in the carriage, it appeared that there was one place too few. I motioned Herwegh to a seat. Herwegh, who was by no means distinguished for such delicacy, would not take it. I looked at him, shut the carriage door and said to the coachman, 'Off!'

We remained together, the two of us, on the seashore in front of the house. I felt as though there was a flagstone weighing on my heart; he was silent, white as a sheet and avoiding my eyes. Why did I not directly begin speaking or push him off the rock into the sea? I was prevented by a kind of nervous impotence. He said something to me about the sufferings of a poet and life's being so abominably arranged that a poet brings misfortune everywhere.

He suffers himself and makes everyone near him suffer. . . . I asked him whether he had read George Sand's *Horace*. He did not remember,[3] and I advised him to read it.

He went to Visconti's for the book. *I never saw him again.*

When we assembled for dinner between six and seven, he was not there. His wife came in, her eyes swollen with tears: she announced that her husband was ill – we all exchanged glances; I felt quite capable of sticking into her the knife that was in my hand. He had shut himself up in his rooms on a floor above. By this *étalage* he did for himself: I was freed from him.

At last all the others had gone, and the children were in bed – *we two were left together*. Natalie sat at the window and wept. I walked about the room, the blood throbbing in my temples; I could not breathe.

'He is going,' she said at last.

'I think that's not at all necessary: it is I who must go.'

'For God's sake . . .'

'I shall go.'

'Alexander, Alexander, don't say what you may regret. Listen to me – save us all. You alone can do it. He is crushed, he is utterly despondent. You know yourself what you have been to him: his frantic love, his frantic friendship and the consciousness that he has caused grief – and worse. *He* wants to go away, to vanish; but for that to happen you must not complicate things; otherwise he is only one step from suicide.'

'Do you believe that?'

'I am certain of it.'

'And he says so himself?'

'He and Emma. He has cleaned his pistol.'

I burst into a roar of laughter and asked:

'Is it his Baden one? He'd have to clean it, for it probably fell in the mud. You can tell Emma, by the way, that I answer for his life. I'll insure it for any sum.'

'Mind that you do not regret your laughter,' said Natalie, shaking her head sombrely.

'If you like I'll go and persuade him . . .'

'What will it all lead to?'

'The consequences,' I said, 'are hard to foresee and still harder to avert.'

'My God! My God! The children, the poor children: what will become of them!'

'About them,' I said, 'you should have thought before!'

And this, of course, was the cruellest thing of all that I said to her. I was too much provoked for a humane understanding of the sense of words. I felt a spasm in my chest and in my head, and perhaps I would have been capable of bloody deeds as well as of cruel words.

She was utterly crushed, and a silence followed.

About half an hour passed.[4] I wanted to drink my cup to the dregs; I put several questions to her, and she answered them. I felt myself bruised; I was intoxicated by savage impulses to vengeance, of jealousy, of outraged *amour-propre*; no trial, no gibbet could have had terrors for me; I no longer valued my life at a farthing. This is one of the first things that makes mad, frightful acts possible. I did not say a word – I stood in front of the big table in the drawing-room, with my arms folded – probably my face was utterly distorted.

The silence continued. Suddenly I looked at her and was frightened: her face was covered with a deathly pallor, a bluish pallor; her lips were white, her mouth half-open and twitching; not saying a word she gazed at me with mad, lack-lustre eyes. This look of infinite suffering, of mute agony instantly stilled the ferment of passion; I felt sorry for her, tears flowed down my cheeks and I was ready to fall at her feet, to ask for forgiveness. . . . I sat down beside her on the sofa, took her hand, put her head on my shoulder and tried to comfort her in a kind, quiet voice.

My conscience gnawed at me – I felt I had been an inquisitor, an executioner. . . . Was this what was needed? Was this a friend's help, was this sympathy? And so with all my culture, with all my humanity, in a fit of fury and jealousy I could torment an unhappy woman, could play the part of a Raoul Bluebeard?

A few minutes passed before she said anything, before she could say anything, and then she suddenly threw herself sobbing on my neck. I let her sink on the sofa; utterly exhausted; all she could say was: 'Don't be afraid, my dear; these are good tears, tears of tenderness; no, no, I will never part from you.'

Her agitation and spasmodic sobbing were too much: she closed her eyes, she was fainting. I poured eau de Cologne on her head and moistened her temples; she grew calm, opened her eyes, pressed my hand, and sank into semi-consciousness which lasted for more than an hour. I remained on my knees beside her; when she opened her eyes, she met my sad, mild look – the tears were still rolling down my cheeks – and she smiled at me. . . .

This was the crisis. From that minute the evil spell began to lift, the poison to have less effect.

'Alexander,' she said, recovering herself a little, 'finish what you have done: swear to me – I must have this, for I can't live without it – swear that it shall all end without bloodshed. Think of the children . . . of what will become of them without you – and without me.'

'I give you my word that I shall do everything possible: I shall reduce any chances of collision, I shall sacrifice a great deal – but for this I must have one thing: that he shall leave tomorrow – if only for Genoa.'

'That shall be as you wish. And we shall begin a new life, and let all bygones be bygones.'

I embraced her tightly.

Next morning Emma came to see me. She was dishevelled, with tear-stained eyes, looking very ugly, in a blouse girt round her waist with a cord. She came towards me with tragic slowness. At any other time I should have burst out laughing at this German theatricality, but now I was in no mood for laughing. I received her standing and made no attempt to conceal that her visit was unpleasant to me.

'What do you want?' I asked.

'I have come to you from *him*.'

'Your husband,' I said, 'might come himself if he wants anything – or has he already shot himself?'

She folded her arms across her bosom.

'And is this you speaking – you, his friend? . . . I don't recognise you! Surely you must understand the tragedy of what is taking place before your eyes . . . his tender constitution will not endure a separation from her, nor a rupture with you. Yes, yes, with you! . . . he is weeping over the grief he has caused you – he bade me tell you that his life is in your hands; he asks you to kill him . . .'

'This is a farce,' I said, interrupting her. 'Come, who invites people to murder him like this, and through his wife too? I find this persistence in vulgar melodramatic antics repulsive. I am not a German.'

'Herr Herzen! . . .'

'Mme Herwegh, why do you undertake such difficult commissions? You could hardly expect to hear anything pleasant from me.'

'This is a fatal calamity,' she said after a pause; 'it has struck you and me alike . . . but look at the difference between your exasperation and my devotion. . . .'

'Madame,' I said, 'our parts have not been alike; I ask you not to compare them, for fear you should have to blush.'

'Never!' she said vehemently. 'You do not know what you are saying.' And then she added:

'I shall take him away: he must not remain in this situation: your will shall be done. But you are no longer in my eyes the man I have so much respected and looked upon as Georg's best friend. No, if you were that man, you would have parted with Natalie, let her go, let him go; I would have remained here with you and the children.'

I laughed aloud.

She went crimson in the face, and with a voice shaking with anger and exasperation she asked me:

'What does that mean?'

'Why do you jest about serious matters?' I said to her. 'But that's enough; here is my ultimatum: go at once to Natalie, yourself alone, speak to her – if she wants to go, she shall go – I shall put no obstacle in the way of anyone or anything except (excuse me) except your staying here; I can deal with the housekeeping myself somehow. But listen: if she does not want to go, this is the last night I will spend under the same roof as your husband, we shall not both spend another night here alive!'

An hour later Emma came back and gloomily announced to me – in a tone as though to say: See the fruits of your wickedness! –

'Natalie is not going – she has destroyed a great being out of vanity – I shall save him!'

'And so?'

'And so we are going in a day or two . . .'

'A day or two? What do you mean? Tomorrow morning! Have you forgotten the alternative?' (When I repeated this I did not at all break my promise to Natalie: I was absolutely certain she would take him away.)

'I don't know you. How bitterly I've been mistaken in you!' observed the crazy woman, and she went out again.

This time her diplomatic mission was an easy one. She came back twenty minutes later saying that he agreed to everything, both to going away and to a duel, but as well as this he had told her to tell me that he had taken a vow not to lift a pistol to my breast, but was ready to accept death at my hands.

'You see how he keeps on joking with you. Why, even the French king was put to death simply by a public executioner, and not by a friend. And so you set off tomorrow?',

'I really don't know how to do it. We have nothing ready.'

'Everything can be got ready during the night.'

'We must get a *visa* for our passport.'

I rang the bell, and Rocca, our cook, came in. I told him that Mme Emma asked him to get their passports *viséd* for Genoa at once.

'But we haven't any money for the journey.'

'Do you need much to get to Genoa?'

'600 francs.'

'Allow me to hand them to you.'

'We owe bills at the shops here.'

'Approximately?'

'500 francs.'

'Don't worry – and *bon voyage*.'

This tone she could not endure: vanity was almost her prime passion.

'Why,' she said, 'why this behaviour to *me*? Me you have no right to hate or to despise.'

'So him I have?'

'No,' she said, choking with tears, 'no, I only meant to say that I liked you sincerely as a sister, that I don't like to leave you without shaking hands with you. I respect you; you may be right, but you are a cruel man. If you knew what I have been through! . . .'

'But why have you been a slave all your life?' I said to her, giving her my hand; at that moment I was not capable of sympathy. 'You have earned your fate.'

She went out covering her face.

At ten o'clock next morning, in a hired carriage loaded with various baskets and trunks, the poet set off – *mit Weib und Kind* – for Genoa. I was standing at the open window; he managed to whisk into the carriage so rapidly that I did not even catch sight of him. She held out her hand to the cook and the maid, and took her seat beside him. Anything more ignominious than this *bourgeois* departure I cannot imagine. Natalie was badly shaken. We went out alone together into the country; the walk was a sorrowful one, for the blood was streaming from fresh, raw wounds. On our return home the first person to meet us was Herwegh's son Horace, a boy of nine, mischievous and thievish.

'Where have you come from?'

'From Mentone.'

'What has happened?'

'Here is a note from Maman for you.'

'*Lieber* Herzen,' she wrote, as if nothing had happened between us, 'we are stopping for a couple of days at Mentone, the room at the hotel is small – Horace is in Georg's way – allow us to leave him with you for a few days.'

This lack of tact amazed me. In addition to this Emma had written to Karl Vogt asking him to come for a consultation – so outsiders were to be involved. I asked Vogt to take Horace and to say that we had no room.

'However,' she sent word to me through Vogt, 'we still have the upper rooms for three full months, and I can make use of them.'

That was perfectly true – only it was I who paid the rent for those rooms.

Yes, in this tragedy, as in Shakespeare, the coarse laughter, vulgar oaths, and rascality of the market-place are side by side with sounds that tear the heart, with the moan with which life goes out, the last spark dies and thought is extinguished.

Emma had a maid-servant, a handsome, very honourable French girl from Provence, called Jeannette; she remained behind for a couple of days, and was to go with their things by steamer to Genoa. On the next morning Jeannette softly opened my door and asked whether she might come in and speak to me alone. This had never happened before; I thought that she wanted to ask for money, and was prepared to give her some. Blushing up to her ears and with tears in her eyes, the good Provençale handed me various unpaid bills of Emma's from different shops, and added:

'Madame ordered me, but I really can't do it without asking

you — you see, she ordered me to buy various different things at the shops and have them put down on these bills — I couldn't do that without telling you.'

'You were quite right to tell me. What did she tell you to buy?'

'Here is the list.'

On the list were several pieces of linen, several dozen handkerchiefs and a complete outfit of children's underclothes.

It is said that Caesar could read, write, and dictate at the same time: but what versatility is here! — to be able to think of thrifty means of obtaining linen and children's stockings when a family is being destroyed, and people are feeling the cold blade of Saturn's scythe. The Germans are a glorious people!

CHAPTER V

THE FAST THAT FAILED

AGAIN we were alone, but this was not the past come back: every-thing bore the traces of the storm. Faith and doubt, weariness and exasperation, the feeling of annoyance and indignation were a torture; but what was most tormenting of all was that the thread of life was broken, that blessed heedlessness which had made life so easy was gone, and nothing sacred was left. If all that had been had been, nothing was impossible. Thoughts of the past frightened me for the future. How many times we came in to dinner alone in the evening and, neither of us touching anything, and not utter-ing a word, got up from the table wiping away tears, and seeing the good Rocca, with an angry face, shake his head as he took out the dishes.

Idle days, sleepless nights . . . misery – misery. I drank whatever came to hand, Schiedam, cognac, old Bellet, drank at night alone and drank by day with Engelson[1] – and this in the climate of Nice. The Russian weakness of drowning one's sorrows in drink is by no means so bad as is said. Heavy sleep is better than heavy sleeplessness, and the headache of the morning after is better than mortal sorrow on an empty stomach.

Herwegh sent me a letter. I sent it back without reading it. He started writing to Natalie, letter after letter. He wrote once more to me – I sent the letter back. I looked sadly on at this. This time should have been a time of profound experience, of peace and of freedom from external influences. What sort of peace, what sort of freedom could there be with letters coming from a man who posed as being frantic, and threatened not only suicide but the most fearful crimes? Thus he wrote, for instance, that he had moments of frenzy when he wished to cut his children's throats, to fling their corpses out of the window and to appear before us bathed in their blood. In another he wrote that he should come to kill himself at my feet, and say: 'See to what you have brought the man who loved you so!' Side by side with this he kept implor-ing Natalie to reconcile him with me, to take everything on herself, and to suggest that he should come as Sasha's tutor.

A dozen times he wrote of his loaded pistol and Natalie still believed it all. He only asked her blessing on his death. I persuaded her to write to him that at last she agreed, that she was convinced there was no way out but death. He answered that her words had come *too late*, that he was now in a different mood and felt he had not strength enough to do it, but that, forsaken by all, he was going away to Egypt. This letter did him great damage in Natalie's eyes.

After that Orsini arrived from Genoa, and he told us with laughter, how the husband and wife had tried to commit suicide. . . . Learning that the Herweghs were in Genoa Orsini went to see them and met Herwegh walking along the marble sea-wall. Hearing from him that his wife was at home Orsini went there. She immediately explained to him that they had made up their minds to starve themselves to death, that this form of suicide had been chosen by him for himself, but that she wished to share his fate, and she besought Orsini not to abandon Horace and Ada.

Orsini was stupefied.

'We have not eaten for thirty hours,' Emma went on; 'do persuade him to eat something: save the great poet for humanity,' and she sobbed.

Orsini went out on to the verandah and returned at once with the joyful tidings that Herwegh was standing in the corner eating *salami*. Emma, overjoyed, rang the bell and ordered a basin of soup. At this moment her husband came back gloomily, without a word about the *salami*, but the tell-tale basin was standing there.

'Georg,' said Emma, 'I was so delighted to hear from Orsini that you were eating, that I decided to ask for some soup myself.'

'I felt so sick, I took a little piece of *salami* – but anyhow it's all nonsense; death by starvation is most painful, I shall poison myself.' And he started on the soup.

His wife raised her eyes to the ceiling and glanced at Orsini as though to say: 'You see, there is no saving him.'

Orsini is dead, but there are several witnesses of his story living, Karl Vogt, for instance, Mordini,[2] and Charles Edmond. These antics made it very hard for Natalie. She was humiliated through him, I was humiliated through him, and she felt this poignantly.

In the spring Herwegh went away to Zürich and sent his wife to Nice (another outrageous indelicacy). I wished, after all that had happened, to rest. I took advantage of my Swiss naturalisation and went to Paris and Switzerland with Engelson.

Natalie's letters were calm; her soul seemed to be more at ease.

On the way back I met Sazonov in Geneva. Over a bottle of wine he asked me, with the most absolute nonchalance, how things were going now between my wife and me.

'As always.'

'Well, I know the whole story, and I am asking you from friendly sympathy.'

I looked at him with a shudder of horror; he noticed nothing. What was the meaning of it? I had reckoned on its all being a secret, and suddenly a man speaks to me of it over a glass of wine as though it were the most ordinary, everyday affair.

'What have you heard, and from whom?'

'I have heard the whole story from Herwegh himself – and I shall tell you frankly I don't think you are right at all. Why don't you let your wife go, or else leave her yourself? – upon my soul, it's such weakness – you could begin a new, fresh life . . .'

'But what makes you imagine that she wants to go? You can't believe that I can let her go or not let her?'

'You coerce her, but morally, of course, not physically, by your words, your distress. I am very glad, though, to find you much more serene than I expected: and I don't care to be only half open with you. Herwegh left your house, in the first place because he's a coward, and fears you as he fears the fire, and in the second because your wife gave him her word to come to Switzerland as soon as you calmed down.'

'That's a foul lie!' I shouted.

'That's what he told me, I give you my word of honour on it.'

When I got back to my hotel I flung myself, sick and shattered, on the bed, without undressing, in a state bordering on madness or death. Did I believe it or not? I do not know, but I cannot say that I entirely disbelieved what Sazonov had told me.

'And so,' I repeated to myself, 'this is how our poetical life is ending, in deceit, and incidentally in European gossip . . . ha, ha, ha! They pity me, they spare me out of mercy, they give me time to breathe, like a soldier who is taken to the hospital, half-way through his flogging, when his pulse beats feebly, and zealously tended in order to make up the other half when he is stronger.' I was insulted, outraged, humiliated.

In this state of mind I wrote a letter that night – my letter must have borne traces of frenzy, despair, and distrust. I repent, I deeply repent that insult 'by correspondence' – that wicked letter.

Natalie answered in words of sombre grief.

'I should do better to die,' she said; 'your faith is destroyed, and every word now will call back all the past to you. What am I to do, and how can I ever convince you? I weep and weep.'

Herwegh had lied.

Her following letters were meekly sorrowful; she was sorry for me, she wanted to heal my wounds, and what must she herself have suffered! . . . Why did that man turn up who repeated to me that lie . . . and why was there not someone else there to stop that letter of mine, written in a fit of criminal fury?

CHAPTER VI

OCEANO NOX (1851)[1]

1

ON the night between the 7th and 8th of July, after one o'clock in the morning, I was sitting on the steps of the Carignano Palace in Turin; the square was completely empty: at a little distance from me a beggar was dozing, and a sentinel was walking slowly up and down, whistling a tune from some opera and jingling his musket. ... It was a night parching and hot, saturated with the *scent* of the sirocco.

I felt unusually happy, as I had not for a long time; I felt again that I was still young and full of energy, that I had friends and beliefs, that I was as full of love as I had been thirteen years before. My heart was beating as I had lately been unused to feeling it; it was beating as on that March day in 1838, when I had stood wrapped in my cloak under the lamp-post waiting for Ketscher in Povarsky Street.

Now, too, I was awaiting a meeting, a meeting with the same woman, and waiting perhaps with even more love, though sad, black notes were mingled with it; but on that night they were hardly audible. After the frantic crisis of bitterness and despair that had come over me on my way through Geneva, I felt better. Natalie's gentle letters, filled with sadness, tears, pain and love, completed my cure. She wrote that she was coming from Nice to Turin to meet me, and that she wanted to spend a few days in Turin. She was right: we needed to look into each other once more alone, to sponge the blood from each other's wounds, to wipe away our tears and lastly to find out once for all whether there was still happiness for us together; and all this *tête-à-tête*, even without the children, and in another place too, not in the surroundings where the furniture, the very walls, might at the wrong moment recall something, whisper some half-forgotten word.

The post-chaise was to arrive between one and two from the direction of the Col di Tenda; I was waiting for it at the gloomy Carignano Palace, not far from which it was to turn the corner.

I arrived on the morning of the same day from Paris,[2] over Mont

Cenis; at the Hotel Feder I was given a large, lofty, quite hand-somely furnished sitting-room and bedroom. I liked this holiday air, for it was in keeping. I ordered a light supper for us and went to wander about the town, waiting for the night.

When the chaise drove up to the posting-station Natalie recognised me.

'You're here!' she said, nodding to me at the window. I opened the door and she threw herself on my neck with such rapturous joy, with such an expression of love and thankfulness, that words from her letter flashed like lightning into my mind: 'I come back like a ship to its own haven, after storms, shipwrecks and mis-fortunes – broken but safe.'

One look, two or three words, were more than enough ... all was explained and understood; I took her little travelling-bag, threw it on a stick over my back and gave her my arm, and we walked gaily along the empty streets to the hotel. There everyone was asleep except the porter. On the table (laid for supper) stood two unlighted candles, bread, fruit, and a decanter of wine. I did not want to wake anyone; we lit the candles and, sitting down at the unoccupied table, looked at each other and at once both remembered our days at Vladimir.

She had on a white muslin dress or blouse, put on for the journey because of the blazing heat; and at that first interview of ours, when I had arrived from my place of exile, she had been all in white, too, and her wedding dress had been white. Even her face, though it bore clear-cut traces of profound shocks, anxieties, medi-tations and sufferings, recalled by its expression her features at that time.

And we ourselves were the same, only now we gave each other our hands, not as arrogant young creatures, self-confident and proud in our faith in ourselves, in our faith in each other, and in our fate's being somehow exceptional, but like veterans tempered in the battle of life, who had experienced not only our strength but our weakness too, and had hardly survived painful blows and irretrievable mistakes. ... Starting on the road anew, we shared the heavy burden of the past, with no settling of accounts.

With that burden we had to move forward at a more modest pace, but within our aching hearts was preserved all that was needed for settled, mature happiness. Through our horror and dull agony we had seen the more clearly how inseparably years and circumstances, children and life in foreign lands, had welded us together.

At this meeting all was ended: the broken threads grew together more firmly than ever, as sometimes parts of a broken bone will grow together, not without leaving a scar. The tears of sorrow, not yet dry in our eyes, united us again with a new bond – the feeling of deep compassion for each other. I saw her conflict, her agony, I saw how exhausted she was. She saw me weak, unhappy, affronted and affronting, ready for sacrifice and ready for crime.

We had paid too dear a price for each other not to understand what we were worth, and what we had cost each other. 'In Turin,' I wrote at the beginning of 1852, 'was our second wedding! it meant perhaps something deeper and more significant than the first, for it was accompanied with full consciousness of all the responsibility which we were taking up anew in relation to each other, it was accomplished within sight of fearful events. . . .'

Love by some miracle survived the blow which should have destroyed it.

The last dark clouds passed further and further away. We said much, we talked for a long time . . . as though we had been parted for several years; bright streaks of daylight had long been filtering through the lowered blinds when we got up from the empty table. . . .

Three days later we started home together along the Riviera to Nice. We caught a glimpse of Genoa, a glimpse of Mentone, where we had so often been, and in such a different frame of mind, a glimpse of Monaco, engraved upon a sea of velvety grass; everything greeted us gaily, like old friends after a falling out; and here were the vineyards, the rose-bushes, the orange-trees, and the sea lying spread before the house, and the children playing on the beach. . . . Now they had recognised us and rushed to meet us. We were at home.

I am thankful to destiny for those days and for the four months which followed them. They made a triumphant ending to my personal life. I thank her, the eternal pagan, for crowning the doomed victims with a sumptuous wreath of autumn flowers and strewing their path, if only for a time, with her poppy and her fragrance!

The gulfs that had parted us vanished, their sides had come together. Was not this the hand which had lain in mine through my whole life, were not these the same eyes, though sometimes dimmed with tears? 'Be comforted, my sister, friend, companion, for all is over – and we are the same as in our young, bright, holy years!'

'After sufferings you, perhaps, know how great, have come some minutes full of bliss; all the cherished beliefs of childhood, of youth, have not only come true, but have passed through fearful ordeals, losing neither their freshness nor their fragrance, and have blossomed with new splendour and new strength. I have never been so happy as now,'[3] she wrote to her friend in Russia.

Of course, there was a sediment left from the past which could not be touched with impunity, something broken within, some lightly sleeping fear and pain.

The past is not a proof that can be corrected, but a guillotine knife; after it has once fallen there is much that does not grow together again, and not everything can be set right. It remains as though cast in metal, exact in detail, unalterable, dark as bronze. Men forget on the whole only what is not worth remembering, or what they do not understand. If only a man could forget two or three incidents, such and such traits, such a day, such a word, he would be young, strong and bold; but being burdened with them he sinks like a key to the bottom. One need not be a Macbeth to meet the ghost of Banquo; ghosts are not criminal judges, nor pangs of conscience, but the happenings that can never be obliterated from the memory.

But one ought not to forget; that is a weakness, that is a lie of a sort; the past has its rights, it is fact; one must come to terms with it and not forget it – and we moved towards this, keeping step.

Sometimes it would happen that a trivial word, uttered by outsiders, or some object that caught the eye, would pass like a razor over the heart, and blood would flow, and the pain would be insufferable; but at the same instant I would meet eyes full of alarm, looking at me with infinite distress and saying, 'Yes, you are right, it can't be helped, but . . .' and I would try to chase away the lowering clouds.

Sacred time of reconciliation: I think of it through my tears. . . .

No, not reconciliation, that is not the word. Words are like ready-made clothes that fit all people of the same height to a certain degree and are a bad fit for each one individually. We had no need to make things up, for we had never quarrelled; we had each suffered because of the other, but we had never parted. At the darkest moments a sort of indissoluble unity, of which neither had any doubt, and a deep respect for each other were inherent in us. Rather than reconciled, we were like people recovering after a high fever; the delirium had passed, we recognised each other again with eyes that were somewhat dim and enfeebled. The pain we had all suffered was still remembered, and we were still conscious of exhaustion, but we knew that all the evil was past and that we were on shore again.

An idea, that had several times occurred to Natalie before, now absorbed her more and more. She wanted to write down her confession. She was dissatisfied with the beginning and burnt the pages; only one long letter and one brief page have survived. . . . By these one can judge what has been lost. . . . It is painful to read them: one feels that one is touching with one's hand a warm, suffering heart, one seems to hear the whisper of those unvoiced secrets, for ever hidden, and scarcely awakening in the consciousness. In these lines one can trace the racking struggle being translated into a new temper and pain into reflection. If this work had not been rudely interrupted, it would have set a great precedent for filling the gap left by the submissive silence of woman and the haughty patronage of her by man. But the most senseless blow broke over our heads and shattered everything for ever.

2

Dans une mer sans fond, par une nuit sans lune,
Sous l'aveugle océan à jamais enfouis.
                                        V. HUGO: *Oceano Nox.*

So ended the summer of 1851. We were quite alone. My mother, with Kolya and Spielmann, had gone to Paris to stay with Maria Kasparovna [Reichel]. We spent our days quietly with the children. It seemed as though all the storms were behind us.

In November we received a letter from my mother to tell us she was starting soon, and then another from Marseilles in which she wrote that on the next day, the 15th of November, they would take the steamer and come to us. During her absence we had moved into another house, also on the sea, in the suburb of Ste Hélène. In this house, which had a large garden, there was accommodation for my mother; we decorated her room with flowers, our cook went with Sasha and got some Chinese lanterns, and we hung them about the walls and trees. Everything was ready. From three o'clock in the afternoon the children did not move from the verandah; at last, between five and six, a dark streak of smoke could be discerned against the sea on the horizon and a few minutes later the steamer, too, could be seen, a motionless point growing larger and larger. All was bustle in the house. François went down to the jetty and I got into the carriage and drove there, too.

When I reached the jetty the steamer had already arrived, and boats were waiting round her for the permission of the quarantine authorities for the passengers to land. One of these came up to the landing-stage. François was standing in it.

'How is this?' I asked, 'are you coming back already?'

He did not answer. I looked at him and my heart sank; he was green and trembling all over.

'What is it?' I asked, 'are you ill?'

'No,' he answered, avoiding my eyes, 'but our people have not come.'

'Haven't come?'

'Something happened to the steamer there, so not all the passengers have come.'

I jumped into the boat and told them to push off at once.

On the steamer I was met with an ominous consideration and complete silence. The captain himself was waiting for me; all this was utterly unusual, and I began to expect something awful. The captain told me that between the island of Hyères and the mainland the steamer on which my mother was sailing had collided with another vessel and sunk to the bottom, and that most of the passengers had been picked up by him and by another steamer that was passing. 'I have only two young girls of your party,' he said, and he led me on to the forecastle – everyone stepped aside with the same sombre silence. I walked vacantly without even asking a question. My mother's niece, who had been staying with her, a tall, graceful girl, was lying on the deck with wet, dishevelled hair; beside her was the maid who looked after Kolya. Seeing me the young girl wanted to sit up in order to say something, but could not; she turned away sobbing.

'What has happened, anyhow? Where are they?' I asked, taking the maid's hand with a sick dread.

'We know nothing,' she answered. 'The steamer sank, and we were pulled out of the water half dead. An Englishwoman gave us her clothes to change into.'

The captain looked sorrowfully at me, shook my hand and said:

'Don't despair; go to Hyères: perhaps you will find some of them there.'

Leaving Engelson and François to look after the girls, I drove home in a state of stupefaction; everything in my head was confused and shuddering, and I wished that our house had been a thousand miles away. But soon I caught a gleam between the trees, and again and again; it was the lanterns lighted by the children. At our gate stood our servants and Tata and Natalie, with Olga in her arms.

'What, you alone?' Natalie asked me calmly. 'Why, you might have brought Kolya, at least.'

'They are not there,' I said, 'something happened to their steamer. They had to change on to another, which could not take them all. Luiza is here.'

'They are not!'[4] cried Natalie; 'it's only now I can see your face properly: your eyes are dim; all your features are distorted. For God's sake, what has happened?'

'I am going to look for them in Hyères.'

She shook her head and added: 'They are not! They are not!' Then in silence she pressed her forehead against my shoulder. We walked along the avenue without saying a word. I led her into the dining-room; as we passed him I whispered to Rocca: 'For God's sake, the lanterns'; he understood me and rushed to put them out. In the dining-room everything was ready – a bottle of wine was on ice, a nosegay of flowers before my mother's place and some new toys before Kolya's.

The dreadful news quickly spread about the town, and our house began to be filled with friends, such as Vogt, Tessier,[5] Chojecki, and Orsini, and even complete strangers; some wanted to find out what had happened, others to show their sympathy, others still to give all sorts of advice, mostly rubbish. But I shall not be ungrateful; I was deeply touched by the sympathy shown me then at Nice. In the face of these senseless blows of fate people wake up and feel their common bonds.

I decided to go to Hyères that same night. Natalie wanted to go with me, but I persuaded her to stay behind; besides, there had been an abrupt change in the weather: the mistral was blowing cold as ice, and it was raining hard. I had to get a pass to go into France across the Var bridge; I went to Léon Pilet, the French Consul; he was at the opera, and I went to his box with Chojecki. Pilet, who had already heard something of what had happened, said to me:

'I have no right to give you permission, but there are circumstances in which it would be a crime to refuse. I shall give you a permit to let you cross the frontier on my own responsibility. Come to the consulate for it in half an hour.'

At the entrance to the theatre about a dozen of the people who had been at my house were waiting for me. I told them that Léon Pilet was giving a permit.

'You go home and don't bother about anything,' I was told from all sides. 'The rest shall be done; we'll get the permit, we'll get it *viséd* at the office, and we'll order post-horses.'

My landlord, who was present, ran to get a carriage; an hotel-keeper offered his own for nothing.

At eleven o'clock in the evening I set off in pouring rain. It was an awful night; the gusts of wind were sometimes so violent that the horses stopped; the sea, in which there had so recently been a burial, beat and roared almost unseen in the darkness. We mounted the Esterel and the rain turned into snow; the horses slipped and almost fell on the ice. Several times the driver, absolutely exhausted, set about warming himself; I handed him my flask of brandy and, promising him double the fare, besought him to hurry.

What for? Did I believe in the possibility of finding any of them? that anyone had been saved? It was hard to suppose so after all I had heard – but to search, to look at the very spot, to find something, a rag, to see an eyewitness, in short ... one had to be convinced that there was no hope, to be doing something, to be away from home, to gather my wits.

While the horses were being changed at Esterel I got out of the carriage; my heart ached, and I almost sobbed as I looked about me. This was near the very tavern where we had spent a night in 1847. I remembered the huge trees which formed a canopy over it; the same view lay spread before it, only then it had been lit up by the rising sun, and now it was hidden behind grey clouds, unlike Italy, and in places was white with snow.

I vividly recalled that time in every minute detail; I remembered how our hostess had regaled us with a hare, the highness of which was quenched by a fearful quantity of garlic; how bats had flown about the bedroom, how with the help of our Luiza I had chased them with a towel, and how for the first time we felt the breath of the warm southern air. . . .

At the time I wrote: 'From Avignon one feels and sees the South. For a man who has always lived in the north the first encounter with nature in the south is filled with solemn joy: one feels younger, one wants to sing, dance and weep; everything is so clear, so bright, so gay, so luxuriant. After Avignon we had to cross the Maritime Alps. We went up the Esterel by moonlight; when we began to go down again the sun was rising, the mountain chains stood out from the morning mist, the ray of the sun cast a red light upon the dazzling, snowy peaks; round us was vivid green, flowers, sharply cut shadows, immense trees and sombre rocks, scarcely covered with sparse, stiff vegetation; the air was intoxicating, unusually limpid, refreshing and resonant; our words and the singing of the birds sounded louder than usual, and suddenly, at a little curve in the road there was a fringe of brilliance round the mountains and there, quivering with silvery light, lay the Mediterranean.'[6]

And now, after four years, I was at the same place again.

We had not been able to reach Hyères before night; I went at once to the *Commissaire* of Police; with him and a brigadier of gendarmes I went first to the *Commissaire* of the Port. He had various things that had been salvaged; I found nothing among them. Then we went to the hospital. One of the victims was dying; the others told me they had seen an elderly lady, a child of about five, and with him a young man with a fair, bushy beard . . . that they had seen them at the very last minute, and that therefore they, too, had gone to the bottom like everyone else. But here again one asked oneself how it was that these people who were telling their story were alive, though they too, like Luiza Tsabel and the maid, had no clear idea of how they had been saved.

The bodies that had been found were lying in the crypt of a nunnery. We went there from the hospital; the Sisters of Mercy met us and conducted us, lighting our way with church candles.

In the crypt stood a row of boxes newly knocked together, and in each box was one body. The *Commissaire* ordered them to be opened, for it appeared that they had been nailed up. The brigadier sent a gendarme for a cold chisel and then told him to break open one lid after the other.

This examination of the bodies was insufferably painful. The *Commissaire* had a little book in his hand, and at the opening of each box he asked in an official tone: 'Do you testify in our presence that this body is not known to you?' I nodded. The *Commissaire* made a mark with his pencil, turned to the gendarme and ordered him to shut it again. We passed on to the next. The gendarme lifted the lid; with a sort of horror I looked at the dead and was, as it were, relieved when I met unfamiliar features, though in reality it was still more fearful to think that all three had vanished like this without a trace, and were lying abandoned at the bottom of the sea, at the mercy of the waves. A body without a coffin, without a tomb, is more dreadful than any burial, and here we had not even our dead.

I found no one. I was struck by one body: a beautiful woman of about twenty in smart Provençal dress; her bosom was bare (she had had a baby with her, carried away, of course, by the waves) and a stream of milk was still oozing out and trickling down her breast. Her face was utterly unchanged: the brown sunburn made her look absolutely alive.

The brigadier could not help saying: 'What a beauty!' The *Commissaire* added nothing to that; the gendarme, as he covered her, remarked to the brigadier: 'I knew her. She was a peasant woman living here not far from the town; she was going to her husband at Grasse. Well, he'll have to wait.'

My mother, my Kolya, and our good Spielmann had vanished without a trace. Nothing was left of them; among the things salvaged there was not a scrap belonging to them. There could be no doubt of their death. All the survivors were either at Hyères or on the steamer which had brought Luiza. The captain had made up a tale to calm me down.

I was told something else at Hyères, of an elderly man who had lost his whole family, and would not stay in the hospital, but went off somewhere on foot, with no money, in a state bordering on madness; and of two Englishwomen who had gone to see the English Consul: they had lost father, mother, and brother.

It was approaching daybreak; I ordered the horses to be brought. Before I left, a waiter took me to a part of the shore that jutted out into the sea, and showed me from there the place of the shipwreck. The sea was still boiling and tossing, grey and turbid after the storm of the day before; in the distance, in one place, there undulated a peculiar patch, as though of a thicker transparent liquid.

'The steamer was taking a cargo of oil, you see, it has stayed on the surface. That's where the disaster happened.'

That floating patch was *all*.

'And is it deep there?'

'It'll be about a hundred and eighty metres.'

I stood there for a little; the morning was very cold, particularly by the sea. The mistral was still blowing, the sky was covered with the clouds of a Russian autumn. Farewell! ... A hundred and eighty metres deep, and a floating patch of oil!

> 'Nul ne sait votre sort, pauvres têtes perdues!
> Vous roul[er]ez à travers les sombres étendues,
> Heurtant de vos fronts des écueils inconnus. ...'[7]

I came back to a fearful certainty. Natalie, who had only just recovered, had not the strength to bear this blow. From the day of the loss of my mother and Kolya, she never recovered her health again. The fright, the pain remained: they entered into her blood. Sometimes, in the evening or the night, she would say to me, as though asking for my help:

'Kolya, Kolya is always with me; poor Kolya, how frightened he must have been, how cold he was! and then the fishes, the lobsters!'

She would take out his little glove, which had been safely in the maid's pocket – and a silence would follow, that silence in which life flows away as though a sluice has been raised. At the sight of these sufferings, which passed into nervous illness, at the sight of her glittering eyes and increasing thinness, I doubted for the first time whether I should save her. ... The days dragged by in an anguish of uncertainty, in something like the existence men lead between sentence and execution, when a man at once hopes and knows for certain that he will not escape the axe.

CHAPTER VII

THE YEAR 1852

AGAIN the New Year came on; we met it round Natalie's bed: at last her strength had failed, and she had taken to her bed.

The Engelsons, Vogt, and a couple of intimate friends were at our house. We were all melancholy. The 2nd of December in Paris lay like a stone on our hearts. . . . Everything – public and private – had started rushing towards the abyss, and had already gone so far downhill that there was no stopping or changing course; one could but wait, dully, passively, until it all jerked itself off the rails and went flying into darkness.

The usual glass was handed at twelve o'clock: we forced a smile; within was death and horror; we were all ashamed to add any wish for the New Year. To look ahead was more frightening than to look back.

The symptoms became more definite; pleurisy set in in the left lung.

She spent a terrifying fortnight between life and death, but for this time life was the winner. At one of the most painful moments I asked Dr Bonfils whether his patient would live through the night.

'Probably,' said Bonfils.

'Are you telling me the truth? Please don't deceive me.'

'I give my word of honour. I guarantee. . . .' He stopped. 'I guarantee *three days*. Ask Vogt if you don't believe me.'

The reverse of Hudson Lowe's *on en plantera* would have been a good thing.[1]

Convalescence slowly followed and with it a last ray of hope shed its pale radiance over our troubled life. The powers of her spirit came back first . . . there were marvellous moments: the last chords of a music that was falling silent for ever.

A few days after the crisis of the illness I went into my study early one morning and fell asleep on the sofa. I must have slept soundly, for I did not hear the manservant come in. When I woke up I found a letter on the table. Herwegh's handwriting. What was the purpose of his writing, and how dared he write to me after all that had happened? I had given him no cause whatever. I picked up the letter in order to send it back, but seeing on the back of the envelope the words, '*In re* an honourable challenge,' I opened it. The letter was revolting, infamous. He said that by my *calumnies* I had bewildered Natalie, that I had taken advantage of her weakness and my influence over her, that she had *betrayed him*. In conclusion he *informed against her*, and said that fate was deciding between me and him: that 'it drowns in the sea your offspring (*votre progéniture*) and your family. You wanted to end this affair in blood, when I supposed it was possible to finish it humanely. Now I am ready and demand *satisfaction*.'*

This letter was the first insult I had ever received in my life. I leapt up like a wounded beast with a moan of fury. Why was not that blackguard at Nice. . . . Why was a dying woman lying across the passage?

After sousing my head two or three times in cold water, I went in to Engelson (after my mother's death he occupied her rooms), and after waiting till his wife had gone out I told him that I had received a letter from Herwegh.

'So you have actually received it?' said Engelson.

'Why, did you know? Did you expect it?'

'Yes,' he said; 'I heard about it yesterday.'

'From whom?'

'From Karl Vogt.'

I put my hand to my head: it seemed to me that I had gone mad. Our silence had been so absolute that neither my mother nor Maria Kasparovna Reichel had once spoken to me about what had happened. I was more intimate with Engelson than with other people, but to him I had spoken only once, as we were walking in the environs of Paris, spoken in a brief reply to a question he put to me about the cause of my rupture with Herwegh. I had been astounded at Geneva when I heard from Sazonov of the scoundrel's gossip, but how could I have thought that round us, close at hand, the other side of the door, everyone knew, everyone was talking of what I regarded as a secret buried among a few people . . . that they even knew of letters which I had not yet received.

We went in to Vogt. He avowed to me that two days before

---

* I never read this letter again, and only once opened it afterwards. In 1853, on Natalie's birthday, the 23rd of October (O.S.), I burnt it without reading it.

Emma had shown him a letter from her husband, in which he said he was going to send me a *fearful letter*, that he would cast me down from the height on which Natalie had set me, that he would cover us 'with disgrace, if he had to walk over the dead bodies of children, and bring us all, and himself too, into the dock of the accused at the criminal court'. Finally he wrote to his wife (and she had shown all this to Vogt, to Charles Edmond and to Orsini!): 'you alone are pure and innocent, you ought to be the avenging angel', that is, I suppose, to cut our throats. There were people who said he had been driven out of his mind by love, by his rupture with me, by humiliated vanity: that is nonsense. The man never took one *step* that was dangerous or incautious; his madness was only in *words*, he raged rhetorically. His *amour-propre* had been stung, silence was for him harder to bear than any scandal, and the peace that had been restored to our life gave him no rest. A *petit bourgeois*, like George Sand's Horace, he babbled to avenge himself upon the woman whom he had loved and the man whom he had called brother and father, and, a German *bourgeois*, he threatened in melodramatic phrases, composed in a pseudo-Schilleresque key.

At the time when he was writing his letter to me, and a string of insane letters to his wife, at that very time he was living with and being kept by an old, forsaken mistress of Louis Napoleon's, a dissolute woman,[2] who was notorious all over Zürich. With her he spent his days and nights, at her expense he lived in luxury, drove about with her in her carriage and revelled at big hotels . . . no, that is not madness!

'What do you intend to do?' Engelson finally asked me.

'Go and kill him like a dog. That he is the greatest coward you know and everyone knows: the chances are all in my favour.'

'But how can you go . . . ?'

'That's the whole point. Write to him meanwhile that it is not for him to demand satisfaction from me, but for me to punish him, and I shall choose the means and the time to do it myself; that I will not leave a sick woman for that, and that I spit on his boorishness.'

To that effect I wrote to Sazonov, and asked whether he would help in the matter. Engelson, Sazonov, and Vogt accepted my proposal with fervour. My letter was a great mistake, and gave him a reason for saying later that I had accepted his challenge, and that I had refused it only afterwards.

To refuse a challenge is a difficult thing and requires much firmness of spirit or much weakness.

The feudal duel has a firm standing in modern society, demonstrating that it is not at all as modern as it seems. Rarely does anyone venture to touch this holy thing set up by aristocratic honour and military *amour-propre* – and, indeed, it is rare for anyone to be in a position so independent as to be able to insult the bloody idol with impunity and to accept the slur of cowardice.

It is not worth while to try to prove the absurdity of the duel: no one tries to justify it theoretically except a few bullies and fencing masters; but in practice everyone submits to it in order to prove – the devil knows to whom – his courage.

The worst thing about the duel is that it justifies any blackguard, either by giving him an honourable death, or by making him an honourable man-killer.

A man is accused of cheating at cards: he insists on a duel, as though one cannot cheat at cards and not be afraid of a pistol. And how disgraceful to put the card-sharper on a level with the man who denounces him!

A duel may sometimes be accepted as a means of escaping the gallows or the guillotine, although even here the logic is not clear, and I still do not understand how it is that a man is obliged, on pain of general contempt, not to fear his antagonist's sword, but may be afraid of the knife of the guillotine.

The death penalty has this advantage, that it is preceded by a trial, which may condemn a man to death, but cannot deprive him of the right of showing up his enemy, dead or alive. In a duel everything remains sewn up and concealed. It is an institution belonging to that pugnacious social circle on whose hands the blood is still so far from being dry that the wearing of deadly weapons is looked on as a sign of nobility and practice in the art of killing as an official duty.

While the world is governed by military men duelling will not be abolished; but we may boldly demand that the decision should be left to ourselves when we are to bow the head to an idol we do not believe in, and when to appear in our full stature as free men and, after battling with the god and the authorities, to dare to throw down the gauntlet to the medieval tribunal. . . . How many men have passed with proud and triumphant faces through all the misfortunes of life, prison and poverty, sacrifice and toil, inquisitions and I do not know what, and have been lopped off by the impudent challenge of some mischievous scoundrel? There must be no more such victims. The principle that regulates a man's actions should lie within himself, in his own reason; if it lies outside him, he is a slave, however valiant he may be.

I neither accepted nor refused the challenge; the punishment of Herwegh was for me a moral necessity, a physical necessity — I sought in my mind for some sure means of revenge and one, moreover, which could not redound to his glory. But whether I was to achieve it by a duel or simply with a knife, was a matter of indifference to me. He himself suggested the solution. He wrote to his wife — and she, as usual, showed the letter to her acquaintances — that in spite of everything 'I was head and shoulders above all the riff-raff that surrounded me; that I was led astray by men like Vogt, Engelson, and Golovin (!);[3] that if he could see me for one minute all would be set right' — 'he (that is, I) alone can understand me'; and this was written after his letter to me!

'And for this reason,' the poet concluded, 'what I should like best of all is that Herzen would accept a duel without witnesses. I am convinced that at the first word we should fall on each other's breast, and all would be forgotten.'

And so a duel was being proposed as a means of theatrical reconciliation. If I could at that time have got away for five days or a week, I should certainly have gone to Zürich and appeared to him, alone, in fulfilment of his desire — and he would not have remained alive.

A few days after the letter arrived, Orsini came to see me one morning at nine o'clock. Orsini, by some physiological absurdity, had a passionate attachment to Emma; what there was in common between this fiery, handsome, young Southerner and the ugly, lymphatic German woman, I could never understand. I was surprised at his early arrival. He told me simply and without speechifying that the news of Herwegh's letter had upset his whole circle of friends, and that many common acquaintances had suggested that a *jury d'honneur* should be formed. Then he began to defend Emma, saying that she was not at fault except in her senseless love for her husband and her slavish submission to him. He said that he himself had been a witness of how much all this cost her.

'You,' he said, 'ought to hold out a hand to her; you ought to punish the guilty man, but you ought also to rehabilitate an innocent woman.'

I refused decidedly and absolutely. Orsini was too penetrating not to understand that I should not change my mind, so did not insist.

Among other things, speaking of a *jury d'honneur*, he told me that he had already written to Mazzini about the whole affair and asked his opinion. There was no court of honour, but I received later on a letter that amounted to a verdict against Herwegh, signed by names that were dear to me, among others the heroic martyr Pisacane, Mordini, Orsini, Bertani, Medici, Mezzacapo, and Cosenz.[4] Was not this odd again? Parties were formed, verdicts were passed, Mazzini was written to – all this without my knowledge, and all this concerning events at which a week before no one had dared to hint in my presence.

After seeing Orsini out I took a sheet of paper and began a letter to Mazzini. Here there was revealed to me what amounted to a *Vehmgericht*, a court that had forced its services on me. I wrote to him that Orsini had told me of his letter and that, fearing that he had not conveyed the affair entirely correctly, since he had never heard a word about it from me, I wanted to tell him about the business and to ask his advice. Mazzini replied at once.

'It would have been better,' he wrote, 'to bury the whole thing in silence, but I doubt whether that is possible for you now, and therefore you had better come boldly forward as the accuser and leave us to form a court.'

That I believed in the possibility of this court was, perhaps, my last illusion. I was mistaken and paid dearly for my mistake.

Together with Mazzini's letter I received a letter from Haug, to whom Mazzini, knowing that he was well acquainted with me, had shown Orsini's letter and mine. After my first meeting with Haug in Paris he had served under Garibaldi and had fought with distinction near Rome.[5] In this man there was much that was good and a vast deal that was immature and absurd. He had slept the inexplicable barrack-sleep of an Austrian lieutenant, till he was suddenly woken up by the alarm bell of the Hungarian rising and the Vienna barricades. He snatched up a weapon, not to slaughter the people, but to take his place in their ranks. The transition had been too violent and left some traces of angularity and incompleteness. A dreamy and rather impetuous man, generous and devoted, and vain to the point of insolence, a *Bursch*, a Cadet, a student and a lieutenant, he had a sincere affection for me.

Haug wrote that he was going to Nice, and besought me to take no steps without him. 'You have left your country and have come to us as a brother; do not think that we shall permit any of our people to go unpunished after adding calumny to all his previous treacheries, and then to cover all this with an insolent challenge. No, we have a very different conception of our solidarity. It is enough that a Russian poet[6] fell by the bullet of a Western adventurer — a Russian revolutionary shall not fall!'

In reply I wrote a long letter to Haug. This was my first confession. I told him everything that had happened — and started to wait for him. . . .

Meanwhile in the bedroom a great life was feebly flickering out in a desperate struggle with bodily infirmity and fearful forebodings. I spent day and night beside her sick-bed — she liked me alone to give her her medicine and to prepare her orangeade; at night I kept up the fire, and when she slept quietly I again had a hope of saving her. But there were moments when the burden was intolerable. I felt her thin, feverish hand, and I saw her sombre, anxious eyes fixed upon me with entreaty, with hope, and the fearful words: 'The children will be left alone . . . they will be orphans . . . all will be lost; you are only waiting for . . . for the children's sake drop it all, don't defend yourself from the filth . . . let me, me, defend you — you shall come out clean, if only I can get a little stronger physically. But no, no: strength does not come. Do not leave the children!' And for the hundredth time I repeated my promise.

In one of such conversations Natalie suddenly said to me:

'He's written to you?'*

'Yes.'

'Show me the letter.'

'What for?'

'I want to see what he could say to you.'

I was almost glad that she had spoken of the letter; I passion-

---

* Rumours of what had happened had reached her, and I suppose that this was not by chance. There was a hint about his letter in a letter from Maria Kasparovna, who had heard all this in Paris from N. A. Melgunov.

ately wished to know whether there was a grain of truth in one of his charges. I could never have brought myself to ask, but now she had spoken of the letter herself and I could not restrain myself, I was horrified at the thought that the doubt would always remain, and perhaps grow stronger when her lips were closed.

'I am not going to show you the letter, but tell me, did you say something like it . . . ?'

'How can you think such a thing!'

'He writes it.'

'It's almost incredible: he writes that with his own hand . . . ?'

I folded the letter back at the passage and showed her[7] – she looked at it, was silent for a little and then said sadly: 'The scoundrel!'

From that minute her contempt passed into hatred, and never by one word or one hint did she forgive him or express pity for him.

A few days after this conversation she wrote him the following letter:

'Your persecution and your abominable conduct force me once more to repeat, and in the presence of a witness, what I have written to you several times already; yes, my passion was great and blind, but your character, perfidious and Jewish in the worst sense and your unbridled egoism, were revealed in all their hideous nakedness when you were leaving here, and have been since, while Alexander's dignity and devotion have grown with every day. My unhappy infatuation has served only as a new pedestal on which to raise my love for him. You have wanted to pelt that pedestal with dirt, but you will not succeed in doing anything to break our union, which is more indissoluble and more unshakable now than ever it was. Your charges, your slanders against a woman only fill Alexander with contempt for you. You have dishonoured yourself by this baseness. What has become of your everlasting protestations of religious respect for my will, of your love for the children? It is not long since you swore rather to vanish from the face of the earth than cause Alexander a moment's distress. Have I not always told you that I could not for one day survive parting with him? that if he left me, even if he died – I would remain alone to the end of my life . . . ?

'As for my promise to see you again some day, it is true that I made it – I was sorry for you then and I wanted us to say goodbye like human beings; but you have made it impossible for me to keep that promise.

'From the very first time you went away you began tormenting me, demanding first one promise, then another. You meant to vanish for years, to go away to Egypt, if only you could take with you the faintest hope. When you saw that that did not answer, you suggested one absurd thing after another – impossible, ludicrous – and ended by threatening *publicity*; you wanted to make me quarrel finally with Alexander, tried to make him kill you, fight you, and eventually threatened to commit the most frightful crimes. These threats had no more effect on me, for you had repeated them too often.

'I'll tell you again what I wrote in my last letter: "I remain in my home, and my home is Alexander and my children," and if I cannot remain there as mother and wife, I shall remain as a nurse and a servant. "There is no bridge between me and you." You have made the very past loathsome to me.

'N.H.

'NICE, 18 February 1852.'

A few days later the letter returned from Zürich. Herwegh had sent it back unopened; the letter had been sent registered with three seals, and was returned with the address on the same envelope.

'If it's like that,' Natalie remarked, 'it shall be read to him.'

She sent for Haug, Tessier, Engelson, Orsini, and Vogt, and said to them:

'You know how I have wanted to exculpate Alexander, but what can I do, chained to my bed? Perhaps I shall not get over this illness – let me die in peace, trusting that you will carry out my last request. That man sent me back my letter; let one of you read it to him, and in the presence of witnesses.'

Haug took her hand and said:

'Your letter shall be read to him: I stake my life on that!'

This simple, vigorous action affected us all, and the sceptical Vogt went out as much stirred as the fanatical Orsini, who preserved a fervent respect for Natalie to the end of his days. The last time I saw him before he went away to Paris – at the end of 1857 – he spoke with tenderness of Natalie, and perhaps with secret reproach. Of us two Orsini certainly is not the one who could be accused of moral unsoundness – of dualism of word and deed. . . .

Once, late in the evening, or rather night, I had been having a long, melancholy discussion with Engelson. At last he went to his room, and I went upstairs. Natalie was sleeping quietly; I stayed a few minutes in her bedroom and went out into the garden. Engelson's window was open. Standing disconsolately at the window he was smoking a cigar.

'Such is fate, it seems,' he said and came towards me.

'Why aren't you asleep? Why have you come?' he asked, and there was a nervous quiver in his voice. Then he seized my hand and went on, 'Do you believe in my boundless love for you? Do you believe there is no one in the world dearer to me than you? Leave Herwegh to me – there is no need of a court, or of Haug: Haug's a German. Give me the right to avenge you: I am a *Russian* . . . I've thought out a complete plan; I need your confidence, your blessing.'

He stood with a pale face before me, his arms folded, in the light of the dawning day. I was deeply touched and felt ready to fall on his neck with tears.

'Do you believe or not that I would sooner perish, vanish from the face of the earth, than compromise an affair in which so much is involved that is sacred to me? But without your confidence my hands are tied. Tell me frankly: yes or no? If it's no – goodbye and the devil take everything, and the devil take you and me too! I shall go away tomorrow, and you will not hear of me again.'

'I believe in your friendship and in your sincerity, but I am afraid of your imagination and your nerves, and I have no great faith in your practical sense. You are nearer to me than any of the others here, but I must own, I think you will do harm and destroy yourself.'

'So you think General Haug has practical talents?'

'I didn't say that, but I think Haug is a more practical man, just as I think Orsini is more practical than Haug.'

Engelson would hear no more; he began dancing about on one leg, singing, and at last, calming down somewhat, he said to me:

'There you are: you're caught flat aback.'

He put his hand on my shoulder and added in an undertone:

'It was with Orsini that I thought out the whole plan, the most practical man in the world. Come, give us your blessing, father.'

'But do you give me your word not to undertake anything without telling me?'

'Yes.'

'Tell me your plan, then.'

'I can't do that, or at any rate I can't do it now.'

A silence followed. It was not hard to understand what he meant to do.

'Good-night,' I said, 'let me think a little,' and added involuntarily: 'What possessed you to tell me about it?'

Engelson understood me.

'My damned weakness, but anyhow nobody will ever find out that I've told you.'

'But I know,' said I in answer, and we parted.

Anxiety over Engelson and fear of some catastrophe that would be a fatal shock to Natalie's ailing constitution compelled me to prevent his scheme from being carried out. Orsini shook his head, and watched this with regret ... and so instead of punishing Herwegh, I saved him, though of course it was not for his sake nor for my own. There was no sentimentality or magnanimity about it.

And, indeed, how was magnanimity or sympathy possible with this hero turned inside out? Emma, alarmed by something, quarrelled with Vogt for speaking impertinently of her Georg, and asked Charles Edmond to write to Herwegh advising him to stay quietly at Zürich and give up every kind of provocation – or it would be the worse for him. I do not know what Charles Edmond wrote: his task was not an easy one; but Herwegh's answer was remarkable. To begin with he said that 'it was not for the Vogts nor for Charles Edmond to judge him,' then that it was I who had broken the bond between him and me, and so let all the consequences fall upon my head. After going into everything and defending himself even in the double-faced part he had played, he concluded thus: 'I do not even know whether there is treachery in this. Those blockheads are still talking about money: to put a stop to that paltry accusation once for all, I say openly that Herzen did not with his few thousand francs pay too dearly for the moments of distraction and enjoyment which we spent together at that grievous time!' 'C'est grand, c'est sublime,' said Charles Edmond, 'mais c'est niederträchtig.' To which Chojecki[8] replied that the stick was the only answer to such letters, and that he would give it him the next time they met.

Herwegh relapsed into silence.

CHAPTER VIII

THE LAST SCENE

WITH the coming of spring Natalie's health improved; she was already sitting up for the greater part of the day in an easy-chair, she was able to do her hair for the first time during her illness, and finally she could listen without exhaustion when I read aloud to her. We were intending to go, as soon as she should be still better, to Seville or Cadiz. She was eager to recover, eager to live, eager to go to Italy.

After the return of the letter there was complete silence, as though both husband and wife felt in their conscience that they had reached a limit rarely attained, had over-stepped it and were weary.

Natalie had not yet gone downstairs and was in no haste to do so: she meant to go down for the first time on the 25th of March, my birthday. She was having a white merino blouse made for that day, and I had ordered an ermine cloak from Paris. Natalie herself wrote or dictated to me the list of the guests she wanted to invite besides the Engelsons: Orsini, Vogt, Mordini, and Paccelli[1] and his wife.

Two days before my birthday Olga developed a cold and cough; in the town there was influenza. Natalie got up twice in the night and went through another room to the nursery. It was a warm night, but stormy. She woke up in the morning with an acute attack of influenza – she had an agonising cough and by the evening she was feverish.

There could be no thought of her getting up next day; a feverish night was followed by a terrible prostration and aggravated symptoms. All the faint, reviving hopes to which we clung were punctured. The unnatural sound of the cough was a threatening omen.

Natalie would not hear of putting off our visitors. Sorrowful and alarmed, at two o'clock we sat down at table without her.

Mme Paccelli brought with her a song her husband had composed in my honour. She was a silent, melancholy and very kindhearted woman. It seemed as though some sorrow weighed upon her: either the curse of poverty was too great a burden, or perhaps life had promised her something more than everlasting music-lessons and the devotion of a feeble, colourless man, conscious of his subjection to her.

In our house she met with more simplicity and a warmer welcome than with her other pupils, and she came to love Natalie with a southern fervour.

After lunch she sat a little with the invalid, and came away from her as pale as a sheet. The guests begged her to sing the song she had brought with her; she sat down at the piano, struck a few chords, began singing and suddenly, with a frightened look at me, burst into tears, let her head fall on the instrument and sobbed convulsively. This put an end to our fête. Almost without saying a word the guests dispersed. With a weight like a great stone on my heart I went upstairs. The same fearful cough continued.

This was the prelude to a funeral.

And there were two of them.

Two months after my birthday Mme Paccelli was buried too. She had started for Mentone or Roccabrunn on a donkey. The donkeys in Italy are accustomed to go up mountains at night without missing their footing. On this occasion the donkey stumbled in broad daylight – the hapless woman fell off, rolled down on to some sharp rocks and died on the spot in terrible agonies. . . .

I was at Lugano when I received the news. So she was no more. Nur zu – what grotesque absurdity was coming next . . . ?

After that everything is overcast . . . there comes a dull, sombre night, indistinct in my memory; it is useless or impossible to describe it . . . a time of pain, alarm, sleeplessness, a stupefying feeling of fear, of moral insignificance and fearful physical strength.

Everything in the house had declined. There was a peculiar disorder and chaos, a hurry and bustle, the servants were run off their legs; and together with the approach of death there came fresh scandal, fresh nastiness. . . . Fate did not gild the pill for me, nor did men spare me either; his shoulders are broad, they seemed to say: let him bear it.

Three days before Natalie's end, Orsini brought me a note for her from Emma. She besought Natalie 'to be forgiving for all that had been done against her, to forgive *everyone*'. I told Orsini that it was impossible to give the sick woman the note, but that I fully appreciated the feeling that had made Emma write these lines and I accepted them. I did more: in one of her last calm moments I said quietly to Natalie: 'Emma asks for your forgiveness.' She smiled ironically and did not answer a word. She knew the woman better than I did.

In the evening I heard a loud conversation in the billiard-room, which was where our intimate friends usually gathered. I went in and found a heated conversation going on. Vogt was shouting and Orsini was explaining something and was paler than usual. The argument stopped at my entrance.

'What is it?' I asked, feeling certain that some fresh nastiness had come up.

'It's this,' Engelson answered: 'what's the use of making a secret of it? it's such a beauty, such a German gem, that I should be ready to stand on my head if it had happened at any other time. ... The chivalrous Emma has commissioned Orsini to pass on to you that, since you forgive her, she asks that in proof of it you should give her back the I.O.U. for 10,000 francs which she gave you when you paid off their creditors. *Stupendisch teuer, stupendisch teuer!*'

Orsini was embarrassed and added: 'I think she has gone out of her mind.'

I took out her note and offered it to Orsini, saying to him: 'Tell the woman that she is asking too much; and that even if I did value her feeling of repentance it was not at 10,000 francs!'

Orsini did not take the note.

This was the sort of filth that I had to walk through when she was dying. What was it? Madness or vice, depravity or denseness? This is as hard to settle as the question: where was that family unearthed, in a madhouse or a gaol?

On the evening of the 29th of April Maria Kasparovna arrived. Natalie had been expecting her from day to day. She had sent for her several times, fearing that Madame Engelson would take the upbringing of the children into her hands. Every hour she looked for her to come, and when we got a letter from her she sent Haug and Sasha to meet her at the bridge over the Var. But in spite of this, seeing Maria Kasparovna was a violent shock to her. I remember the faint cry, almost a moan, with which she said, 'Masha', and could say nothing more.

The illness had come on Natalie half-way through her pregnancy. Bonfils and Vogt thought that her condition had assisted her recovery from the pleurisy.

Maria Kasparovna's arrival brought on the birth, which was easier than had been expected; the baby was born alive, but her strength was exhausted, and a fearful weakness ensued.

The baby was born towards morning.[2] Towards the evening she told us to give her the new-born babe and call the children. The doctor had prescribed absolute quiet, and I asked her not to do this. She looked at me gently.

'And are you heeding what they say, too, Alexander?' she said. 'Mind that you are not very sorry later on for depriving me of this moment. I am a little easier just now. I want to show the baby to the children myself.' I called the children.

Not having the strength to hold the little one, she laid him beside her, and with a bright, joyful face said to Sasha and Tata:

'Here is another little brother for you. You must love him.'

The children rushed happily to kiss her and the baby. I remembered that not long before Natalie had repeated, looking at the children:

> 'And may the young life sport and play
> About the entrance to the grave.'[3]

Stunned by grief, I looked at this apotheosis of the dying mother. When the children had gone I besought her to rest and not to speak. She wished to rest but could not: tears rolled down her cheeks.

'Remember your promise. ... Ah, how fearful to think that they will be left alone, quite alone ... and in a foreign land! ... But can it be that there is no hope? ...'

She fixed on me a look of entreaty and despair. These transitions from a fearful hopelessness to illusive dreams of recovery tore my heart inexpressibly during those last days. At the moments when I had least faith, she would take my hand and say:

'No, Alexander: it can't be so, it's too stupid, we shall live yet, if only this weakness would pass.'

The rays of hope slipped away; they vanished of themselves and were replaced by unutterably mournful, quiet despair.

'When I am not here,' she would say, 'everything will be arranged somehow; now I can't imagine how you'll live without me, for it seems I am so necessary to the children; but when one thinks – they will grow up just the same even without me, and everything will go on its way as though it had always been so.'

She added a few words more about the children and about Sasha's health; she was glad that he had grown stronger at Nice, and said that Vogt, too, agreed that he had.

'Look after Tata: you must be very careful with her, for she has a deep, reserved nature. Ah,' she added, 'if only I could live till my Natalie⁴ comes! ... Well, are the children asleep?' she asked a little later.

'Yes,' I said.

There was the sound of a child's voice in the distance.

'That is Olenka,' she said, and smiled (for the last time). 'See what she wants.'

Towards night she became extremely restless; without speaking she signed that her pillows were uncomfortable. But however I arranged them she was still restless, and with distress and even vexation kept changing the position of her head; then she fell into a heavy sleep.

In the middle of the night she made a movement with her hand as though she was thirsty; I gave her a spoonful of orange juice mixed with sugar and water, but her teeth were firmly clenched; she was unconscious. I was numb with horror. The day was breaking. I drew back the curtain and, with a crazy feeling of despair, saw that in a few hours not only her lips but even her teeth had turned black.

Why was there this, too? Why this awful unconsciousness, why this black colour?

Dr Bonfils and Karl Vogt had been sitting all night in the drawing-room. I went and told Vogt what I had observed. He avoided my eye and went upstairs without answering. No answer was needed. Her pulse was scarcely beating.

About midday she came to herself; again she sent for the children, but did not utter a word. She thought that it was dark in the room. It was the second time that this had happened. The day before she had asked me why there were no candles (two candles were burning on the table). I lit another candle, but she did not notice it and said that it was dark:

'Oh, my dear, how my head aches!' she said, and two or three words more.

She took my hand – hers was no longer like a living hand – and covered her face with it. I said something to her, and she answered indistinctly — consciousness was lost again and never came back. . . .

One more word . . . one word . . . or let there be an end of everything! In that condition she remained till the following morning, from midday or one o'clock on the 1st of May till seven o'clock in the morning of the 2nd. Those frightful, inhuman nineteen hours!

At moments she returned to half-consciousness, said distinctly that she wanted to take off the flannel or her dressing-jacket, or asked for a handkerchief, but said nothing more.

Several times I began to speak: it seemed to me that she heard but could not utter a word; a look of bitter pain seemed to flash across her face. Twice she pressed my hand, not convulsively but intentionally. At six o'clock in the morning I asked the doctor how much time was left.

'Not more than an hour.'

I went into the garden to call Sasha. I wanted him to keep his mother's last minutes in his memory for ever. As I went upstairs with him, I told him what misfortune was in store for us. He had had no suspicion of the whole danger. Pale and almost fainting, he went into the room with me.

'Let us kneel here side by side,' I said to him, pointing to a rug by the head of the bed.

The sweat of imminent death covered her face and her hand caught spasmodically at her dressing-jacket, as though trying to take it off. A few moans, a few sounds that reminded me of Vadim's death agony — then those, too, ceased. The doctor took her hand and let it drop: it fell like an inanimate thing.

The boy was sobbing. I do not properly remember what happened in the next few minutes. I rushed away into the *salon*, met Charles Edmond and tried to say something to him, but, instead of words, a sound such as I had never heard burst from my chest. I stood before the window, stupefied, with no clear understanding, and stared at the senselessly moving, sparkling sea.

Then I remembered the words: 'Look after Tata.' I felt alarmed that the child might be frightened. Before this I had forbidden her being told, but could I rely upon it? I had her fetched, and shutting myself up with her in the study I set her on my knee; preparing her little by little, I eventually told her that 'Mama' was dead. She trembled all over, her face flushed in patches and tears started into her eyes. . . .

I took her upstairs. Everything there had been changed. Natalie was lying as though alive on a bed decked with flowers, beside the baby, who had died the same night. The room was draped with white and full of flowers. The taste of the Italians, artistic in everything, knows how to bring something gentle into the heart-rending sorrow of death. The frightened child was struck by the exquisite setting.

'Mama is here,' she said; but when I lifted her up and she touched the cold face with her lips, she broke into hysterical weeping. I could bear no more, and went out. . . .

An hour and a half later I was sitting alone again by the same window and again staring blankly at the sea and the sky. The door opened and Tata came in alone. She came up to me and caressed me, and said in a frightened whisper:

'Papa, I behaved sensibly: I didn't cry much.'

I looked with deep distress at the motherless child. 'Yes,' I thought, 'you will need to be sensible. You will know no mother's caresses, no mother's love. Nothing can take their place. There will be a void in your heart; you will not know the best, purest devotion, the only disinterested devotion in the world. You will feel it yourself, perhaps, but no one will feel it for you. What is a father's love compared with a mother's anguish of love?'

Natalie lay all surrounded with flowers. The blinds had been lowered. I was sitting on the chair, the same chair as usual, beside the bed; all was quiet except the splash of the sea under the window. The muslin of her nightgown, it seemed, was lifted by her faint, very faint breathing.

Grief and distress gently froze still, as though suffering had ended without leaving a trace, effaced by the careless serenity of a monument that knows not what it stands for. And I watched and watched all night: what if she really should wake?

She did not wake; it was not sleep, it was death.

So it was true.

On the floor and along the staircase masses of orange-red geranium had been flung. That fragrance even now convulses me like an electric shock, and I recall all the details, every minute, and see the room draped in white, with the veiled looking-glass; beside her, also covered with flowers, the yellow body of the baby,

fallen asleep without waking; and her cold, her fearfully cold
forehead. ... I go with rapid steps, with no thought, no object,
into the garden. Our François is lying on the grass sobbing like a
child. I want to say something to him, and I have no voice at all; I
run back *there* again. A lady I do not know, dressed all in black,
with two children, softly opens the door: she asks leave to say a
Catholic prayer – I am ready to pray with her myself. She kneels
down and whispers a prayer in Latin, and the children softly
repeat it after her. Then she says to me:

'They, too, have no mother, and their father is far away. You
were at the burial of their grandmother.'

They were the children of Garibaldi.

Within twenty-four hours crowds of exiles had gathered in the
courtyard and the garden. They had come to accompany her to
the grave. Vogt and I laid her in her coffin, and it was carried out.
I walked firmly after it, holding Sasha by the hand and thinking:
this is how people look at the crowd when they are being led out
to be hanged. Two Frenchmen (one of them, I remember, was the
Comte de Vogué[5]) in the street pointed with hatred and derision
at the absence of a priest. Tessier would have called a reproach to
them, but I was alarmed, and made a sign to him with my hand;
silence was essential.

An immense wreath of small, dark-red roses lay on the coffin.
We each picked a rose; it was like a drop of blood on everyone.
When we were going up the hillside the moon rose and there was
a glitter on the sea that had had its share in killing her. We buried
her on the hillside that juts out into the sea in sight of Esterel on
the one hand and the Corniche on the other. There was a garden
all round, so she was still among flowers as she had been on her
bed....

A fortnight later Haug recalled her last wish and his promise,
and he and Tessier prepared to go to Zürich.

It was time for Maria Kasparovna to return to Paris. Everyone
insisted that I should let Tata and Olga go with her, and that
Sasha and I should go to Genoa. It was painful for me to part with
them, but I had no confidence in myself. Perhaps, I thought, this
really will be better for them; well, if it is better, so let it be. I only
asked her not to take the children before the 9th of May, for I
wanted to spend with them the fourteenth anniversary of my
wedding.

The day after that I saw them as far as the bridge over the Var. Haug went with them to Paris. We watched the customs-house authorities, the gendarmes and police of all sorts harassing the passengers. Haug lost his stick, a present from me, searched for it and grew angry and Tata was crying. The guard, in a uniform jacket, took his seat beside the driver. The *diligence* drove off along the Draguignan road, and we, Tessier, Sasha and I, went back across the bridge, got into the carriage and drove to the place where I lived.

A home I had no longer. With the departure of the children the last trace of family life vanished. Everything had assumed the appearance of bachelorhood. Engelson and his wife went away two days later. Half the rooms were shut up. Tessier and Edmond moved into the house. The feminine element was excluded. Only Sasha's age and features reminded one that there had been something else . . . recalled someone who was no longer there.

Post scriptum . . . Five days after the funeral Herwegh wrote to his wife: 'This news has grieved me deeply. I am full of gloomy thoughts. Send me by the first post I *Sepolcri* of Ugo Foscolo.' And in a subsequent letter:* 'Now has come the time for a reconciliation with Herzen. *The cause of our quarrel exists no longer* . . . if only I could see him face to face – he is the only man capable of understanding me!'

Yes, I did understand him.

*

* Both letters were passed from hand to hand at Nice.

CHAPTER I
N. I. SAZONOV

SAZONOV, Bakunin, Paris. These names, these men, this city, tug one back ... far back into time and space, to the days of youthful conspiracies, to the days of the cult of philosophy and the idolatry of revolution.

My youth with each is too dear for me not to dwell on it for a little. ... With Sazonov early in the 'thirties I shared my boyish fancies of a plot *à la* Rienzi ... with Bakunin ten years later, in the sweat of my brains, I mastered Hegel.

Of Bakunin I have spoken already and shall have much more yet to say. His striking personality, his eccentric and vigorous appearance, everywhere – in the youthful circle of Moscow, in a lecture-room at Berlin University, among Weitling's communists, and the Montagnards of Caussidière, his speeches in Prague, his leadership in Dresden, his trial, imprisonment, sentence to death, torture in Austria, extradition to Russia, where he vanished behind the frightful walls of the Alexeyevsky Ravelin – make of him one of those individual figures which neither the contemporary world, nor history can overlook.

That man had within him the germ of a colossal activity for which there was no demand. Bakunin bore within himself the ability to become an agitator, a tribune, a preacher, the head of a party or of a sect, a heresiarch or a fighter. Put him down in any position you like, so long as it is an extreme one – let him be an Anabaptist, a Jacobin, a comrade of Anacharsis Clootz or a friend of Gracchus Babeuf – and he would have won over the masses and shaken the destinies of nations.

'But here under the yoke of Tsars,'[1]

a Columbus without an America or a ship, after serving for two years in the artillery against his will and for two more in the ranks of Moscow Hegelianism, he made haste to leave the country in which an idea was persecuted as an evil intention, and an independent word as an offence against social morality.

After tearing himself from Russia in 1840, he did not return there until a picket of Austrian dragoons handed him over to a Russian officer of gendarmes in 1849.

The worshippers of expediency, the nice fatalists of rationalism, are still surprised at the all-wise appropriateness with which great talents and leaders appear as soon as there is a need for them; but they forget how many embryos perish, are stifled without seeing the light, how many capacities and aptitudes wither away because they are not needed.

Sazonov's example is still more striking. Sazonov has passed without leaving a trace, and his death has been as unnoticed as the whole of his life. He died without carrying out one of the hopes that his friends built upon him. It is easy to say he was to blame for his fate; but how can we weigh or appraise how much of the blame rests on the man and how much on his environment?

The age of Nicholas was a soul-destroying age; it murdered not only by means of labour in the mines and the 'white straps', but with its stifling, degrading atmosphere, with its, so to speak, negative blows.

To deliver the funeral oration over the beings of that period, whose life has dragged on worn out with striving to drag our ship off the shoal where it has grounded so deeply, is my speciality. For them I play the part of Domazhirov, the old retired orderly of Prozorovsky's,[2] now forgotten by everybody, but at one time a familiar figure in Moscow. With a powdered head, wearing a light green uniform of the days of Paul, he used to appear at all the funerals at which a bishop officiated and, taking the foremost place, led the procession, imagining that he was doing his job.

... In our second year at the university – that is in the autumn, of 1831 – in the lecture-room of the Faculty of Physics and Mathematics, Ogarëv and I met among our new comrades two with whom we became particularly intimate.

Our friendships, our sympathies and antipathies, were all derived from the same source. We were young men and we were fanatics: learning, art, connections, home, and social position, everything was subordinated to one idea and one religion. Wherever there was an opening to convert, to preach, there we were on the spot with all our heart and mind, persistent, importunate, unsparing of time, work, and even blandishments.

We went into the lecture-room with the firm purpose of founding in it the nucleus of a society in the image and semblance of the
Decembrists, and therefore we sought proselytes and adherents.
The first of our comrades to understand us clearly was Sazonov;
we found him completely ready, and at once made friends with
him. He gave us his hand with consciousness of what he was
doing, and on the next day brought us another student.[3]

Sazonov had conspicuous gifts and conspicuous egoism. He was
eighteen or rather less, but in spite of that he had many interests
and read everything in the world. He tried to dominate his comrades, and put no one on a level with himself. That was why they
respected him more than they liked him. His friend, on the contrary, handsome and as delicate as a girl, sought for someone to
take refuge with; full of love and devotion, fresh from the shelter
of his mother's wing, with noble aspirations and half-childish
dreams, he longed for warmth and tenderness, he clung to us and
gave himself up entirely to us and our cause – he had the nature
of a Vladimir Lensky,[4] of a Venevitinov.[5]

... The day on which we sat side by side on a bench in the
amphitheatre, looked at each other with the consciousness of our
dedication, our league, our secret, our readiness to perish, our faith
in the sacredness of our cause – and looked with loving pride at the
multitude of handsome young heads about us, as at a band of
brothers – was a great day in our lives. We gave each other our
hands and *à la lettre* went out to preach freedom and struggle in
all the four quarters of our youthful 'universe', like the four
deacons who go on Easter Day with the Four Gospels in their
hands.

We preached in every place at all times ... exactly what it was
we preached it is hard to say. Our ideas were vague: we preached
the Decembrists and the French Revolution, then we preached
Saint-Simonism and the same Revolution; we preached a constitution and a republic, the reading of political books and the concen-
tration of forces in one society. Most of all we preached hatred for
every form of violence, for every sort of arbitrary tyranny
practised by governments.

Our society in reality was never formed; but our propaganda
sent down deep roots in all the faculties, and extended far beyond
the walls of the university.

Since those days our propaganda has gone on uninterrupted

during our whole lives, from the university lecture-room to our London printing-press. Our whole life has been the carrying out of our boyish programme as far as lay in our power. It is not hard to follow the connecting thread through the questions we have touched upon, through the interests we have aroused, in periodicals, in lectures, in literary circles. ... Though it took different forms and developed itself, our propaganda remained true to itself and imported its individual character into every environment. Punishment lifted us up and gave us on its own account the prestige of *prison and exile*. We came back to Moscow, 'authorities' at five-and-twenty. We were joined by Belinsky, Granovsky, and Bakunin, while through our articles in *Notes of the Fatherland* we ourselves joined the Petersburg movement of the Lycée students and the young literary men. The Petrashevsky[6] group were our younger brothers as the Decembrists were our elder ones.

To be silent about the importance of our circle because I belonged to it would be hypocritical or stupid. Quite the contrary: whenever in my memoirs I come upon those days, on old friends of the 'thirties and 'forties, I purposely pause and speak, regardless of repetition, if only I can make the younger generation better acquainted with them. It does not know them, it has forgotten them, it does not care for them, and disowns them as people who were less practical and less businesslike, who did not know so well where they were going; it is angry with them, and rejects them wholesale as out of date, as idle and superfluous men, as fantastic dreamers, forgetting that the evaluation of men of the past, their significance and the assaying of them, depends less on the comparison of the sum of knowledge, and the manner of formulating problems of the old period and of the new, than on the energy and strength they brought to their solution. I should like to save the younger generation from the ingratitude of history, and even from the mistakes of history. It is time for the fathers not to devour their children like Saturn, but it is time for the children, too, to cease following the example of those natives of Kamchatka, who kill off their old people.

Boldly, and with full consciousness, I shall say once more of our comrades of those days 'that they were a wonderful set of young men, that such a circle of talented, pure-hearted, cultured, intelligent, and devoted men I have never met', and I have wandered pretty widely about this world, and about the revolutionary world. I am not speaking only of my own circle of intimate friends; I am bound to say the same thing as emphatically of Stankevich's circle and the Slavophils. Young men, frightened by horrors of reality, amid gloom and oppressive misery, gave up all and went in search of a way out. They sacrificed everything that others strive after — social position, wealth, everything which their traditional life offered them, to which environment and example drew them, to which their family urged them — for the sake of their convictions, and they remained true to them. Such men must not be simply put into the archives and forgotten.

They are persecuted, tried, put under police supervision, exiled, dragged from place to place, overwhelmed with insults and humiliations — they remain the same: ten years pass — they are still the same: twenty, thirty years pass — they are still the same.

I demand that recognition shall be accorded and justice done to them.

To this simple demand I have heard a strange objection, and more than once, too: 'You, and even more the Decembrists, were the dilettanti of revolutionary ideas; participation in the cause was for you a luxury, something romantic; you say yourselves that you all *sacrificed* social position; you had means, so for you the revolution was not a question of bread and butter and of human existence, a question of life and death. . . .'

'I suppose,' I answered once, 'that for those who were executed it was. . . .'

'At least they were not fateful, inevitable questions for you. You liked being revolutionaries, and that of course is better than if you had liked to be senators or governors; for us the struggle with the existing order is not a matter of choice, it is due to *our* social position. Between you and us there is the difference between the man who has fallen into the water and the man who is bathing; both have to swim, but one does it from necessity and the other for pleasure.'

To refuse recognition to men because they have done from inner impulse what others *are going* to do from necessity is remarkably like the monastic asceticism which only attaches value to duties the fulfilment of which is very repugnant.

Extreme views of this sort easily take root among us; and though the roots do not go deep, they are as hard to eradicate as horse-radish.

We are greatly given to theoretical pedantry and argumenta-tiveness. This German propensity is associated in us with our national element – which we might call the Arakcheyev element – a ruthlessness, a passionate rigidity, and an eagerness to despatch our victims. To satisfy his ideal of a grenadier guardsman Arak-cheyev flogged living peasants to death; we flog to death ideas, arts, humanity, past leaders, anything we please. In dauntless array we advance pace by pace to the limit and overshoot it, never falling out of step as regards logic but only as regards *truth*; unaware, we go on further and further, forgetting that real sense and real understanding of life are manifested precisely in stopping short before extremes . . . that is the *halte* of moderation, of truth, of beauty, that is the eternal balance of the organism.

The oligarchic pretension of the have-nots to be the exclusive sufferers from the social system and to possess a monopoly of the feeling of social injustice is as unjust as all forms of exclusiveness and monopoly. Neither through evangelical mercy nor through democratic envy will you ever get beyond charity and violent spoliation, the redistribution of property and general poverty. In the Church it has remained a theme for rhetoric and a sentimental exercise in compassion; in the ultra-democrats, as Proudhon has observed, it is confined to the feeling of envy and hatred; and in neither case has it gone on to any constructive idea, to any practical result.

In what way are men to blame who understood the pain of the sufferers before they themselves did, and showed it to them and, what was more, the way of escape too? It was not through starva-tion that either Saint-Simon the descendant of Charlemagne, or Robert Owen the manufacturer, became apostles of socialism.

This view will not persist; it lacks warmth, kindness, breadth. I should not have referred to it if these critics had not included on their black lists not only our names, but those of the men who sowed the first seeds of all that has come up and is coming up – the Decembrists whom we honour so deeply.

This digression is hardly in place here.

Sazonov was, in fact, an idle man, and wasted his immense abilities; frittering his life away in all sorts of trivialities abroad, he was lost like a soldier who is taken prisoner in his first battle and never comes home again.

When we were arrested in 1834 and clapped into prison,

Sazonov and Ketscher, by some miracle, were not touched. They both lived almost uninterruptedly in Moscow, and talked a great deal but wrote little, and none of us had any letters of theirs. We were sent into exile; Sazonov's mother succeeded in getting a passport for him to go to Italy. His lot, which was separated from ours, may have given a start to his subsequent life, which was that of a star with no fixed orbit, falling and leaving no trace.

A year later he returned to Moscow; it was just at one of the most stifling and oppressive periods of the last reign. In Moscow he was met by a dead *calme plat*, nowhere a shade of sympathy or a lively word. We, in the rear ranks of exile, among the *reserves*, were cherishing our past life, were living on hope and memory, were working and learning something of the crude reality of provincial existence.

In Moscow everything reminded Sazonov of our absence. Of his old friends the only one on the spot was Ketscher, with whom Sazonov, a man of stiff and aristocratic manners, was less able to be intimate than with any of the rest. Ketscher, as I have said, was a conscious savage — one of Fenimore Cooper's civilised pioneers who had intentionally gone back to the primitive state of the human race, coarse on principle, slovenly by theory, a student of five-and-thirty in the part of a Schilleresque youth.

Sazonov struggled on and on in Moscow — he was overcome by boredom and he had no motive for work, for activity. He tried moving to Petersburg; that was even worse: *à la longue* he could not stand it, and went to Paris without any definite plan. That was still the time when France and Paris had all their enchanting power over us. Our tourists glided over the veneered surface of French life, knowing nothing of its rough side, and were in raptures over everything — over the liberal speeches, over the songs of Béranger and the caricatures of Philipon. It was the same with Sazonov; but he found nothing to do there either. Noisy, lively idleness succeeded to his life of dumbness and oppression. In Russia he had been bound hand and foot, and here he was a stranger to everyone and everything. Another long series of years of aimless excitement and over-stimulated nerves began for him in Paris. He was incapable of concentrating on himself, of devoting himself to intellectual work without waiting for some impulse from outside; it was not in his nature. The objective interest of science was not strong enough in him; he was looking for some activity, and would have been ready for any amount of work so long as it was conspicuous, so long as it could be rapidly applied and realised in practice – and moreover in a noisy *milieu*, amidst applause and the outcry of his enemies. Not finding such work he flung himself into the dissipations of Paris.

... Yet his eyes too, glowed and filled with tears at the memories of our dreams as students. In the recesses of his deeply wounded vanity there was still preserved a faith in the imminence of revolution in Russia, and that he was called to play a great part in it. It seemed as though he was carousing only *meanwhile*, in the wearisome suspense of waiting for the great work before him, and was convinced that one fine evening he would be summoned from his table in the Café Anglais and borne off to govern Russia. ... He kept intent watch on what was being done, and impatiently awaited the moment when he would have to take part in earnest and utter the final, conclusive word.

After the first, noisy days in Paris more serious conversation began, and at once it became evident that we were tuned to very different keys. Sazonov and Bakunin (like Wysocki and the members of the Polish Central Committee later on)[7] were displeased that the news I brought referred more to the literary and university world than to political spheres. They expected to be told about parties, secret societies, ministerial crises (under Nicholas!) and the opposition (in 1847!), but I told them about professorships, Granovsky's public lectures, Belinsky's articles, and the mood of the students, and even of the seminarists. They had been too much separated from Russian life, and had entered too thoroughly into the interests of the 'all-world' revolution and French problems, to remember that among us the appearance of *Dead Souls* was more important than the appointment of a couple of Paskeviches[8] as field-marshals and a couple of Filarets as metropolitans. Without regular means of communication and with no Russian books and periodicals, they were related to Russia theoretically and from memory, which throws an artificial light on everything far away.

The difference in our views almost led to a breach between us; it happened like this: on the day before Belinsky left Paris we saw him home in the evening and went for a walk in the Champs-Élysées. I saw fearfully clearly that for Belinsky all was over and that I was pressing his hand for the last time. The mighty, passionate fighter had burnt himself out; death had already carved its

nearness in big characters on his face that was exhausted with
suffering; he was in acute consumption, but still full of holy
energy and holy indignation, still full of his agonising, angry love
for Russia. I had a lump in my throat and for a long time had been
walking in silence, when the unlucky argument, which had been
sur le tapis a dozen times already, was renewed once more.

'It is a pity,' observed Sazonov, 'that Belinsky has had no career
but journalistic work, and under the censorship, too.'

'I think it is hard to reproach him, of all people, for doing too
little,' I answered.

'Well, with abilities like his he might in other circumstances
and in another field have done rather more. . . .'

I felt vexed and wounded.

'But do tell me, please: you now, who are not under the censorship, who are so full of faith in yourselves, so full of strength and
talent, what have you done? Or what are you doing? Surely you
don't imagine that walking from one end of Paris to the other
every day to talk over the boundaries of Poland and Russia once
more with Sluzhalski or Chodkewicz[9] is doing something? Or that
your talks in cafés and at home, where five fools listen to you and
understand nothing, while another five understand nothing and
talk, is doing something?'

'Wait a bit, wait a bit,' said Sazonov, by now considerably
nettled: 'you forget our situation.'

'What situation? You have been living here for years in freedom, in no dire extremity: what more do you want? Situations are
created. Strong men make themselves acknowledged and force
themselves in. Enough of that: one critical article of Belinsky's is
of more value for the younger generation than playing at being
conspirators and statesmen. You are living in a delirium, walking
in your sleep; you're in a perpetual optical illusion with which
you deceive your own eyes. . . .'

I was particularly annoyed at the time by the two different standards which not only Sazonov but Russians in general applied in
their appreciation of people. The severe criticism of their own
people was transformed into slavish worship of French celebrities.
It was vexatious to see our friends give in before those champion
babblers, who flung them a word, a phrase, a commonplace,
uttered with *vitesse accélérée*; and the more meekly the Russians
behaved, the more they blushed and tried to conceal their idols'

ignorance (as tender parents and self-satisfied husbands do), the more the latter gave themselves airs and swaggered before their hyperborean Anacharses.[10]

Even as a student in Russia Sazonov had been fond of surrounding himself with a retinue of various mediocrities, who listened to him and followed his lead; and here, too, he was surrounded by various *lazzaroni* of the literary Chiaia,[11] barren in mind and body, penny-a-liners, journalistic scavengers such as the gaunt Jules Vécourt, the half-crazy Tardif de Melot, the unknown but great poet Bouilhet;[12] in his chorus, there were also the most narrow-minded Poles – some of the followers of Towjanski[13] – and dull-witted German atheists. How it was they did not bore him is his secret. He almost always brought one or two authorised witnesses from his retinue even when he came to see me, although I was always bored by them and did not conceal the fact. It struck one as particularly odd, too, that he himself was in the position of a Jules Vécourt in his relation to the Marrasts, the Ribeyrolles[14] and to even lesser celebrities.

All this is not entirely intelligible to contemporary visitors to Paris. It must never be forgotten that the present Paris is not the *real* Paris, but a new one.

Having become a gathering-place for the whole world Paris has ceased to be a pre-eminently French city. Formerly the whole of France was in Paris, and nothing besides; now the whole of Europe is there, and the two Americas besides, but there is less of Paris itself: it has become merged in its function of a world-hotel, a caravanserai, and has lost its individual personality, which once inspired ardent love and burning hate, boundless respect and unlimited aversion.

I need hardly say that the attitude of foreigners to modern Paris has changed. The Allied troops who bivouacked in the Place de la Révolution knew that they had taken a *foreign* town. The nomadic tourist now regards Paris as his own; he buys it, enjoys it and knows very well that he is necessary to Paris, and that the old Babylon has rigged herself out, and rouged and powdered herself, not for her own sake but for his.

In 1847 I still found the old Paris – a Paris, moreover, with a quickened pulse, that had been singing Béranger's songs, with the chorus 'Vive la réforme!' unexpectedly changed into 'Vive la République!' Russians in those days continued to live in Paris with an ever-present, conscious sense of thankfulness to Providence (and to the regular despatch of remittances) that they were living in it, that they were strolling in the Palais Royal and going *aux Français*. They frankly worshipped lions and lionesses of every kind – celebrated doctors and dancing-girls, the dentist Désirabode and the mad Ma-Pa,[15] and all the literary charlatans and political jugglers of the day.

I hate the systematic, *prémédité* insolence which is the fashion among us. I recognise in it the family traits of the old dashing arrogance of our officers and landowners, like that of D'Anthès, adapted to the manners of Vasilevsky Island and its streets. But it must not be forgotten that our servility to West European authorities also has come out of the same barracks, the same government offices, the same antechambers, though it has come out of the other door and is addressed to the grand gentleman, the office chief or the commanding officer. In our lack of anything whatever to which to do homage, except brute force and its symbols, stars and ranks in the service, the demand to have a moral Table of Ranks is easy to understand; but, to make up for that, to what men have not the best of our contemporaries bowed down with tender devotion? Even before Werder[16] and Ruge, those mighty dullards of Hegelianism. From this reverence for Germans it may easily be gathered how far they went in their attitude to Frenchmen, to men who are really remarkable – to Pierre Leroux, for instance, or George Sand herself. ...

I am ashamed that I was at first carried away, and thought that to talk in a café with the historian of the *Ten Years*,[17] or at Bakunin's with Proudhon, was something like a promotion, an honour; but in me all attempts at idolatry and fetish-worship do not last, and very soon give place to complete negation.

Three months after I arrived in Paris I began strenuously attacking this respect for authority, and it was at the very peak of my opposition to it that the quarrel about Belinsky took place. Bakunin, with his usual good-heartedness, half assented and laughed; but Sazonov resented it, and continued to regard me as a profane outsider in questions of practical politics. I soon confirmed him still more in this conviction.

The Revolution of February was a complete triumph for him; his journalistic friends received posts in the government, thrones were tottering and leaning for support on poets and doctors. German princelings were asking advice and help from professors and journalists, who only the day before had been persecuted. The Liberals taught them how to fit their narrow crowns on more firmly, that they might not be carried away by the rising storm. Sazonov wrote letter after letter to me in Rome, and summoned me to come *home*, to Paris, to the one and indivisible republic.

On my return from Italy I found Sazonov preoccupied, Bakunin was not there; he had already gone off to stir up the Western Slavs.

'You don't mean to say,' Sazonov said to me at our first inter-view, 'that you don't see that our *time has come?*'

'How do you mean?'

'The Russian Government is in an *impasse*.'

'Why! what has happened? A republic has not been proclaimed in the Peter-Paul Fortress, has it?'

'*Entendons-nous*: I don't think that tomorrow we shall have a 24th of February in Russia. No, but the state of public opinion, the influx of liberal ideas, Austria broken to pieces, Prussia with a constitution, will force the men about the Winter Palace to think a little. They cannot do less than grant some sort of constitution, *un simulacre de charte*: well, and with that,' he added with some solemnity, 'they must have a liberal, cultured ministry who can speak the language of today. Have you thought of that?'

'No!'

'You queer fellow! Where are they going to get cultured ministers?'

'Oh, they'd find them right enough if they wanted them; but I fancy they won't look for them.'

'This scepticism is quite out of place now; *history is being made*, and very rapidly too. Think a minute: the government will have no choice but to appeal to *us*.'

I looked at him, trying to make out what he meant: whether he was joking or not. His face was quite serious; it looked a little flushed and nervous with excitement.

'You mean literally to *us*?'

'Whether to us personally or to our circle does not matter. But just think again: to whom else can they turn?'

'Which portfolio shall you take?'

'You're wrong to laugh. It's our misfortune that we don't know how to take advantage of circumstances *ni se faire valoir*. You keep thinking about your little articles: articles are all very well, but times are changed now; one day in power is worth more than a whole volume of them.'

Sazonov looked with compassion on my unpracticalness, and eventually found less sceptical people who put faith in his coming advent to power. At the end of 1848 two or three German refugees were very regular visitors at the little evening gatherings that were held at Sazonov's. Among them was an Austrian lieutenant[18] who had distinguished himself as chief of staff to Messenhauser. Once, as he was going out at two o'clock in the morning in a heavy downpour of rain, the officer complained of his hard lot, reflecting on the considerable distance between the Rue Blanche and the Quartier Latin.

'Why were you forced to trudge all that way in such weather?'

'Of course I was not forced; but, you know, Herr von Sessanoff is vexed if one does not turn up, and I believe that we ought to maintain good relations with him. You know better than I do that with his talent and intellect ... with the position he occupies in his party, he will go far in the coming revolution in Russia. ...'

'Well, Sazonov,' I said to him next day, 'you have found Archimedes' fulcrum; there is a man who believes in your future portfolio, and that man is Lieutenant So-and-So.'

Time passed, the revolution in Russia did not come off, and no one sent envoys to fetch us home. The menacing days of June passed, too; Sazonov set to work on a leading article not for a periodical but for an Epoch. He spent a long time working at it; read aloud a few fragments, made corrections and alterations, and only just finished it by the winter. He thought it essential 'to explain the last revolution to Russia'. 'Do not expect me,' he wrote at the beginning, 'to describe the events; others will do that better than I. I am giving you the significance, the idea of the revolution which has taken place.' Mere work was not enough for him; whenever he did take up the pen, he wanted to do something unusual, something thunderous; his mind was always haunted by Chaadayev's letter. The article reached Petersburg, was read in friendly circles, and made no impression.

As early as the summer of 1848, Sazonov founded an International Club. He brought to it all his Tardifs, Germans, and Messianists. With a beaming face he walked up and down the empty room in a dark blue dress-suit. He opened the International Club with a speech addressed to five or six listeners, of whom I was one, by way of audience, the rest of the little party being on the platform in the capacity of committee. Sazonov was followed by Tardif de Melot, a dishevelled figure looking half-asleep, who stood up and boomed out a poem in honour of the Club.

Sazonov frowned, but it was too late to stop the poet.

> 'Worcel, Sassonoff, Olinski[19] Del Balzo, Léonard . . .
>   Et vous tous . . .'

Tardif de Melot bawled with enraptured frenzy, unaware of the laughter.

Two or three days afterwards Sazonov sent me *a thousand* copies of the programme of the opening ceremony; with that the Club ended. It was only subsequently that we heard that one of the representatives of humanity, in fact the representative at that congress of Spain, who had delivered a speech in which he called the executive power *potence ehécoutive*, supposing this to be French, nearly finished up in England on a real gallows and was sentenced to penal servitude for forging some document.

The failure to become a minister and the collapse of the Club were followed by more modest but far more practical attempts to become a journalist. When *La Tribune des Peuples* was established, with Mickiewicz as chief editor, Sazonov occupied one of the chief positions on the paper, wrote two or three very good articles . . . and was silent, and before the failure of the *Tribune* – that is, before the 13th of June, 1849 – he was already at odds with the whole staff. To him it all seemed petty and poor, *il se sentait dérogé*, was vexed at this, finished nothing, dropped what he had begun and cast aside what was half done.

In 1849 I suggested to Proudhon that the post of foreign editor of the *Voix du Peuple* should be given to Sazonov. With his knowledge of four languages, of the literature, politics, and history of all the European nations, and his acquaintance with political parties, he might have done wonders for the French with this part of the paper. Proudhon had nothing to do with the internal arrangements of the foreign news department; it was in my hands, but I could do nothing from Geneva. A month later Sazonov transferred the foreign editorship to Chojecki and severed his connection with the paper. 'I have a profound respect for Proudhon,' he wrote to me in Geneva, 'but there is not room on one journal for two such personalities as his and mine.'

A year later Sazonov joined *La Réforme*, which was then being revived by the followers of Mazzini. Lamennais was the chief editor. But here too there was no room for two great men. Sazonov worked on it for three months, and then threw up *La Réforme*. With Proudhon he had fortunately parted peacefully, but with Lamennais he quarrelled. Sazonov accused the niggardly old man with using the funds of the paper for his personal ends. Lamennais, recalling the habits of his clerical youth, resorted to what is the *ultima ratio* in Western Europe, and spread about Sazonov the suggestion that he might be an agent of the Russian Government.

The last time I saw Sazonov was in Switzerland in 1851. He had been deported from France, and was living in Geneva. This was the very greyest, most oppressive period; a brutal reaction was triumphant everywhere. Sazonov's faith in France and in the coming change of ministry in Petersburg was shaken. He was bored and tormented by his idle life, his work was not successful, he caught at anything, but did not keep it up, he grew irascible and drank. Moreover the life of petty anxieties, the everlasting war with his creditors, the effort to obtain money, together with the talent for throwing it away and the inability to put his affairs in order, brought a great deal of nervous irritability and dismal prose into Sazonov's daily existence; there was no longer any merriment in his dissipation: it was a habit; but once upon a time he had been an expert at it.

. . .

Between him and me 'a cat ran': he did not treat me openly in a matter that was very dear to my heart. I could not get over this.

Meantime a new epoch was beginning for Russia. Sazonov longed to take part in it: wrote articles* that were unsuccessful, wanted to return to Russia and did not return, and finally left Paris. For a long time nothing was heard of him.

One day a Russian who had recently come to London from Switzerland said to me: 'An old friend of yours was buried the day before I left Geneva.'

---

* His article on 'The Place of Russia in the All-World Exhibition' was published in Vol. II of *The Pole Star*.

'Who was that?'

'Sazonov; and think of it: there was not one Russian at the funeral.'

I felt the shock of it: I suppose I regretted having dropped him for so long. . . .

(Written in 1863.)

THE ENGELSONS

THEY are both dead. He was not more than thirty-five, and she was younger.

He died about ten years ago in Jersey: his coffin was followed to the grave by his widow, his child and a sturdily built, dishevelled-looking old man with large, marked, but undefined features; in his face were jumbled at random genius and insanity, fanaticism and irony, and the wrath of an Old Testament prophet and of a Jacobin of the year 1793. This old man was Pierre Leroux.

She died in Spain at the beginning of 1865. I heard of her death a few months later.

I have not heard where the child is.

The man of whom I am speaking was once near and dear to me; he first dressed deep wounds when they were fresh; he was a brother, a sister to me. She, scarcely knowing what she was doing, estranged him from me. He became my enemy. . . .

The news of her death called them back to my mind again. . . .

I took up the manuscript I had written about them in 1859, and read it through by way of reading the Psalter over the dead.

For a long time I considered whether I should print it or not, and recently I decided to do so. My intention is good, and my story is true. I do not want to cast reproaches on their grave, but together with the reader to trace once again, and once again, using fresh persons as examples, all the intricate, morbid warping of people's characters in the last generation under Nicholas.

Château Boissière, 31 December 1865

1

At the end of 1850 a Russian arrived at Nice with his wife. They were pointed out to me on the parade. They both belonged to the number of those who were waiting for the troubling of the waters: he was thin, pale, and consumptive, with reddish fair hair;

she was a beauty who had faded early, worn-out, half-shattered, harassed.

A doctor living in the household of a Russian lady told me that the fair-headed gentleman had been a Lycée[1] student, that he was reading *Vom andern Ufer*, that he had been mixed up in the Petrashevsky affair, and that for all these reasons he wished to make my acquaintance. I answered that I was always glad to meet a good Russian, especially a Lycée student, and even more one who had had a hand in an affair of which I knew little, but which had been for me like the olive branch brought by the dove to Noah's ark.

Some days passed without my seeing either the doctor or the new Russian. Suddenly between nine and ten one evening a card was brought to me; it was he. We were sitting with Karl Vogt in the dining-room. I told the servant to ask the visitor upstairs to the drawing-room, and I went up there before the others. There I found him, pale and trembling with some kind of fever. He could scarcely tell me his name; when he was a little calmer, he jumped up from his chair, rushed at me, kissed me effusively and, before I also could recover myself, with the words, 'So at last I am really seeing you,' he kissed my hand. 'What are you about? Upon my soul!' I said, but by then he was in tears.

I looked at him in perplexity; was this nervous indiscipline or plain madness?

Apologising and showering me with compliments, he told me with unusual rapidity and vehement gesticulation that I had saved his life, and that this was just how it had been: Sunk in acute depression in Petersburg, expelled from the Lycée for some nonsense or other, abominating civil service,[2] in which he had been obliged to accept an appointment, and seeing no way out for himself personally or for things in general, he had made up his mind to poison himself; and a few hours before carrying out his intention went wandering aimlessly about the streets: he went into Izler's and picked up a volume of the *Notes of the Fatherland*. It contained my article, 'Concerning a certain Drama'.[3] The reading of it gradually absorbed his attention; he felt better, he felt ashamed of having so given in to sorrow and despair when public interests were gaining strength on all sides and calling on all who were young, for all who had strength, and instead of poison Engelson asked for half a bottle of Madeira, read the article over again, and from that time became my ardent admirer.

He sat on till late at night, and went away asking leave to come again soon. Through his tangled talk, interspersed with anecdotes and digressions, one could discern an intensely well organised brain, pronounced dialectical ability and, still more clearly, something warped and distorted that hurled him from one extreme to the other, from an indignation shocked by affliction, and made poignant by misfortune, to ironical clowning, from tears to grimaces.

He left me with a strange impression. At first I did not quite trust him, then I was tired by him − he seemed to have too powerful an effect on one's nerves; but by degrees I grew used to his oddities and was glad of an out-of-the-way person to break the monotonous boredom induced by the vast majority of Western Europeans.

Engelson had read a vast amount and had done a vast amount of study; he was a linguist and a philologist, and brought into everything the scepticism with which we are so familiar, and which exacts so high a price for the pain it leaves. In old days they would have said of him that he had read himself silly. His over-stimulated intellectual activity was too much for the strength of his frail organism. Liquor, with which he conquered fatigue and stimulated himself, fanned his thoughts and imagination into long, bright tongues of fire, and was rapidly consuming his sick body.

His disorderly living and drinking, his perpetual, irritable mental activity, his conspicuous many-sidedness and his conspicuous futility, his utter idleness, his extreme passionateness and extreme apathy, vividly recalled the past to me, in spite of the immense difference between all this and our old ways in Moscow. Again I heard the sounds not only of my native language but of my native way of thinking. He had been a witness of the reign of terror in Petersburg after 1848, and he knew the literary circles. Entirely cut off from Russia as I was at that time, I listened greedily to his accounts.

We took to seeing each other often, and then every evening.

His wife, too, was a strange creature. Her face, beautiful by nature, was distorted by neuralgic pains and a sort of restless anxiety. She was a Russified Norwegian, and spoke Russian with a slight accent which suited her. In general she was more silent and reserved than he. Their home life was not serene: it seemed to be nervous, *unheimlich*, strained; there was something lacking in their life, and something superfluous in it, and one felt this continually like electricity, unseen and menacing, in the air.

I often found them in the large room which served them as bedroom and sitting-room in the hotel, in a state of utter prostration. She, spent and with tear-stained eyes, in one corner; he pale as death, with white lips, distraught and silent in the other. . . . So they would sit at times for whole hours, whole days together, and that a few steps from the dark blue Mediterranean, from groves of orange-trees, to which everything – the sapphire sky and the bright, noisy gaiety of southern life – invited one. They did not actually quarrel; it was not a case of jealousy nor estrangement, nor any tangible cause, indeed. . . . He would suddenly get up, go to her, fall on his knees and, sometimes with sobs, repeat: 'I have been your ruin, my child, your ruin!' and she would weep and believe that he had been her ruin. 'When shall I die at last and leave him in freedom?' she used to say to me.

All this was new to me, and I felt so sorry for them that I wanted to cry with them, and most of all to say to them: 'Oh, come, that's enough; you are not at all so miserable or so bad; you are both splendid people. Let us take a boat and drown sorrow in the dark blue sea.' I did do this sometimes, and succeeded in drawing them out of themselves. But by next morning the paroxysms would have returned. . . . They were somehow so on each other's nerves, and had reached such an hysterical *impasse*, that the slightest word destroyed their harmony and once more called up furies or something from the bottom of their hearts.

I sometimes fancied that, in continually exacerbating their own wounds, they found in this pain a smarting enjoyment; that this gnawing at each other had become essential to them, like vodka or pickles. But unfortunately the physique of both was unmistakably beginning to be exhausted; they were on the high road to the lunatic asylum or the grave.

Her character, by no means without talents, but unfinished and at the same time corrupted, was far more complex, and in a certain sense she had far more fortitude and strength than he had. Moreover she had not a shadow of the unity, the consistency, that unfortunate consistency which he retained even in the most violent extremes and the sharpest contradictions. In her, side by side with her despair, her desire to die, her habit of moaning and groaning, there was a thirst for worldly pleasures and a concealed coquetry, a love for finery and elegance, denied as it were intentionally, to spite herself. She was always dressed becomingly and with taste. She wanted to be an emancipated woman according to the notions of the time, and the victim of a vast, egregious, psychic unhappiness, like George Sand's heroines ... but her former, customary, traditional life dragged her like a heavy weight towards quite a different sphere.

What gave poetic charm to Engelson, and did much to make up for his defects, and what served as a safety-valve for himself, she could not understand. She could not follow his racing thought, his rapid transitions from despair to witticisms and laughter, from candid mirth to candid tears. She lagged behind, losing the thread, perplexed. ... His caricatures of his own gloomy thoughts were beyond her comprehension. When Engelson, after a perfect feast of puns and jokes, mockery and teasing, got more and more excited and began acting regular dramatic scenes at which one could only laugh helplessly, she took offence and would go out of the room; she was offended at 'his unseemly behaviour before outsiders'. He usually noticed this and, since nothing could stop him when once he had taken the bit between his teeth, he would play the fool more extravagantly than ever, and then waltz up to her and ask her with glowing cheeks and perspiring brow: '*Ach, mein lieber Gott, Alexandra Christianovna, war es denn nicht respektabel?*' She would weep more than ever, and he would suddenly change, grow gloomy and morose, drink glass after glass of brandy, and go home, or simply fall asleep on the sofa.

On the next day I had to mediate and make things up ... and he so earnestly kissed her hands and so drolly begged forgiveness for his sins, that even she sometimes could not restrain herself and laughed with us.

I must explain in what these performances consisted, that caused such sorrow to poor Alexandra Christianovna. Engelson's comic talent was indubitable and very great; such biting satire was never equalled by Levassor,[4] hardly by Grasso at his best, or by Gorbunov[5] in certain of his stories. Moreover half of it was improvised; he would bring in additions and variations while preserving the same framework. If he had cared to train and develop this capability, he would certainly have been in the foremost ranks of *spiteful* comedians, but Engelson never trained nor developed anything in himself. Talents shot up like vigorous wild suckers and were choked in his unstable soul, both by domestic cares which took away half his time, and by his habit of catching at everything in the world, from philology and chemistry to political economy and philosophy. In this respect Engelson was a pure Russian, although his father was of Finnish extraction.

He acted everything in the world – officials and Russian gentlemen, priests and police-constables; but the best of his performances were concerned with Nicholas, for whom he had a profound, heartfelt and active hatred. He would take a chair *à la* Napoleon, sit astride it, and ride sternly up to a corps on parade . . . epaulettes, helmets, and casques quaking all round him . . . this is Nicholas at a review; he is moved to wrath and, turning his horse, says to the corps commander, 'Bad'; the commanding officer listens with reverent awe, looks after Nicholas, and then, dropping his voice and gasping with fury, whispers to a divisional general: 'You appear, your Excellency, to be busy about something else and not the service. What a wretched division! what regimental commanders! I'll show them.'

The divisional general turns redder and redder, and pounces on the first colonel he comes across, and so from one rank to the next, with hardly discernible, lifelike nuances, the Imperial 'bad' passes down to the sergeant-major, at whom the squadron commander swears like a trooper, and who, without answering, pokes the hilt of his sword with all his might into the ribs of a flank-marker, who has done nothing.

Engelson would portray with amazing fidelity not only the characteristics of each rank, but also each rider's movements as he tugged at his horse in his fury and then raged at it for not standing still.

Another performance was of a more peaceful kind. The Emperor Nicholas is dancing the French quadrille. *Vis-à-vis* to him is a foreign diplomatist, and on one side a general, stiff as on parade, on the other a civilian grandee. This was a complete *chef-d'œuvre* of its own. Engelson would take one of us for his partner. The flower of it all was Nicholas – playing the autocratic Tsar over the quadrille, the conscious firmness of each step, the brilliant perfection of every movement, together with the indulgent and gracious gaze at his partner, which is transformed at once into an order to the general, and a word of advice to the civilian gentleman not to forget himself. To convey this in words is impossible. The general who, rigidly erect, his elbows turned out a little, marches in time, with strained attention, to assault the figures under the stern observation of His Majesty the Emperor; and the distracted civilian with his legs giving way under him from fright, with a smile on his face and almost a tear in his eye – all this was performed in such a way that a man who had never seen Nicholas could thoroughly grasp the agonising ordeal of an Imperial quadrille, and the danger of having His Majesty as a vis-à-vis. I have forgotten to say that the foreign diplomatist was the only one who danced with studied negligence and great finish, concealing the uncomfortable feeling of uneasiness of which the most valiant is conscious with a lighted cigar close to a barrel of gunpowder.

But although Engelson's grimacing and foolery roused his wife's indignation, it does not follow that there was any more unison or harmony in herself; quite the contrary, in her head there was real confusion, that was destructive of all order, of all consistency, and made her impossible to grasp. In her I learnt for the first time how little can be done with logic in discussion with a woman, especially when the discussion relates to practical affairs. In Engelson his disorder recalled the confusion after a fire, after a funeral, or perhaps after a crime; but in her it was like a room that has not been tidied up, in which everything has been thrown about at random – children's dolls, a wedding dress, a prayer-book, a novel by George Sand, slippers, flowers, plates. In her half-conscious ideas and half-undermined beliefs, in her pretensions to an impossible liberty and her independence of the customary external bonds, there was something suggestive of a child of eight, a girl of eighteen, and an old woman of eighty. Many times I told her this herself; and, strange to say, even her face was prematurely faded; it looked old because some of her teeth were missing, and at the same time it retained a childish expression.

For the chaos in her mind it was Engelson who was entirely to blame.

His wife was the spoilt child of a mother who had been infatuated with her. An elderly, phlegmatic civil servant of Swedish origin sought her in marriage when she was eighteen. In a moment of childish caprice and vexation with her mother she agreed to marry him. She wanted to be her own mistress and sit at the head of the table.

When the honeymoon of freedom, visits, and fine clothes was over, the bride was insufferably bored; although her husband behaved with strict propriety, took her to the theatre and arranged evening tea-parties for her, she had an aversion for him; she struggled with him for three or four years, grew tired of it, and went back to her mother. They were divorced. Her mother died, and she was left alone, suffering and melancholy, with her health prematurely broken in the struggle with her absurd marriage, with emptiness and hunger in her heart and an idle brain.

It was just at this time that Engelson was expelled from the Lycée. He was nervous, irritable and, with a passionate yearning for love and a morbid lack of confidence in himself, was consumed by *amour-propre*. ... He had made her acquaintance while her mother was living, and they became great friends after her death. It would have been strange if he had not fallen in love with her. Whether the feeling were likely to be lasting or not, he was bound to love her passionately; everything helped to bring this about ... the fact that she was a woman without a husband, a widow and not a widow, a bride and not a bride, and that she was pining for something, was in love with another man, and made miserable by her love. This other was an energetic young fellow, an officer and a literary man, but a desperate gambler. They quarrelled over this frantic passion for play; later on he shot himself.

Engelson never left her side; he comforted her, amused her, interested her. It was his first and last love. She wanted to study, or rather to learn without studying; he undertook to be her Mentor – she asked for books.

The first book Engelson gave her was Feuerbach's *Das Wesen des Christenthums*. He took the place of commentator, and day by day he pulled from under the feet of his Héloïse, who could not step on firm ground for the cramping Chinese shoes of her early Christian training, the ladder on which she might somehow have kept her balance. ...

Emancipation from the traditional morality, said Goethe, never leads to good unless the mind has grown strong; indeed only reason is worthy to replace the religion of duty. Here was a woman sleeping the deep slumber of moral unconcernedness, lulled by traditions and full of the dreams natural to a patriarchal soul, tinged with Christianity, tinged with romantic and moral notions; and Engelson tried to educate her at one blow by the method of English nurses who, when the baby screams from stomach-ache, pour a glass of gin into its mouth. He flung into her immature, childish conceptions a corrosive ferment with which men are rarely equal to coping, which he himself could not cope with but only understand.

Dizzied by the overthrow of all her moral conceptions and all her religious convictions, and finding in Engelson himself nothing but doubt, nothing but irony and denial of the old, she lost the only compass, the only rudder she had left, and was like a boat adrift at sea, wheeling about at random without a helm. The equilibrium arrived at by life itself, which is kept – as by the compensating strips in a pendulum – by absurdities which exclude each other and are maintained by so doing, was broken.

She flung herself violently into reading, understanding and not understanding, and mixing up the philosophy of her nurses with the philosophy of Hegel, sentimental socialism with the economic conceptions of conventional housekeeping. With all that, her health grew worse, her boredom and misery continued; she pined and grew thin, longed desperately to go abroad, and dreaded persecutions and enemies.

After a prolonged struggle Engelson rallied all his forces and said to her:

'You want to travel; how will you go alone? . . . You will meet with a vast number of unpleasantnesses; you will be lost without a friend, without a protector with the right to protect you. You know that I would lay down my life for you . . . give me your hand – I will care for you, soothe you, watch over you. . . . I will be your father, your mother, your nurse, and your husband, but it must be legally. I will be with you, near you. . . .'

This was said by a man under thirty and passionately in love. She was touched, and accepted him as her husband unconditionally. A short time afterwards they went abroad.

Such had been the past of my new acquaintances. When Engelson told me all this, when he complained bitterly that this marriage had been the ruin of them both, and I saw for myself how they were languishing in the fumes of a moral brazier which they deliberately fanned, I was convinced that their unhappiness was due to their having known too little of each other beforehand, to their being too close to each other now, their having built their life too much on personal emotion, and their putting too much faith

in being husband and wife. If they could have parted, each might
have sighed in freedom, have grown calm, and perhaps begun to
blossom afresh. Time would have shown whether they were really
so necessary to each other; in any case the delirium would have
been broken for a time without catastrophe. I did not conceal my
opinion from Engelson; he agreed with me, but all this was a
mirage; in reality he had not the strength to leave her, nor she to
take the plunge. . . . They secretly *wanted* to hover on the brink
of these resolutions without carrying them into execution.

My opinion was too sane and simple to be right in regard to
such intricately pathological subjects and such sick nerves.

2

The type to which Engelson belonged was at that time rather
new to me. At the beginning of the 'forties I had seen such a type
only in embryo. It developed in Petersburg towards the end of
Belinsky's career, and was formed after I left and before Chernyshevsky[6] appeared. It was the type of the Petrashevsky group and
their friends. That group was made up of young, gifted men,
extremely intelligent and extremely well educated, but nervous,
morbid and hard to please. Among them there was no instance of
a crying lack of talent, no one who wrote ungrammatically: these
were phenomena of a quite different period; but in them there was
something degenerate, deranged.

The followers of Petrashevsky made a bold and ardent dash into
activity, and astonished all Russia with the *Dictionary of Foreign
Words*. The intense mental activity of the 'forties was their heritage, and they passed straight from German philosophy into
Fourier's phalanx,[7] into becoming followers of Kant.

Surrounded by petty, worthless people, proud of the attentions
of the police, and conscious of their own superiority at the very
time they left school, they prized too highly their negative
achievement, or rather their potential achievement. This led to
immoderate vanity – not that youthful, healthy vanity that is
becoming in a lad who dreams of a great future, and becoming in
a man in the fullness of his power and activity, not that which in
days gone by made men perform miracles of daring and endure
chains and death for the sake of glory, but, on the contrary, a
morbid vanity, preventing any act with its vast pretensions,

irritable, ready to take offence, conceited to the point of insolence, and diffident at the same time.

Between their overrating of themselves and their appreciation by their neighbours the disproportion was very great. Society will not accept notes of hand for the future, but demands completed work in return for the ready money of recognition. They had too little power of hard work and perseverance; they only had enough of each to understand and assimilate what had been worked out by others. They wanted to have harvests in return for the intention to sow, and laurels for having their seedbins full. 'The insulting way in which they were overlooked by society' tormented them, made them unjust to others, and reduced them to despair and *Fratzenhaftigkeit*.

In the person of Engelson I studied the difference between that generation and our own. Later on I met many men not so talented, not so cultured, but with the same visible, morbid fracture in their whole composition.

A fearful sin lies at the door of the reign of Nicholas in this moral murder of a generation, in this spiritual damage done to its children. The wonder is that the strong and healthy, though warped, still survived. Everyone knows the celebrated orders to instructors in the Cadet Corps.[8] At the Lycée things were better, but of late years it, too, had incurred the hatred of Nicholas. The whole system of government education lay in instilling the religion of blind obedience, leading to power as its reward. The feelings of the young, naturally radiant, were rudely driven inwards, and replaced by ambition and jealous, envious rivalry. What did not perish came out sick or mad.... Together with burning egoism they were inoculated with a sort of spiritlessness, a sense of impotence, of fatigue before beginning work. Young men became hypochondriacal, suspicious, tired before they were twenty. They were all tainted with the passion for introspection, self-analysis, self-accusation; they scrupulously believed in their psychical phenomena, and loved making endless confessions and giving descriptions of neurotic incidents in their lives. In later years I often happened to receive the confessions not only of men but of women belonging to this category. After watching with sympathy their remorse, their neurotic self-castigation, which extended to gross calumny upon themselves, I came at last to the conviction that this was only one of the forms of that same egoism. One had but to cease protesting and sympathising and to agree with the repentant sinners, to see how readily malignant and how mercilessly vindictive were these Magdalens of both sexes. With them, like the Christian priest before the mighty of this world, you have only the right solemnly to remit their sins and to be silent.

These nervous people though excessively ready to take offence, shuddering like a sensitive plant at the slightest clumsy touch, were, on their side, incredibly harsh in their language. As a rule, when it came to avenging themselves, they did not measure their terms – a fearful defect of taste, which betrays a profound contempt for the person addressed and a scandalous indulgence for oneself. This unbridled speech among Russians comes from the houses of landed gentry, from government offices and army barracks; but how is it that it has survived and developed in the younger generation after skipping ours? That is a psychological problem.

In our old student circles we cursed each other roundly, argued roughly and passionately, but in the most violent fray something stayed outside the lists. . . . For our nervous friends of Engelson's generation this place of covenanted truce did not exist: they did not think it necessary to restrain themselves; for the sake of an empty and momentary revenge, for the sake of getting the upper hand in a dispute, they spared nothing, and I have often, with horror and amazement, seen them – led by Engelson himself – without a trace of pity, fling the most precious pearls into the corrosive fluid of their bitterness, *and afterwards weep*. With the change of the nervous current, remorse would follow, and entreaties for forgiveness from the outraged idol. They were not fastidious, and poured filth into the very cup from which they drank.

Their repentances were sincere, but did not prevent repetitions of the offence. In these people some spring, which regulates the action of the wheels and directs them, is broken; the wheels turn with tenfold swiftness, doing no work, but injuring the machine; harmonious combination is broken, the aesthetic mean is lost; there is no living with them, and with this defect they cannot live themselves.

Happiness did not exist for them; they were not able to look after it. The slightest cause provoked them to ruthless resistance and made them behave churlishly to everyone near them. With their irony they have ruined and spoilt as much in life as the Germans have by mawkish sentimentality. Strange to say, these people are avidly desirous of being loved; they seek enjoyment, and when they lift the cup to their lips some evil spirit jogs their arm, the wine is spilt upon the ground, and the cup, passionately flung down, rolls in the mud.

3

The Engelsons soon went away to Rome and Naples; they meant to stay there for six months, and returned in six weeks. Having seen nothing they trailed their boredom about Italy, were wretched in Rome and melancholy in Naples, and eventually made up their minds to come back to Nice – 'to you for treatment,' he wrote to me from Genoa.

Their sombre mood had grown worse while they were away. In addition to their nervous disorder there were now quarrels which assumed a more and more exasperated and splenetic nature. Engelson was to blame for his unbridled words and harsh expressions, but it was always she who provoked them, provoked them intentionally, with secret spite and peculiar success in his most good-natured moments; he was never allowed to forget himself for an instant.

Engelson was incapable of holding his tongue; talking to me was a relief to him, and so he used to tell me everything, even more than he ought, which was awkward for me. I felt that I could not be as frank with them as they were with me. Talking came easy to him: voicing a grievance calmed him for a time – but me it did not.

One day, sitting in a little tavern with me, Engelson said that he was worn out by the daily struggle, that there was no way of escape from it, that once again the thought of cutting short his life seemed to him the last way of salvation left. . . . With his nervous indiscipline it might have been expected that if a pistol or a phial of poison did come in his way he would sooner or later try one or the other.

I was sorry for him. And both of them were to be pitied. She might have been a happy woman if her husband had been a man of serene temper who would have known how to draw her out gently, to be light-hearted in his merriment, and in case of need to influence her not merely by persuasion but also by authority – by sober authority, without irony. There are immature natures which cannot guide themselves, just as there are persons of lymphatic constitution who need a corset to avoid curvature of the spine.

While I was thinking about this Engelson, going on with his tale, had come to the same conclusion himself. 'The woman does not love me,' he said, 'and cannot love me; what she does understand and looks for in me is bad, and what is good in me is so much Chinese to her. She is corrupted by *bourgeois* ideas; with our outward *Respektabilität*, our petty domesticity. We shall torment each other to death; I see that clearly.'

It seemed to me that, if a man could talk in that way of the woman nearest to him, then the chief bond between them had been broken; and so I admitted to him that, having watched their life together for a long time past with deep sympathy, I had often asked myself why they went on living together.

'Your wife is pining for Petersburg, for her brothers and her old nurse; why don't you arrange that she shall go home, and you remain here?'

'I've thought of that a thousand times; it's the one thing I wish for. But in the first place she has no one to go with; and in the second she is bored to death in Petersburg.'

'Well, but she's bored to death here, too. As for having no one to send her with, that's a relic of our old lordly caprices. You can see your wife to the steamer at Stettin, and the steamer will find its way by itself. If you haven't the money, I'll lend it you.'

'You're right, and that's what I shall certainly do. I am sorry for her, my heart aches for her; all the love I have in me I have concentrated on her. I sought in her not only a wife, but a being whom I wished to draw out and educate according to my own notions. I thought that she would be my child – the task was beyond my strength. But who could have known that I should find such resistance, such stubbornness?' He paused, and then added: 'If I'm to tell you all I think – she needs a different husband ... if a man turned up who was worthy of her, with whom she fell in love, I would put her hand in his, and we should both recover – that's more important than Petersburg.'

I accepted all this *au pied de la lettre*. That he was sincere there is no doubt. That is just the difficulty with these impulsive, un-controlled natures; they can, like good actors, enter so thoroughly into different parts, and so identify themselves with them, that a cardboard dagger seems to them a real one, and they shed genuine tears over Hecuba.

We were then living together at Ste Hélène. Two days after my conversation with Engelson, late in the evening, Mme Engelson, with a tear-stained face, came into the drawing-room, a candle in her hand; she set the candle on the table, and said she wanted to have a talk with me. We sat down ... after a brief and obscure prelude about the fate that pursued her, on Engelson's unfortunate nature and her own, she announced that she had made up her mind to return to Petersburg, and did not know how to do it. 'Only you alone have influence over him: persuade him *really* to let me go. I know that in moments of vexation he is ready in words to put me in the post-chaise at once, but all that is only words. Persuade him, save us both, and give me your word to look after him just at first, comfort him ... it will be hard for him, he is ill and nervous.' She sobbed again and hid her face in her handker-chief.

I did not believe in the depth of her woe, but I understood very well what a false move I had made by speaking openly to Engelson; it was clear to me that he had repeated our conversation to her.

I had no choice left; I repeated my own words, softening the form of them. She got up, thanked me and added that if she did not go she would throw herself into the sea; that she had that evening been burning a great many papers, and wished to put some others in a sealed packet in my keeping. It was clear to me that she was by no means so passionately anxious to go away, but through some self-indulgent caprice wanted to drag on and pine away in melancholy. Moreover I saw that, if she was wavering without any settled plan, he was not even wavering but distinctly did not want her to go. She had great power over him; she knew this and, building upon it, allowed him to rage, to rear, to foam at the bit, knowing that, however he might jib, things would go not as he willed but as she did.

She never forgave me for my advice; she feared my influence, though she had unmistakable proof of my powerlessness.

For ten days there was no talk about going away. Then periodical skirmishes started up. Once or twice a week she would come to me with tear-stained eyes and announce that now all was over, and that next day she would get ready to go to Petersburg or to the bottom of the sea. Engelson would come out of his room, twitching convulsively, with a green face and trembling hands; he would vanish for some ten hours, and would come back covered with dust, exhausted and very drunk, would take a passport to be viséd, or obtain a permit for Genoa; then it would all calm down and flow into the usual channel again.

Externally Mme Engelson was completely reconciled with me, but from that time she began to conceive for me something like hatred. Before that she had differed with me and been angry without concealing it ... now she became unusually amiable. She was annoyed that I had seen through something; that I had not been touched by her tragic destiny or taken her for an unhappy victim, but had looked on her as a capricious invalid; that, far from shedding tears of Platonic sympathy with her, I doubted whether she did not find enjoyment rather than distress in tears, heart-rending scenes, explanations lasting several hours, and so on and so on.

Time passed, and by degrees much was altered. With the rapidity which only occurs in nervous invalids she regained her health, became gayer and even more careful of her dress. And although the most nonsensical causes would lead once more to the old scenes between her and Engelson, to a farewell *à la* Socrates before the hemlock, and to a readiness to follow in Sappho's footsteps to the bottom of the sea, yet on the whole things went better. The woman who had been for ever half reclining from weakness, for ever exhausted, drew herself up as erect as Sixtus V, and began to grow so stout that one day poor Kolya, sitting at dinner and looking at her full bosom, shook his head and said: '*Sehr viel Milch!*'

It was evident that some new interest was occupying her, that something had awakened her from her morbid lethargy. From the time that I had spoken plainly to her she had begun a persistent game, thinking over every move, no less carefully than the gamblers *du Café Régent*, and patiently correcting her mistakes. Sometimes she betrayed herself and made a blunder, was lured too far in one direction or another; but she constantly returned to her original plan. This plan now went beyond tightening her grip over Engelson, and beyond avenging herself on me; it was to get us all, the whole household, into her power, and taking advantage of

Natalie's being more and more seriously ill to control the education of the children and our whole life: *si non – non*, that is, if she
failed, to break off my relationship with Engelson at all costs.

But before she could attain her final result there were many
moves in the game to be made, very difficult, painful concessions,
cat-like tactics, and much patient waiting: she accomplished a
great deal, but not everything, Engelson's incessant chatter
hindered her as much as my opened eyes.

She might have made better use of the energy, the force and the
persistence which she wasted on her craftily interwoven schemes
... but personal feeling and vanity intoxicate people and, once
entering upon the dark intrigues of passion, it is hard to stop and
hard to see anything clearly. Usually light is brought into the
room at the noise of a crime that has already been committed; that
is how it is that both the catastrophe and the sting of conscience
are irremediable.

4

... Of the misfortunes that descended upon me in 1851 and
1852 I speak in another place. Engelson brought me much comfort
in my sorrow. He and I would have lived for a long time near the
graveyard, but the restless egoism of his wife had no pity even on
mourning.

Some weeks after the funeral[9] Engelson, sorrowful and alarmed,
with evident reluctance and evidently not of his own initiative,
asked me whether I was not thinking of entrusting the upbringing
of my children to his wife.

I answered that the children, except my son Sasha, were going
to Paris with Maria Kasparovna Reichel, and that I admitted
frankly that I could not accept his suggestion.

My answer wounded him, and it hurt me to wound him. 'Tell
me,' I said, 'speaking honestly, do you think your wife competent
to bring up children?'

'No,' Engelson answered, 'but ... but perhaps it would be a
planche de salut for her; she is just as wretched as ever, and this
would mean your trusting her, and a new duty.'

'Yes, but if the experiment didn't answer?'

'You are right; let us say no more about it, but it is
sad.'

Engelson really agreed with me, and said no more. But she had not expected so simple an answer; on this question I could not give way and she would not and, beside herself with vexation, she immediately made up her mind to take Engelson away from Nice. Three days later he told me he was going to Genoa.

'What is the matter?' I asked, 'and why are you going so soon?'

'Well, you see for yourself my wife does not get on with you, nor with your friends, so I've made up my mind ... and perhaps it is for the best.'

And the next day they went away.

Afterwards I left Nice. On my way through Genoa we met peaceably. Surrounded by our friends, among whom were Medici, Pisacane, Cosenz, and Mordini, she seemed calmer and better in health. Nevertheless she could not let slip any chance for having a spiteful dig at me. I went away, said nothing; that was no use. Even when I had gone to Lugano she kept up her poisoned *petits points*, and this in the rare postscripts to her husband's letters, as though with his 'visa'.

Finally these pin-pricks, at a time when I was utterly crushed by grief and distress, drove me out of all patience. I had done nothing to deserve them, nothing to provoke them. On getting one of her spiteful postscripts, which said that Engelson would still have to pay dearly for devoting himself heart and soul to his friends, not knowing that they would do nothing for him, I wrote to Engelson that it was time to put a stop to this.

'I do not understand,' I wrote, 'why your wife is angry with me. If it is because I did not give my children into her keeping, surely that is no justification for it?' I reminded him of our last conversation, and added: 'We know that Saturn ate his children, but for anyone to show his gratitude to his friends for their sympathy by bestowing his children's upbringing on them is something unheard of.'

She never forgave me that sally but, what is far more remarkable, he never forgave me for it either, though at first he showed no sign of resenting it ... but he reproached me with those words years later. ...

I went to London;[10] Engelson settled for the winter in Genoa, and afterwards moved to Paris.

5

The saying, 'He who has not been at sea has not prayed to God,' may be rephrased in this way: the woman who has not had children does not know what disinterested devotion is, and this refers particularly to married women; in them childlessness almost always develops a crude egoism – if, that is, some impersonal interest does not incidentally rescue them. The old maid has some belated yearnings that soften her, she is still seeking and still hoping: the childless woman with a husband has reached harbour safely; at first she has grieved instinctively at having no children, then she has comforted herself and lives for her own satisfaction and, if she is not successful in that, for *her own sorrow*, or for somebody else's dissatisfaction, somebody else's sorrow, if it is only her maid's. The birth of a child may save her. A child trains its mother in sacrifice, in subordinating her will, in eagerly spending her time *not on herself*, and trains her to indifference to all external reward, recognition, gratitude. A mother does not keep an account with a baby; she requires nothing from it but to be well, to be hungry, to sleep – and to smile. Without taking the woman out of the home, the baby transforms her into a citizen.

It is quite a different thing when another woman's child comes for any reason whatever into the house of a childless woman, and especially when this is inevitable. She will dress it up, perhaps, and play with it, but only when she wants to; she will spoil it, in her own fashion: at all other times the child will knock in vain at the doors of a heart that has grown numbed or fatty from self-indulgence. In short, the child can be certain of all the indulgence and pampering which are given to a pug or a canary, but nothing more.

One of our friends had a daughter[11] whose mother was a young widow when she was born. With a view to the mother's marrying again an attempt was made to abduct the child, and she was kidnapped at a time when the father was away. After a prolonged search the little girl was found; but the father, having been expelled from France, could not come to Paris to fetch her, and besides he had not the money. Not knowing what to do with her he asked Engelson to take her to begin with. Engelson consented, but very quickly regretted it. The child was naughty – indeed, considering the irregular way in which she had been brought up, it is quite likely she probably was very naughty; but all the same her naughtiness was that of a child of five years old, and Engelson was too humane and understanding to be capable of turning against a child for naughtiness. And indeed the trouble was not that she was naughty; the child was in the way, not of him so much as of his wife, though she had never done anything. Engelson, somewhat harshly, wrote to me complaining of the child!

By the way, in regard to her father Engelson wrote to me: 'Is it not strange that Chojecki, who once agreed with you that my wife was *not a suitable person to bring up your children*, has entrusted his *own daughter to her?*'

He knew perfectly well that the father had not chosen Mme Engelson to bring up his little girl, but had been forced by financial necessity to have recourse to her assistance. There was something so harsh, so ungenerous in this remark that it sent a pang to my heart. I could not get used to this lack of mercy, this temerity of language which did not hesitate at anything! Intensely malignant allusions which may in a moment of irritation occur to anyone's mind, but which our lips refuse to utter, are spoken by people like Engelson with readiness and enjoyment at the slightest misunderstanding.

Giving vent to his irritation, Engelson in his letter incidentally attacked Tessier too and other friends, and even Proudhon, for whom he had great respect. Together with Engelson's letter came one from Tessier, who was also in Paris; he made some friendly jests about Engelson's 'tempers and tantrums', without suspecting that the latter had been writing about him. I disliked the rôle of a sort of negative traitor, and I wrote to Engelson that it was shameful to talk in that abusive way of men with whom life had brought us into friendly contact; that in spite of their failings they were still good people, as he knew himself. In conclusion I told him that it was shameful to exaggerate everything so, and to be sighing and groaning and reduced to despair over the naughtiness of a child of five.

That was enough. My fervent admirer, the friend who had kissed my hand in a burst of enthusiasm, who had come to me to share every grief and had offered to shed his blood and lay down his life for me, not in word but in deed[12] ... this man, bound to me by his own confession and by my misfortunes, of which he was the witness, by the coffin which we had followed together, forgot it all. His vanity was wounded . . . he had to avenge himself, and avenge himself he did.

Four days later I received from him the following reply:

2 February 1853

'There are rumours that you have decided to come here. Maria Kasparovna is, I believe, recovering (at least in the past week she has been in better spirits, gets up for five minutes and has an appetite). Concerning the commission you gave me for T., all I have to tell you is that the things the General[13] asks him to get ready are not at T.'s, but were left by him at Vogt's at Geneva, and that Mme T. thinks your silence *peu gracieux*, and adds that a correspondence with you could not cause them any unpleasantness.

'In short, I need not have written to you before you came if it had not occurred to me that silence may often be taken as a sign of assent. I do not wish to mislead you or keep you in error in regard to me: I do not agree with what you said in your last letter to me (of 28 January).

'These were your words: "Come now, was it worth while to part like that? – And oh, the baby – the child – and oh dear, oh dear – and good God! Just think: is this worthy of you? Surely, it's nothing new! You have seen life and know what people are. Every day I grow more indulgent and more aloof from others."

'To this I answer, without for the present going off into a dissertation on respectability in general, and without even congratulating you on your satisfaction with yourself, that of course a man is ridiculous who falls into a rage and a frenzy when he is plagued by mosquitoes or bugs, but the man is even more ridiculous who, suffering from the attacks of such insects, forces himself to assume an air of stoical indifference.

'You perhaps do not agree with this, for *you put the part you play above everything*. Don't be angry! Wait a minute! Let me finish. In the first chapter of your *Vom andern Ufer* in the Russian and German versions these are your words: "Man likes to produce an effect, to play a part, especially a tragic one; to suffer is good and noble, it presupposes unhappiness; suffering is a distraction, a comfort . . . yes, yes, it is a comfort." As I have said to you already at Nice, I was at first inclined to take this *dictum* of yours for a slip of the tongue, though not a happy one. At the time you answered that you did not remember the words.

'Though by no means applying those words exclusively to you – that is, not assuming that you judged in this case of men in general by yourself – I had thought hitherto that this *dictum* of yours, like most of the *Réflexions de La Rochefoucauld*, which it greatly resembles, like the description of the talented men of our period, once drawn in a masterly fashion by Belinsky, was "a hyperbole, a jest". And therefore, when I learnt that C. in Switzerland was indignant with the General for the way he behaved in your affair,[14] I took his indignation, not for a part that he was playing, but for real feeling, and wrote to you: "Yes, I see C. is a brother to me." When T., in the presence of a witness declared that he had been sentenced "for life plus two years", I believed this too, and even repeated it to several people. Yesterday Mme T. told me her husband had never been sentenced at all. *Ergo*, in the eyes of the persons to whom I repeated his lie I am just such a *blagueur* as he. I do not like this. Who is to blame? I am, of course, because I was "young and credulous"; but they are to blame too, because they told a lie. I have never seen in Russia or anywhere else such *blagueurs* as I have seen at Nice. In my letter to you of the 19th of January I told you that I want without *esclandre* to get away from these people; they are antipathetic to me. I wrote this to you because I want to be open with you. But *absorbed in yourself* you could not grasp this very simple idea. Otherwise you would hardly, I suppose, have given me even a most trivial commission to T. You too say that you are holding yourself aloof from people, but along with that you ask them to write to you. I cannot manage that sort of aloofness.

'Assuming that in serious matters to be frank is an essential condition of honesty, I have to tell you this, too, without loss of time. You write to me that when you have despatched the General to Australia, and given indefinite leave to everyone else, you will be left with me and with your enemies – and that if, moreover, I were a little more stable, and less dependent upon my own and other people's nervous alarms and caprices you would make *un bout de chemin* with me. To this I am obliged to reply that, feeling in myself neither a taste nor a talent for playing parts, and especially tragic ones, I am ready, if you please, to serve you with advice, but not with actions.'

Of course I had not supposed that a man who with tears and sobs had invited me to make confidences that were difficult to utter, a man who had become so intimate with me and on whom I had leaned as on a brother in moments of weakness and impotence, when my pain was beyond human endurance – that the eyewitness of all that had happened could take my misfortunes for stage trappings and scenery, which I was making use of to play a tragic part. In his ecstasies over my book he had been picking out stones in it and laying them up in his bosom to throw them at me when the chance should come. It was not enough for him to tear the present to pieces; he defiled and vulgarised the past: when he broke with me, he could not show it the respect of a melancholy silence, but covered it with merciless abuse and ironical jeering.

This letter wounded me, wounded me very much.

I answered him sadly, with suppressed tears; I said goodbye to him, and asked him to break off our correspondence.

This was followed by a complete silence between us. . . .

With Engelson once more something had snapped within me. I became even poorer, more isolated; there was coldness all about me, nothing near me. . . . At times a hand seemed held out to me more warmly; some fanatic of no understanding, not discerning at first that we were not of the same religion, would approach hurriedly, and as hurriedly turn away. Besides, I did not myself seek closer intimacy with people; I grew accustomed to men coming and going, to various anonymities of whom I asked nothing, and to whom I gave nothing except cigars, wine and sometimes money. My one salvation lay in work: I was writing *My Past and Thoughts*, and was setting up a Russian printing-press in London.

6

A year passed: the printing-press was in full swing, and was being noticed in London and feared in Russia. In the spring of 1854 I received a short manuscript from Maria Kasparovna. It was not difficult to guess that it had been written by Engelson. I published it at once.

Then came a letter from him asking that an end should be put to our unhappy misunderstanding and that we might be united for the common cause. Of course I held out both hands to him.

Instead of an answer he arrived in London himself[15] for a few days, and stayed with me. Sobbing and laughing, he begged me to forget the past, showered me with friendly words, and again seized my hand and pressed it to his lips. I embraced him, deeply touched and firmly convinced that the quarrel would not be renewed.

But only a few days later clouds appeared on the horizon that foreboded little good. The shade of fatalism, of Bonapartism, which had peeped out in his letters from Geneva, had grown. From hatred of Nicholas and the rank and file of the French Revolution of 1848, he had passed over *armes et bagages* into the enemy's camp. We argued; he was obstinate. Knowing that he always rushed to extremes and came back as quickly, I waited for the turn of the tide, but it did not come.

Unhappily Engelson was busy at that time with an amazing project with which he was passionately in love.

He had made a plan for an air battery, that is a balloon loaded with combustible materials and also with printed proclamations. This was at the beginning of the Crimean War. Engelson proposed letting off such balloons from ships against the Baltic shores. I greatly disliked this scheme; what could one make of propaganda with projectiles? Where was the sense in it for us Russians to burn Finnish villages and help Napoleon and England? Moreover Engelson had discovered no new means of steering balloons. I made little opposition to his plan, supposing he would drop this nonsense himself.

But not at all. He went off with his plan to Mazzini and Worcell. Mazzini said that things of that sort were not in his line, but that he was ready through his friends to send his plans to the Minister of War. The War Office gave an evasive reply, and put the project aside without a definite refusal. He asked me to gather together two or three of the military men among the refugees and put the question of the balloon to them. They were all against it, and I told him over and over again that I was against it, too; that our business, our strength, was in propaganda, nothing but propaganda; that we should lose in moral prestige by siding with Napoleon, and should ruin ourselves in the eyes of Russia *faisant cause commune* with her enemies. Engelson lost his temper and was beside himself. He had come to London confident of a triumph and, meeting with opposition even from me, imperceptibly returned to his hostile attitude.

Soon afterwards he went to fetch his wife, and brought her to London in May. A complete transformation had taken place in their relationship: she was pregnant, and he was delighted at the prospect of a child. Misunderstandings, quarrels, and explanations were all a thing of the past. She with a somnambulistic, half-mad mysticism was turning tables and was absorbed in spiritualism. The spirits foretold many things, and among others the imminent demise of me. He was reading Schopenhauer, and told me with a smile that he was doing all he could to indulge her mystical tendencies, and that this faith and exaltation was bringing peace and calm into her soul.

With me she behaved friendlily, perhaps in expectation of approaching death; she would come to me with her work, and make me read aloud chapters from *My Past and Thoughts* and my newest articles. When a month later differences arose again over Engelson's Bonapartism and air-balloons, *she appeared as the conciliator* – came to me begging me to spare the sick man, and assuring me that every spring Engelson was attacked by a hypochondriacal condition in which he did not know himself what he was doing.

Her placid mildness was the mildness of the conqueror, the clemency of complete triumph. Engelson, imagining that he held her under control by means of the turning tables, lost sight of one thing: that she was not only turning tables, but twisting him round her finger, and that he always gave the answers she wanted better than the tables did.

One evening Engelson again began a dispute over his balloons with a Frenchman, and said various caustic things to him; the latter got out of it by means of irony, and of course that infuriated Engelson more than ever. He snatched up his hat and ran off. In the morning I went round to have it out with him on the subject.

I found him at his writing-table, his face still completely distorted with anger, and a frenzied expression in his eyes. He told me that the Frenchman[16] (a refugee whom I had known for a long time and know still) was a spy, that he would expose him, would kill him; and he gave me a letter he had only just written to a doctor of medicine in Paris; in the letter he implicated persons living in Paris, and slandered refugees in London. I was dumbfounded.

'And do you mean to send that letter?'

'At once.'

'By post?'

'By post.'

'That's informing,' I said; and flung his scrawl on the table. 'If you send that letter....'

'Well, what?' he shouted, interrupting me in a wild, hoarse voice – 'you want to threaten me: with what? I'm not afraid of you nor of your vile friends.' With this he leapt up, opened a big knife, brandished it about and shouted, out of breath though he was: 'Come on, come on: show your mettle ... I'll teach you ... wouldn't you like to try ... you're most welcome!'

I turned to his wife, said to her, 'Has he gone quite out of his mind? You had better get him away somewhere. ...' and went away.

This time, too, Mme Engelson played the peacemaker. She came to me in the morning asking me to forget what had passed the day before. He had torn up the letter; he was ill and melancholy. She took all this for a disaster, for a physical derangement; she was afraid that he was seriously ill, and wept. I gave in to her.

After that we moved to Richmond, and Engelson came too. The birth of a son and the first months of looking after him gave Engelson new life; he was off his head with joy. The minute the baby was born he embraced and effusively kissed first the maid and then his old landlady. Anxiety over the baby's health, the novelty of paternal feeling, the novelty of the baby himself, occupied Engelson for some months, and all went well again.

All at once I got a big envelope from him, with a note asking me to read the enclosed document and tell him my opinion candidly. It was a letter to the French Minister of War. In it he again proposed air-balloons, bombs, and manifestoes. I thought it all bad, from the quarter to which he was addressing himself to the style, which was lacking in dignity, and I told him so.

Engelson answered with an insolent note and began to sulk.

After that he gave me another manuscript to publish. I did not conceal from him that its effect on Russian readers would be a very nasty one, and that I did not advise publishing it. Engelson blamed me for wanting to set up a censorship, and said that I had probably founded the printing-press exclusively to publish my own 'immortal works'. I did publish the manuscript, but my flair was justified: it aroused general indignation in Russia.

All this indicated that a new rupture was not far off. I own that this time I felt no great regret. I was sick of this intermittent fever with its paroxysms of friendship and hatred, of having my hands kissed and of moral boxes on the ear. Engelson had passed the limit beyond which neither memories nor gratitude could save the situation. I liked him less and less, and waited coolly for what was to come.

Then an event occurred so important that for a time all quarrels and dissensions were eclipsed by a single feeling of joy and expectation.

On the morning of the 4th of March I went as usual at eight o'clock into my study, opened *The Times* and read, read a dozen times without understanding, without daring to understand, the grammatical sense of the words at the head of the telegraphed news: *The death of the Emperor of Russia*.[17]

Hardly knowing what I was doing I rushed with *The Times* in my hands into the dining-room; I looked for the children and the servants to tell them the great news, and with tears of joy in my eyes I gave them the newspaper. . . . I felt as though several years had rolled off my shoulders. It was impossible to stay indoors. Engelson was at that time living in Richmond, and I hurriedly put on my coat and hat and was about to go to him, but he anticipated me and was already in the hall; we fell on each other's necks and could say nothing but: 'Well, at last he is dead!' Engelson, as his way was, capered about, kissed everyone in the house all round and sang and danced; and we had hardly recovered ourselves when a carriage suddenly stopped at my front door and someone gave a furious tug at the bell: three Poles had driven at full speed from London to Twickenham, without waiting for a train, to congratulate me.

I ordered champagne; no one reflected that all this was at eleven o'clock in the morning, or earlier. Then, quite needlessly, we all went off to London. In the streets, on the Exchange, in the eating-houses, people were talking of nothing but the death of Nicholas; I did not see one man who did not breathe more easily from knowing that that cataract had been removed from the eye of humanity, and did not rejoice that that oppressive tyrant in the jack-boots had at last returned to clay.

On the Sunday my house was full all day; French and Polish refugees, Germans, Italians, even English acquaintances kept coming and going with beaming faces. It was a clear, warm day; after dinner we went out into the garden.

Some little boys were playing on the bank of the Thames. I called them up to the railing and told them we were celebrating the death of their enemy and ours, and threw them a whole handful of small silver for beer and sweets. 'Hurray! hurray!' shouted the boys. 'Impernikel is dead! Impernikel is dead!' My visitors too began flinging them sixpences and threepenny-bits; the lads fetched ale and tarts and cakes, got hold of a barrel-organ, and began dancing. After that, as long as I lived at Twickenham, the boys used to take off their caps when they met me in the street, and shout: 'Impernikel is dead! hurray!'

The death of Nicholas multiplied our hopes and energies tenfold. I at once wrote the letter to the Emperor Alexander, afterwards published, and made up my mind to bring out *The Pole Star*.[18]

'Long live reason!' broke involuntarily from my tongue at the head of my programme. '*The Pole Star* has been hidden behind the storm-clouds of the reign of Nicholas; Nicholas has gone, and *The Pole Star* shall appear again on the day which is our Good Friday, the day on which five gibbets became for us five crucifixes.'

It was a powerful, stimulating impetus; we set to work with redoubled energy. I announced that I was bringing out *The Pole Star*; Engelson at last set to work on his article on socialism about which he had been talking in Italy. It might have been thought that we should have worked together for a couple of years or more ... but his irritable vanity made any work with him insufferable. His wife encouraged his intoxication with himself. 'My husband's article,' she used to say, 'will be considered as a new epoch in the history of Russian thought. If he writes nothing more, his place in history is assured.' The article, 'What is the State?'[19] was good, but its success did not justify the family's expectations; and besides it appeared at the wrong moment. Awakening Russia required, just at that time, practical advice, and not philosophical treatises *à la* Proudhon and Schopenhauer.

The article had not yet been published in its entirety, when a new quarrel of a different character from all the preceding ones almost definitively severed all relations between us.[20]

One day when I was with them I spoke jestingly of their having sent for the third time for a doctor for their baby, who had a cold in its head and a slight chill.

'So because we are poor,' said Mme Engelson, and all her old spiteful hatred ten times intensified flamed in her face, 'our baby is to die without medical assistance? And you say that? You, a socialist and the friend of my husband, who refuse him fifty pounds, and are exploiting him over his lessons.'

I listened in amazement, and asked Engelson whether he shared this opinion or not. He was embarrassed, his face flushed in patches, he besought her to be silent. . . . She went on. I got up and interrupted her by saying:

'You are ill and are nursing your baby. I am not going to answer you, but I am not going to listen either. . . . You will hardly think it strange that I shall not set foot in your house again.'

Engelson, grieved and distraught, caught up his hat and came out with me into the street:

'Don't take *au pied de la lettre* the unbridled language of a hysterical woman. . . .' He lost himself in a muddle of explanations. 'I shall come to give my lesson tomorrow,' he said. I shook hands with him and went home without a word.

All this calls for explanations, and most painful ones, too, relating not to opinions and public affairs but to the kitchen and account-books. Nevertheless I shall try to clear up this side of our relationship too. Squeamishness, the romanticism of cleanliness, is out of place in pathological investigations.

The Engelsons were scarcely entitled to reckon themselves poor people. They received 10,000 francs a year from Russia, and he could easily earn another 5,000 by translations, reviews, and school-books; Engelson was interested in linguistics. Trübner's, the booksellers, had commissioned from him a lexicon of Russian roots and a grammar; he could, like Pierre Leroux, like Kinkel, like Esquiros, give lessons. But, like a regular Russian, he set about everything — the dictionary, the translations, and the lessons — never finished anything, never put himself out, and never earned a farthing.

Neither husband nor wife was economical or capable of managing their affairs. The continual fever in which they lived prevented them from thinking about running the household. He had come from Russia with no definite plan, and remained in Europe without any object. He had taken no steps to secure his property, and *un beau jour*, in a fright, had made a hasty arrangement of some sort by which he limited his income to 10,000 francs, a sum which he did not receive quite punctually, but did receive.

That Engelson would not make both ends meet with his 10,000 francs was evident; that he would not know how to economise was equally clear; all that was left for him was to work or to borrow. At first, after his arrival in London, he borrowed about forty pounds from me ... some time afterwards he asked for money again. ... I had a serious, friendly talk with him about this, and told him I was ready to advance him money, but that I was determined not to lend him more than ten pounds a month. Engelson frowned. However, he did twice take a ten-pound note; then he suddenly wrote to me that he needed fifty pounds and, if I did not care to lend it him or did not trust him, he begged me to get it for him by pawning some diamonds. All this could hardly be taken seriously; if he really wanted to pawn the diamonds, he ought to have taken them to a pawnbroker and not to me.... Knowing him and being sorry for him, I wrote that I would pawn the diamonds for fifty pounds, if they would give that, and would send him the money. Next day I sent a cheque, but the diamonds, which he would certainly have sold or pawned, I put away to keep for him. He took no notice of the fact that no interest was asked for the fifty pounds, and believed that I had pawned the diamonds.

The second point, relating to the lessons, is even simpler. In London, Savich gave Russian lessons at my house, charging four shillings an hour. At Richmond, Engelson offered to take Savich's place. I asked him about terms; he answered that it was difficult for him to talk of terms with me, but that, as he had no money, he would take what I had paid Savich.

On reaching home I wrote a letter to Engelson: I reminded him that he had himself fixed the terms for the lessons, but I asked him to take double the amount for all the lessons in the past. Then I wrote what had led me to keep his diamonds, and sent them back to him.

He sent a confused answer, thanked me, expressed vexation, and came in the evening himself, and started coming as before. His wife I did not see again.

7

A month later Zeno Swentoslawski,[21] and with him Linton,[22] the English republican, were dining with me. Engelson came in towards the end of dinner.[23] Swentoslawski, the purest-hearted and kindest of men, a fanatic who at over fifty retained the reckless fire of a Pole and the vehemence of a boy of fifteen, was urging the necessity of our returning to Russia and beginning a lively propaganda there in print. He undertook to convey the type, and so on.

After listening to him, I said half in jest to Engelson:

'I say, you know, *on nous accusera de lâcheté* if he goes alone.'

Engelson made a grimace and went away.

Next day I went up to London and came back in the evening; my son, who was lying down with a feverish attack, told me in great excitement that Engelson had come in my absence, that he had abused me fearfully, had said that he would pay me out, that he was not going to put up with my authority any longer, and that he did not need me now *since his article had been published*. I did not know what to think, whether Sasha was delirious from fever or Engelson had come in dead drunk.

From Malwida von Meysenbug[24] I learnt more. She told me with horror of his fury. 'Herzen,' he had shouted in a nervous, gasping voice, 'called me *lâche* yesterday in the presence of two strangers.' Malwida interrupted him, saying that I had not been talking about him at all, that I had said *'on nous taxera de lâcheté'*, speaking of all of us generally. 'If Herzen feels that he is doing something dastardly, let him speak for himself, but I will not allow him to speak like that of me, and in the presence of two blackguards too.'

At the sound of his shouting my elder girl, then ten years old, had run in. Engelson had gone on: 'No, this is the end of it, it is enough. I am not accustomed to it, I will not allow myself to be trifled with, I will show him who I am. . . .' and he pulled a revolver out of his pocket and went on shouting, 'It is loaded, it is loaded: I shall wait for him. . . .'

Malwida got up and told him that she insisted on his leaving her, that she was not obliged to listen to his wild ravings and that she could only put down his behaviour to illness. 'I am going,' he said; 'don't make a fuss; but first I want to ask you to give Herzen this letter.' He opened it and began reading it aloud; the letter was a string of abuse.

Malwida refused the commission, asking him why he expected her to act as an intermediary in passing on such a letter.

'I will find means without your help,' observed Engelson, and went away. He did not send the letter, but a day later he wrote me a note; in it, without saying one word about what had passed, he wrote that he had an attack of haemorrhoids, that he could not come to me, and asked me to send the children to him.

I said that there was no answer, and again all diplomatic relations were broken off; hostile relations remained. Engelson did not let slip a chance of turning them to account.

From Richmond I moved in the autumn of 1855 to St John's Wood. Engelson was forgotten for some months.

Suddenly, in the spring of 1856, I received from Orsini, whom I had seen two days previously, a note which was redolent of a duel. . . .

Coldly and courteously he asked me to let him know whether it was the truth that Saffi and I were spreading a rumour that he was an Austrian spy. He asked me either to give a complete *démenti*, or to indicate from whom I had heard this abominable calumny.

Orsini was right; I should have done the same. Perhaps he ought to have had more confidence in Saffi and in me – but the offence was a great one.

Anyone who knew anything of Orsini's character would understand that such a man, attacked in the very holy of holies of his honour, could not stop short at half measures. The affair could only be settled by our *absolute* innocence or by the death of someone.

From the first minute it was clear to me that the blow came from Engelson. He no doubt reckoned on one side of Orsini's character, but fortunately there was another, which he had overlooked. Orsini combined with indomitable passions a terrific power of self-control; he was prudent among dangers, thought over every step he took, and never plunged at random, because when once he had reached a decision he wasted no time in criticism, in doubt, in reconsideration, but carried it out. We saw this in the Rue Lepelletier.[25] He acted in the same way now. He wished to investigate the matter without haste, to find out who was guilty, and then, if he could manage it, to kill him.

Engelson's second mistake lay in quite unnecessarily bringing in Saffi.

The facts were these. Six months before my rupture with Engelson I was one morning at the house of Mrs Milner-Gibson (the wife of the Minister): there I found Saffi and Pianciani, who were saying something to her about Orsini. As I went away I asked Saffi what they had been talking about. 'Think of it,' he answered: 'Mrs Milner-Gibson had been told in Geneva that Orsini had been bribed by Austria....'

On reaching home at Richmond I had passed this on to Engelson. We were then both dissatisfied with Orsini. 'The devil take him!' observed Engelson, and nothing was said on the subject.

When Orsini made his marvellous escape from Mantua we re-called in our own close circle the accusation heard by Mrs Milner-Gibson. The arrival of Orsini himself, his story, his wounded foot, entirely wiped out this absurd suspicion.

I asked Orsini to give me an appointment, and he invited me to go the following evening. In the morning I went to Saffi and showed him Orsini's note. He at once offered to go with me, as indeed I expected he would. Ogarëv, who had only just arrived in London, was a witness of this interview.

Saffi told of the conversation at Mrs Milner-Gibson's with the simplicity and straightforwardness which are his distinguishing characteristics. I filled in the rest of the story. Orsini thought a minute, and then said:

'Well, may I ask Mrs Milner-Gibson about this?'

'Of course,' answered Saffi.

'Yes, I think I have been too heated; but,' he asked me, 'tell me why did you talk about it to outsiders without warning me?'

'You forget, Orsini, the time when it happened, and that the *outsider* to whom I spoke was at that time not an outsider; you know better than most people what he was to me.'

'I have mentioned no one . . .'

'Let me finish. Why, do you suppose it is easy for a man to repeat such things? If these rumours had spread, perhaps I ought to have warned you – but who speaks about it now? As for your having mentioned no one's name, you are making a great mistake there. Bring me face to face with my accuser, then it will be still more evident what part each has played in these slanders.'

Orsini smiled, got up, came to me and embraced me, embraced

Saffi and said: 'Amici, we will put an end to this business; forgive me, let us forget all about it and talk of something else.'

'That's all very well, and you were perfectly right to ask me for an explanation; but why do you not name my accuser? In the first place, it is impossible to conceal it ... it was Engelson who told you.'

'Give me your word that you will drop the matter?'

'I give you my word, before two witnesses.'

'Well, you have guessed it.'

I had expected this confirmation, but yet it gave me pain as though I had still doubted it.

'Remember what you have promised,' Orsini added after a brief silence.

'You need not worry about that. But now comfort me and Saffi: tell us how it happened; you see, we know all that matters.'

Orsini laughed. 'What curiosity!' he said. 'You know Engelson. He came to me the other day: I was in the dining-room' (Orsini lived in a 'boarding-house'[26]) 'and having dinner alone. He had already dined, and I ordered a decanter of sherry for him; he drank it up, and at once began complaining of you – that you had offended him, that you had broken off all relations with him – and after gossiping about all sorts of things asked how you had received me on my return. I answered that you had given me a very friendly welcome, that I had dined with you, and been to see you in the evening. ... Engelson suddenly shouted: "That's just like them ... I know those gentry. It's not long since he and his friend and admirer Saffi were saying that you were an Austrian spy, but now you're famous again and in the fashion, and he is your friend!" "Engelson," I observed, "do you fully understand the gravity of what you've just said?" "Fully, fully," he repeated. "Will you be ready under any circumstances to repeat your words?" "Under any circumstances!"

'When he had gone I took a sheet of paper and wrote you a letter. That's the whole story.'

We all went out into the street. Orsini, as though guessing what was passing within me, said by way of consolation, 'He's crazy.'

Soon afterwards Orsini went to Paris, and his beautiful classical head rolled bleeding on to the platform of the guillotine.

The first we heard of Engelson after this was the news of his death in Jersey.

No word of reconciliation, no word of remorse reached me. . . . (1858.)

P.S. In 1864 I received a strange letter from Naples. It spoke of the apparition of my wife's soul, and of her having appealed to me to be converted and purify my soul through religion, and to abandon worldly cares. . . .

The writer said that it was all written at the dictation of the spirit; the tone of the letter was warm, friendly, and enthusiastic.

The letter was unsigned, but I recognised the handwriting: it was from Mme Engelson.[27]

*

THE FREE RUSSIAN PRESS
AND 'THE BELL'

CHAPTER I

THE FOGS OF LONDON

WHEN at daybreak on the 25th of August, 1852, I passed along a wet plank on to the shore of England and looked at its dirty white promontories, I was very far from imagining that years would pass before I should leave those chalk cliffs.

Entirely under the influence of the ideas with which I had left Italy, stunned and sick, bewildered by a series of blows which had followed one on the other with such brutal rapidity, I could not look clearly at what I was doing. It seemed as though I had needed to be brought again and again into physical contact with familiar truths in order that I might renew my belief in what I had long known or ought to have known.

I had been false to my own logic and forgotten how different the man of today is in opinions and in actions, how noisily he begins and how modestly he carries out his programmes, how genial are his desires and how feeble his muscles.

Two months had been filled with unnecessary meetings, fruitless seeking, painful and quite useless conversations, and I was still expecting something . . . expecting something. But my real nature could not remain for long in that world of phantoms. I began little by little to grasp that the edifice I was raising had no solid ground beneath it, and that it would inevitably crumble into ruins.

I was humiliated, my pride was outraged and I was angry with myself. My conscience gnawed at me for the sacrilegious deterioration of my grief, for a year of vain anxiety; and I was aware of a fearful, inexpressible weariness. . . . How I needed then the breast of a friend who, without judging and condemning, would have received my confession and shared my unhappiness; but the desert about me extended more and more; there was no one near to me, not one human being . . . and perhaps that was even for the best.

I had not thought of staying longer than a month in London, but little by little I began to perceive that I had absolutely nowhere to go and no reason to go anywhere. Nowhere could I have found the same hermit-like seclusion as in London.

Having made up my mind to remain there, I began by taking a

house in one of the remotest parts of the town, beyond Regent's Park, near Primrose Hill.

The little girls remained in Paris; only Sasha was with me. As the fashion is here, the house was divided into three storeys. The whole middle storey consisted of a huge, cold, uncomfortable 'drawing-room'. I turned it into a study. The owner of the house was a sculptor and had cluttered up the whole of this room with various statuettes and models; a bust of Lola Montes was always before my eyes, together with Victoria.

When on the second or third day after our crossing, having unpacked and settled in, I went into that room in the morning, sat down in a big arm-chair and spent a couple of hours in complete stillness, worried by no one, I felt myself somehow free, for the first time after a long, long time. My heart was not the lighter for this freedom, but yet I looked out of the window with a greeting to the sombre trees in the park, which were hardly visible through the smoky fog, and thanked them for the peacefulness.

For whole mornings I used now to sit utterly alone, often doing nothing, not even reading; Sasha would sometimes run in, but he did not interfere with my solitude. Haug, who lived with me, never came in – without some pressing need – before dinner which was between six and seven. In this leisure I went, fact by fact, over the whole past, words and letters, other people and myself. I found mistakes to the right, mistakes to the left, vacillation, weakness, action hindered by irresolution and over-readiness to be influenced by others. And in the course of this analysis, by degrees, a revolution took place within me ... there were bitter moments and more than once tears rolled down my cheeks; but there were other moments, not of gladness but of courage: I was conscious of power in myself. I no longer relied on anyone else, but my confidence in myself grew stronger; I grew more independent of everyone.

The emptiness about me strengthened me and gave me time to collect myself; I grew unaccustomed to others: that is, I did not seek real intimacy with them: I avoided no one, but people became indifferent to me. I saw that I had no ties that rested on earnest, profound feelings. I was a stranger among outsiders; I had more sympathy for some than for others, but was in no close intimacy with any. It had been so in the past, too, but I had not noticed it, being continually carried away by my own thoughts; now the masquerade was over, the dominoes had been removed, the garlands had fallen from the heads, the masks from the faces, and I saw features different from those that I had surmised. What was I to do? I could help showing that I liked many people less, that is, I knew them better, but I could not help feeling it; and, as I have said, these discoveries did not rob me of my courage, but rather strengthened it.

London life was very favourable for such a break. There is no town in the world which is more adapted for training one away from people and training one into solitude than London. The manner of life, the distances, the climate, the very multitude of the population in which personality vanishes, all this together with the absence of Continental diversions conduces to the same effect. One who knows how to live alone has nothing to fear from the tedium of London. The life here, like the air here, is bad for the weak, for the frail, for one who seeks a prop outside himself, for one who seeks welcome, sympathy, attention; the moral lungs here must be as strong as the physical lungs, whose task it is to separate oxygen from the smoky fog. The masses are saved by battling for their daily bread, the commercial classes by their absorption in heaping up wealth, and all by the bustle of business; but nervous and romantic temperaments, fond of living among people, fond of intellectual sloth and of idly luxuriating in emotion, are bored to death here and fall into despair.

Wandering lonely about London, through its stony lanes and stifling passages, sometimes not seeing a step before me for the thick, opaline fog, and colliding with shadows running – I lived through a great deal.

In the evening, when my son had gone to bed, I usually went out for a walk; I scarcely ever went to see anyone; I read the newspapers and stared in taverns at the alien race, and lingered on the bridges across the Thames.

On one side the stalactites of the Houses of Parliament would loom through the darkness, ready to vanish again; on the other, the inverted bowl of St Paul's ... and street-lamps ... street-lamps ... street-lamps without end in both directions. One city, full-fed, went to sleep: the other, hungry, was not yet awake – the streets were empty and nothing could be heard but the measured tread of the policeman with his lantern. I used to sit and look, and my soul would grow quieter and more peaceful. And so for all this I came to love this fearful ant-heap, where every night a hundred thousand men know not where they will lay their heads, and the police often find women and children dead of hunger beside hotels where one cannot dine for less than two pounds.

But this kind of transition, however quickly it approaches, is not achieved all at once, especially at forty. A long time passed while I was coming to terms with my new ideas. Though I had made up my mind to work, for a long time I did nothing, or did not do what I wanted to do.

The idea with which I had come to London, to seek the tribunal of my own people, was a sound and right one. I repeat this even now, with full, considered conviction. To whom, in fact, are we to appeal for judgment, for the re-establishment of the truth, for the unmasking of falsehood?

It is not for us to litigate in the court of our enemies, who judge by other principles, by laws which we do not recognise.

One can settle one's quarrels for oneself; no doubt one can. To take the law in one's own hands is to snatch back by force what has been taken by force, and so restore the balance; vengeance is just as sound and simple a human feeling as gratitude; but neither revenge nor taking the law into one's own hands explains anything. It may happen that a clear explanation is what matters most to a man. The re-establishment of the truth may be dearer to him than revenge. My own error lay not in the main proposition but in the underlying assumption; in order that there may be a tribunal of one's own people one must first of all have one's own people. Where were mine . . . ?

I had had my own people once in Russia. But I was so completely cut off in a foreign land; I had at all costs to get into communication with my own people; I wanted to tell them of the weight that lay on my heart. Letters were not allowed in, but books would get through of themselves; writing letters was impossible: I would print; and little by little I set to work upon *My Past and Thoughts*, and upon setting up a Russian printing-press.

*

APOGEE AND PERIGEE[1]

1858–1862

1

ABOUT ten o'clock one morning I heard from downstairs a thick, discontented voice:

'*May dee comsa – colonel rioos ver vwar.*'

'*Monsieur ne reçoit jamais le matin et. . . .*'

'*Zhe par deman.*'

'*Et votre nom, monsieur?*'

'*May voo diray colonel rioos*' – and the colonel raised his voice.

Jules was in a very difficult situation. I went to the top of the stairs, and asked:

'*Qu'est-ce qu'il y a?*'

'*Say voo?*' asked the colonel.

'*Oui, c'est moi.*'

'Give orders for me to be admitted, my dear sir. Your man-servant won't let me in.'

'Be good enough to come up.'

The colonel's somewhat testy face became visible and, as he stepped with me into my study, he suddenly assumed an air of some dignity and said:

'I am Colonel So-and-so: I am passing through London and thought it my duty to call.'

I at once felt myself to be a general: I pointed to a chair and added:

'Sit down.'

The colonel sat down.

'Are you here for long?'

'Till tomorrow, sir.'

'Have you been here for a long time?'

'Three days, sir.'

'Why are you staying for such a short time?'

'You see, without speaking the language it's strange here, like being in a forest. I sincerely wanted to see you in person, to thank you from myself and many of my comrades. Your publications are very useful; there's a lot of truth in them, and sometimes they make us split our sides.'

'I'm extremely grateful to you; this is the only acknowledgment we've received abroad. Are many of our issues received in Russia?'

'A great many, sir. And think how many people read each page: they read and re-read them till they're in holes, in rags; there are devotees who even make copies of them. We meet sometimes to read them, and criticise: you know? I hope you will permit the frankness of a military man who has a sincere respect for you?'

'By all means. It hardly becomes *us* to oppose freedom of speech.'

'We often speak so among ourselves: there's much profit in your disclosures. You know yourself how much one can say over there about Sukhozanet,[2] for instance: keep your tongue between your teeth, eh?; or about Adlerberg,[3] let's say? But, you see, you left Russia a long time ago: you've forgotten *too much* about it, and we keep thinking you harp too much on the peasant question . . . it's not ripe yet . . .'

'Isn't it?'

'Yes, indeed, sir . . . I agree with you entirely; good gracious: the same soul, form, image of God . . . and all that, believe me, is seen by many people nowadays, but there mustn't be any hurry *prematurely.*'

'You think not?'

'I'm sure, sir. Our peasant is a fearful slacker, you know. He's a good chap, perhaps, but a drunkard and a slacker. Emancipate him at once, and he'll stop working, won't sow the fields and will simply die of hunger.'

'But why should you worry about that? Nobody has entrusted the feeding of the Russian people to you, Colonel, have they? . . .'

Of all possible and impossible rejoinders, this was the one that the colonel expected least.

'Of course, sir, on the one hand. . . .'

'Well, don't you be afraid about *on the other hand*; he won't really die of hunger, will he, because he will have sown wheat not for his master but for himself?'

'Excuse me: I thought it was my duty to say. . . . Besides, it seems to me I'm taking up too much of your valuable time. . . . Allow me to take my leave.'

'I thank you most humbly for calling.'

'Pray don't trouble. .... *Oo ay mon kab?* You live a good way out, sir.'

'It's not close.'

I wanted, with this splendid scene, to begin the description of the period of our bloom and prosperity. Such scenes and similar ones were continually repeated. Neither the fearful distance at which I lived from the West End – at Putney or Fulham – nor the door that was permanently shut in the mornings – nothing helped. We were the fashion.

Whom indeed did we not see at that time? How many people would now pay dearly to wipe out their visit from the memory, if not of themselves, then of humanity? But then, I repeat, *we were the fashion*, and in a tourist's guide-book I was mentioned as one of the curiosities of Putney.

So it was from 1857 to 1863, but it had not been so before. In proportion as reaction extended and strengthened itself in Europe after 1848, and Nicholas grew more savage not by the day but by the hour, Russians began to be rather frightened and to avoid me. Besides, it became known in 1851 that I had officially refused to go to Russia. At that time there were very few travellers. At long intervals one of my old acquaintances would appear, recount frightful, inconceivable things, speak with dread of his return and disappear, looking round to make sure there was no fellow-Russian there. When I was visited at Nice by A. I. Saburov,[4] in a carriage with a body-servant, I looked on it as a feat of heroism. When I passed secretly through France in 1852 I met some of the Russians in Paris: these were the last. In London there was nobody.* Weeks, months went by. ....

No Russian sound, nor Russian face.[5]

No one wrote me any letters, M. S. Shchepkin was the first who was anything like a friend *from home* that I saw in London. I have told the story of our meeting in another place. His arrival for me was like an All Souls' Day. He and I held a general commemoration of the Muscovite dead, and our very mood was somehow sepulchral. The real dove from the ark with the olive leaf in its mouth was not Shchepkin but Dr Vensky.[6]

* I am not speaking, of course, of two or three emigrants.

He was the first Russian who came to see us, after the death of Nicholas, at Cholmondely Lodge, Richmond, and was perpetually amazed that it should be so spelt, but pronounced Chumly Lodge. The news that Shchepkin brought was gloomy; he was in a mournful state of mind himself. Vensky used to laugh from morning till night, showing his white teeth; his news was full of the hope, the sanguineness, as the English say, that possessed Russia after the death of Nicholas and made a luminous band against the sullen background of Petersburg imperialism. True, he did bring a bad account of the health of Granovsky and Ogarëv, but even this disappeared in the glowing picture of an awakening society, of which he himself was a specimen.

How avidly I listened to his stories, cross-questioned him and ferreted out details. I do not know whether he knew then or appreciated afterwards the immeasurable good he did me.

Three years of life in London had fatigued me. It is a laborious business to work without seeing the fruit from close at hand; and as well as this I was too much cut off from any circle of my kin. Printing sheet after sheet with Chernetski[7] and piling up heaps of printed pamphlets in Trübner's[8] cellars, I had hardly any opportunity to send anything across the frontier of Russia. I could not give up: the Russian printing-press was my life's work, the plank from the paternal home that the ancient Germans used to take with them when they moved; with it I lived in the atmosphere of Russia; with it I was prepared and armed. But with all that, it wore one out that one's work was never heard of: one's hands sank to one's sides. Faith dwindled by the minute and sought after a sign, and not only was there no sign: there was not *one single* word of sympathy from home.

With the Crimean War, with the death of Nicholas, a new time came on; out of the continuous gloom there emerged new masses, new horizons; some movement could be sensed: it was hard to see well from a distance – there had to be an eye-witness. One appeared in the person of Vensky, who confirmed that these horizons were no mirage, but reality, that the boat had moved, was under way. One had only to look at his glowing face to believe him. There had been no such faces at all in recent times in Russia.

Overwhelmed by a feeling so unusual for a Russian, I called to mind Kant taking off his velvet cap at the news of the proclama-͜on of the republic in 1792 and repeating, 'Now lettest Thou Thy servant depart.' Yes, it is good to fall asleep at dawn after a long night of bad weather, fully believing that a marvellous day is coming!

So Granovsky died.[9]

Indeed, the *morning* was drawing near of the day for which I had been yearning since I was thirteen – a boy in a camlet jacket sitting with just such another 'malefactor' (only a year younger) in a little room in the 'old house';[10] in the lecture-room at the university, surrounded by an eager, lively brotherhood; in prison and exile; in a foreign land, making my way through the havoc of revolution and reaction; at the summit of domestic happiness, and shattered, lost on the shores of England with my printed *monologue*. The sun which had set, lighting up Moscow below the Sparrow Hills[11] and carrying with it a boyish vow ... was rising after a twenty-year-long night.

What was the use now of rest and sleep? ... To work! And to work I set myself with redoubled energy. The work no longer went for nothing, no longer vanished in a dark expanse: loud applause and burning sympathy were borne to us from Russia. *The Pole Star* was bought up like hot cakes. The Russian ear, unused to free speech, became reconciled to it, and looked eagerly for its masculine solidity, its fearless frankness.

Ogarëv arrived in the spring of 1856 and a year later (1 July 1857) the first sheet of the *Kolokol* (*Bell*) came out. Without a fairly close periodicity there is no real bond between a publication and its readership. A book remains, a magazine disappears; but the book remains in the library and the magazine disappears in the reader's brain and is so appropriated by him through repetition that it seems his very own thought; and, if the reader begins to forget this thought, a new issue of the magazine, never fearing to be repetitious, will prompt and revive it.

In fact, for one year the influence of *The Bell* far outgrew *The Pole Star*. *The Bell* was accepted in Russia as an answer to the demand for a magazine not mutilated by the censorship. We were fervently greeted by the young generation; there were letters at which tears started to one's eyes ... But it was not only the young generation that supported us ...

'*The Bell* is an authority,' I was told in London[12] by, *horribile dictu*, Katkov, and he added that it lay on the table at Rostovtsev's[13] to be referred to about the peasant question. ... And before him the same thing had been repeated by Turgenev, Aksakov, Samarin, and Kavelin,[14] by generals who were liberals, liberals who were counsellors of state, ladies of the court with a thirst for progress and aides-de-camp of literature; V. P. Botkin himself, constant as a sunflower in his inclination towards any manifestation of power, looked with tenderness on *The Bell* as though it had been stuffed with truffles. All that was wanting for a complete triumph was a sincere enemy. We were before the *Vehmgericht*,[15] and we had not long to wait for him. The year 1858 was not yet over when there appeared the accusatory letter of Chicherin.[16] With the haughty frigidity of an unbending doctrinaire, with the *roideur* of an incorruptible judge he summoned me to a reply and, like Biron, poured a bucket of cold water on my head in the month of December.[17] The behaviour of this Saint-Just of bureaucracy astonished me; but now, after seven years,[18] Chicherin's letter seems to me the flower of politeness after the strong language and strong patriotism of the *Mikhaylovsky* time.[19] Yes, and the temper of society was different in those days; Chicherin's 'indictment' provoked an explosion of indignation and we had to try to calm down our exasperated friends. We received letters, articles, protests by dozens. To the accuser himself his former friends wrote letters singly and collectively, full of reproaches, one of them being signed by common friends of ours (three-quarters of them now are more friendly with Chicherin than with me); with the chivalry of bygone times he sent on this letter himself to be kept in our arsenal.

At the palace *The Bell* had received its rights of citizenship even earlier. Its articles led the Emperor to give orders for a review of the affair of 'Kochubey[20] the marksman' who winged his steward. The Empress wept over a letter[21] to her about the up-bringing of her children; and it is said that Butkov, the bold Secretary of State, repeated in a fit of arrogant self-sufficiency that he was afraid of nothing, 'Complain to the Tsar, do what you like, write to *The Bell*, if you must, it's all the same to me.' An officer passed over for promotion seriously asked us to print the fact, with a particular hint to the Emperor. The story of Shchepkin and Gedeonov I have told in another place[22] – I could tell dozens of such stories.

... Gorchakov[23] pointed with amazement to the account printed in *The Bell* of the secret session of the Council of State[24] to consider the peasant business. 'Now who,' he said, 'can have told him the details so accurately, except one of those present?'

The Council was disquieted and there was a secret conversation once between 'Butkov and the Tsar' about how to muzzle *The Bell*. The *unmercenary* Muravëv[25] advised that I should be bought off; Panin, the giraffe with the ribbon of St Andrew, preferred that I should be inveigled into the Civil Service. Gorchakov, who played between these two 'dead souls' the part of Mizhuyev,[26] had doubts about my venality and asked Panin:[27]

'What position shall you offer him?'

'Assistant Secretary of State.'

'Well, he won't accept an assistant secretaryship of state,' answered Gorchakov, and the fate of *The Bell* was left to the will of God.

But the will of God evinced itself plainly in the flood of letters and correspondence from all parts of Russia. Each one wrote whatever came into his head: one to blow off steam, another to convince himself that he was a dangerous fellow ... but there were letters written in a burst of indignation, passionate cries that revealed the everyday abominations. Letters like this compensated for dozens of 'exercises', just as one visit made up for any number of *colonels rioos*.

Altogether the *bulk* of the letters could be divided into letters with no facts in them but with an abundance of heart and eloquence, letters with magisterial approval or magisterial rebukes, and finally letters with important communications from the provinces.

The important communications, usually written in an elegant court-hand, almost always contained an even more elegant preamble full of lofty sentiments and irresistible flattery. 'You have inaugurated a new era of the Russian tongue and, so to speak, thought; you first, from the eminence of your tribunal in London, have set about publicly stigmatising the people who tyrannise over our good folk, for our folk *are* good; it is not without reason that you love them. You know not how the heart beats with love and gratitude towards you from the far distance of our fatherland. ...'

From sultry Colchis to the northern ice ...

... 'the unassuming river Oka, Klyazma or Such-and-such a Province ... We look upon you as our sole defender. Who but you can expose a monster which through its name and position stands above the law – a monster like the president of our (exchequer, criminal, Crown Lands') court (name, patronymic, surname, rank)? A man of no education, who has crawled to honours from the lowly sphere of a clerkship in chancery, he has retained all the boorishness of a former pettifogger and is by no means averse to gratitude signed by Prince Khovansky[28] (as our elders say). The boorishness of this satrap is well known in all the circumjacent provinces; the officials avoid the exchequer court like a place accursed; he is insolent not only *with us* but also with head clerks. He has left his wife and keeps a kept woman to the general scandal, a widow (name, patronymic, surname, rank of deceased husband); we have nicknamed her the Mina Ivanovna of our province,[29] because everything at the court is done by her. May the resonant voice of *The Bell* rouse and frighten this pasha in the midst of his orgies, in the criminal embraces of his forty-year-old Herodias. If you print anything about him we are prepared to furnish you with abundant information. We have plenty of "pigs in skull-caps" as the immortal author of that work of genius, *The Government Inspector*, put it.[30]

'P.S. Do not forget to mention, with the inimitable graver with which you etch your caustic satire, that on 6 December a lieutenant-colonel of the Home Guard, at the Marshal of the Nobility's ball, to which he had come, somewhat heated, from the Mayor's was so intoxicated by the end of supper that, in the presence of ladies of birth and their daughters, he began to make use of expressions more natural to a public bath or the market-place than to the *salon* of the Marshal of the best educated class of society.'

Along with letters telling the secrets of the behaviour of a president and the president's wife, and the open drunkenness of a lieutenant-colonel, there arrived letters that were quite poetical, disinterested, and absurd. Many of these I destroyed or distributed to my friends, but I have kept some, and I shall certainly share them with my readers.

One of the best was (apparently) from a young officer who had very recently begun to live as his own master: it began with commonplaces and with the words, 'Dear Sir,' very modestly and flatteringly. . . . Little by little his pulse quickened. . . . Advice followed, and admonishment; the heat increased. On the fourth page (of folio) our friendship had gone so far that the unknown called me 'my dear' and *'mon cher'*. 'I write to *thee*,' the gallant officer concluded, 'so frankly because I love thee with all my heart.' As I read the letter I could almost see the young man sitting down after supper with the letter and a bottle of something pretty strong. . . . As the bottle empties his heart fills, our friendship grows and, with the last swallow, the good-natured officer loves me and sets about correcting my faults, loves me and wants to kiss me. . . . Oh officer, officer, only wipe your mouth and I shall have nothing against our rapid friendship *in contumaciam*!

By the way, speaking of officers I should say that the most like-able and rational people of those who came to see us were *officers*. Young civilians were for the most part confused, unsure of themselves, too much swallowed up by the affairs of their own literary circles and unable to get clear of them. The military men were more unassuming and simpler; they felt that they had behind them the inadequate education of the Corps of Cadets and, as if conscious of their own bad reputation, forced themselves forward and tried to learn something for themselves. In essence they were not at all worse trained than the others and, by the great law of moral reactions, under the despotic weight of the Corps they had nourished in themselves a strong love of independence. After the Crimean War an earnest movement began in the world of the officer: it is exhibited equally by those who were executed, like Slivitsky, Arngoldt . . . those who were killed, like Potebnya, and those who were sent into exile with forced labour like Krasovsky, Obruchev and others.[31]

Of course, very many have turned about since those days and have come to their senses and under the articles of war – all that is quite a usual affair.

A *propos* renegades: one young enthusiast, an officer, who was visiting me once with the noble, stainless Serakowski[32] and two other companions, as he was saying goodbye, drew me into the garden and said, as he embraced me warmly:

'If you ever need, for any purpose, a man who is absolutely devoted to you, remember me.'

'Preserve yourself, and in your breast those feelings with which you are filled, and may you never be in the ranks of those who march against the people.'

He drew himself up. 'That's impossible! . . . but if you should ever hear about me anything like that have no mercy on me; write to me: write to me frankly and remind me of this evening.'

Serakowski was strung up on the gallows, wounded as he was; some of the young men who were in London at the time resigned their commissions and went their ways. . . . One name I kept coming across, only because of its owner's *promotions* – the name of my enthusiast. Not long ago he met an old acquaintance at a watering-place. He cursed Poland, praised the government and, seeing that the conversation was not going well, the General recollected himself and he said:

'I think you can't have forgotten our foolish fantasies in London. D'you remember our talks at Alpha Road? What childishness, and what madness!'

I did not write to him – for obvious reasons.

. . .

3

Again it was striking ten o'clock in the morning, and again I heard the voice of a stranger, not a military voice this time, thick and stern, but a woman's, irritable, upset and sounding like tears:

'I must, I absolutely must see him. . . . I shan't go away till I have.'

And after that there came in a young Russian girl, or young lady, whom I had seen twice before.

She stopped in front of me and looked me steadily in the eyes: her features were sad, her cheeks on fire; she hastily excused herself, and then:

'I have only just come back from Russia, from Moscow; friends of yours, people who are fond of you, have commissioned me to tell you . . . to ask you . . .' Her voice failed her and she stopped.

I understood none of this.

'Can it be true that you – you that we were so passionately fond of – *you* . . .?'

'But what is the matter?'

'Tell me, for God's sake, *yes or no* – did you have anything to do with the Petersburg fire?'[33]

'I?'

'Yes, yes, yes! They're accusing you . . . at any rate, they're saying you knew about the wicked scheme.'

'What madness! Can you take it seriously, this accusation?'

'Everyone's saying it!'

'Who's "everyone"? Some Nikolay Filippovich Pavlov?'[34] (my imagination did not go any further at that time!)

'No: people you know well, people who love you dearly; you must clear yourself for their sakes; they're suffering, they're waiting...'

'And do you believe it yourself?'

'*I don't know*. That's why I came, because I don't know: I expect you to explain....'

'Let's begin by you calming yourself, and sitting down and listening to me. If I had secretly participated in this incendiarism, what makes you think that I should tell you so – like that, the first time I'm asked? You've no reason, no basis for believing me. You'd do better to say where in all that I've written there's anything, one single word, that could justify such an absurd accusation. We are not madmen, you know, to try to commend ourselves to the people of Russia by setting fire to the Rag Market.'

'Why do you keep silent? Why don't you clear yourself publicly?' she asked, and in her eyes there was irresolution and doubt. 'Brand these wicked men in print, say you're horrified by them, that you're not with them, or....'

'Or what? Now, that's enough,' I said to her with a smile, 'of playing Charlotte Corday; you've no dagger and I'm not sitting in my bath. It's shameful of you, and twice as shameful of my friends, to believe such rot; but it would be shameful for us to try to clear ourselves of it, all the more if we tried to do so by way of trampling on and doing great harm to people quite unknown to us, who now are in the hands of the secret police and who very likely had as much to do with the fires as you and I.'

'So you're determined not to clear yourself?'

'No, I won't.'

'Then what shall I write to them?'

'Write what you and I have been saying.'

She took the latest issue of *The Bell* out of her pocket and read out: 'What fiery cup of suffering is passing us by? Is it the fire of senseless destruction, or punishment that purifies by flame? What has driven people to this, and what are these people? What painful moments are they for the absent one when gazing where all his love lies, all that a man lives by, he sees only the dull glow of a conflagration.'

'Dark, frightening lines, that say nothing against you and *nothing for you*. Believe me: clear yourself – or remember my words: *Your friends and supporters will abandon you*.'

Just as the *colonel rioos* had been the drum-major of our success, so the unmurderous Charlotte Corday was the prophetess of our collapse in public opinion – on both sides, too. At the same time as the reactionaries lifted their heads and called us monsters and incendiaries, some of the young people bade us farewell, as though we had fallen by the wayside. The former we despised, the latter we pitied, and we waited sadly for the rough waves of life to destroy those who had made too far out to sea, for we knew that only some of them would get back and make fast to the shore.

The slander grew and was quickly caught up by the press and spread over the whole of Russia. It was only then that the denunciatory era of our journalism began. I remember vividly the amazement of people who were simple and honourable, not in the least revolutionaries, before the printed denunciations – it was something quite new to them. The literature of disclosures quickly shifted its weapon and was twisted at once into a literature of police perquisitions and calumniation by informers.

There was a revolution in society itself. Some were sobered by the emancipation of the peasants; others were simply tired by political agitation; they wished for the former repose; they were satiated before a meal which had cost them so much trouble.

It cannot be denied: our breath is short and our endurance is long!

Seven years of liberalism had exhausted the whole reserve of radical aspirations. All that had been amassed and compressed in the mind since 1825 was expended in raptures of joy, in the foretaste of the good things to come. After the truncated emancipation of the peasants people with weak nerves thought that Russia had gone too far, was going too quickly.

At the same time the *radical* party, young, and for that very reason full of theories, began to announce its intentions more and more impulsively, frightening a society that was already frightened even before this. It set forth as its ostensible aim such extreme outcomes, that liberals and the champions of gradual progress crossed themselves and spat, and ran away stopping their ears, to hide under the old, filthy but familiar blanket of the police. The headlong haste of the students and the landowners' want of practice in listening to other people could not help bringing them to blows.

The force of public opinion, hardly called to life, manifested itself as a savage conservatism. It declared its participation in public affairs by elbowing the government into the debauchery of terror and persecution.

Our position became more and more difficult. We could not stand up for the filth of reaction, but our *locus standi* outside it was lost. Like the knights-errant in the stories who have lost their way, we were hesitating at a cross-roads. Go to the right, and you will lose your horse, but you will be safe yourself; go to the left, and your horse will be safe but you will perish; go forward, and everyone will abandon you; go back – that was impossible: for us the road in that direction was overgrown with grass. If only a sorcerer or hermit would appear and relieve us of the burden of irresolution. . . .

. . .

CHAPTER III

BEHIND THE SCENES
1863–1864

WE were left alone, without faith, listening to the far-away roll of shooting, the far-away groan of the wounded. Early in April the news came that Potebnya had been killed in battle at Peskov Rock. In May Padlewski[1] was shot at Plotsk, and so it went on and on.

It was a hard time, unbearably hard. And, to add to all the gloom, one was the involuntary spectator of the stupidity, the senselessness of men, the cursed recklessness destroying every force about one.

V. I. KELSIEV

The name of V. Kelsiev[2] has gained a mournful notoriety of late: the haste of his inward and the speed of his outward transformation, the success of his repentance, the urgent need for a public confession[3] and its strange truncation, the clumsiness of his story, its misplaced jocosity together with the easy levity so unseemly in the penitent and forgiven – all this, among people so unaccustomed as we are to abrupt and public conversions, set the better part of our journalists in arms against him. Kelsiev wanted at all costs to occupy the public attention; he bought a conspicuous place for himself as a target at which one flings a stone without sparing. I am far from condemning the intolerance displayed in this case by our drowsy literature. This indignation proves that among us many luminous, uncorrupted powers have remained whole, in spite of the black spell of moral confusion and immoral talk. The indignation poured upon Kelsiev was the same as that which once was unsparing of Pushkin[4] for one or two poems and turned its back upon Gogol for his *Correspondence with Friends*.

To cast a stone at Kelsiev is superfluous; a whole roadway has been thrown at him already. I want to tell others and to remind him what he was like when he came to us in London, and what he was like when he went to Turkey for the second time.

· · ·

In 1859 I received the first letter from him.

The letter came from Plymouth. Kelsiev had arrived there on the steamer of a North American company, and was leaving for somewhere, Sitka or Unalaska,[5] to work. After spending a little time in Plymouth he gave up the idea of going to the Aleutian Islands, and wrote to me asking whether he could gain a livelihood in London. He had already succeeded in making the acquaintance of some theologians or other in Plymouth, and told me that they had called his attention to remarkable interpretations of prophecies. I warned him off English clergymen, and invited him to London 'if he really wanted to work'. A fortnight later he made his appearance. A rather tall, thin, sickly looking young man with a rectangular skull and a thick crop of hair on his head, he reminded me – not by his hair (for the other was bald) but in his whole being – of Engelson, and he really did resemble him in many ways. At the first glance one could discern in him much that was inharmonious and unstable, but nothing that was vulgar. It was evident that he had escaped to freedom from every form of tutelage and serfdom, but had not yet enrolled himself in the service of any cause or party: he belonged to no guild. He was much younger than Engelson, but yet he did belong to the latest ranks of the Petrashevsky group,[6] and had some of their virtues and all their defects, had studied everything in the world and learnt nothing thoroughly, read everything of every sort, and worried his brains over it all fruitlessly enough. Through continual criticism of every accepted idea, Kelsiev had shaken all his moral conceptions without acquiring any clue to conduct.

What was particularly original about Kelsiev was that in all his sceptical groping there remained an admixture of mystical fancies: he was a Nihilist with the ways and manners of the religious, a Nihilist in the surplice of a deacon. The flavour of the Church, its manner of speech and imagery, were retained in his deportment, his language, his style, and gave his whole life a peculiar character, a peculiar unity, based on a welding together of antithetical metals.

Kelsiev was passing through that stage of revaluation so familiar to us which almost every truly awakened Russian accomplishes within himself, and of which the Western European through practical preoccupations and lack of leisure never thinks at all, drawn as he is by his specialised knowledge into other concerns. Our *elder brothers* never verify traditional beliefs, and that is how it is that their generations succeed each other, building and destroying, rewarding and punishing, investing people with crowns and fetters, always firmly convinced that this is how it should be done and that they are doing their job. Kelsiev, on the contrary, doubted everything and refused to take anyone's word for it that good was good or that evil was evil. This sullen, stubborn spirit that disavows all previous morality and accepted truths was boiled up more than ever in the *mi-carême* of our Lent under Nicholas, and found striking expression as soon as the weight that pressed upon our brains was lifted by one-tenth of an inch. This analysis, so full of life and hardiness, was fiercely attacked by the conservative literary movement – conserving God knows what – and after it by the government.

At the time of our awakening in the din of the Sevastopol cannon, many of our clever fellows began to repeat that Western European conservatism was a thing that had been grafted on to us, that we had been hurriedly thrust into European culture, not that we might share their hereditary diseases and out-of-date prejudices, but that we might compare ourselves with our elder brothers, and that we might have the opportunity of advancing in step with them. But as soon as in actual fact we see that in awakened thought and in mature speech there is no firm principle, 'nothing sacred', nothing but questions and problems, that thought is seeking, that speech is denying, that the 'known' good is tottering together with what is bad, and that the spirit of doubt and experiment is dragging everything indiscriminately into a precipice with no hand-rail – then a cry of horror and frenzy bursts from the breast, and the first-class passengers close their eyes in order not to see, when the train dashes off the rails and the guards try to put on the brakes and stop all movement.

Of course there was no reason to be afraid: the rising force was physically too weak to derail a train carrying sixty million passengers. But it had a programme, perhaps a prophecy.

Kelsiev had developed under the first influence of the period of which we are speaking. He was far from becoming stabilised or reaching any centre of gravity; his moral property was in complete liquidation. All that was old he denied, he let the solid ground slip from him, he shoved off from the shore and set off headlong for the open sea; with equal suspicion and mistrust he regarded belief and disbelief, Russian methods and the methods of Western Europe. The one thing that had sent deep roots into his heart was a passionate and profound recognition of the economic iniquity of the present political order, a hatred of it and an obscure yearning for the social theories in which he saw a way out.

To this consciousness of iniquity and to this hatred of it he had, over and above his understanding of it, an inalienable right.

In London he settled in one of the remotest parts of the town, in an alley in Fulham,[7] inhabited by Irishmen and emaciated workmen of all sorts, with no shine on them, covered with what looked like a layer of ashes. In these damp, stony, unroofed corridors, it is fearfully quiet; there is almost no sound nor light nor colour: people, clothes, and houses all are faded and shrunk. Every line is blackened by smoke and soot till they look like the edges of a mourning envelope. No tradesmen's carts rattle down them with victuals, no cabmen drive that way, no hawkers cry their wares, no dogs bark (there is absolutely nothing for these to feed on); only from time to time a thin, dishevelled, smutty cat emerges, clambers on along a roof and goes up to a chimney to get warm, arching its spine and revealing by its appearance how chilled it has been indoors.

The first time I visited Kelsiev he was not at home. A very young, very plain woman – thin, lymphatic, with tear-stained eyes, was sitting on the floor by a mattress, on which a baby of a year or a year and a half was tossing in a high fever, suffering and dying.

I looked at its face, and remembered the face of another baby on the point of death: it was the same expression. A few days later it died, and another was born.

No poverty could have been more complete. The frail young woman, or rather married child, endured it heroically and with unusual simplicity.

No one looking at her sickly, scrofulous, feeble exterior could have imagined what energy, what force of devotion, resided in that puny body. She might have served as a bitter lesson for our inveterate romanticists. She was, or rather wanted to be, what was afterwards called a Nihilist: she did her hair queerly, was careless in her dress, smoked a great deal, and was not afraid either of bold thoughts or bold words; she was not enthusiastic over the domestic virtues, did not talk of the sacredness of duty and the sweetness of the sacrifice she made daily, or of the lightness of the cross that weighed on her young shoulders. There was no pose or affectation about her struggle with poverty; and she did everything – sewed, washed, fed the baby, cooked the meat and cleaned the room. She was a resolute comrade to her husband, and like a great martyr laid down her life in the Far East, following her husband's restless, wandering flight and losing her two last children in succession.

At first I struggled with Kelsiev, trying to persuade him not to cut himself off at the very start from the path of return before he knew what the life of an exile was like.

I told him that he ought first to learn what poverty in a strange land meant, poverty in England, particularly in London; I told him that in Russia now every vigorous man was precious.

'What are you going to do here?' I asked him. Kelsiev intended to learn everything and to write about everything; most of all he wanted to write about the Woman Question and the organisation of the family.

'Write first,' I told him, 'about the need for the peasants to have land when they are emancipated. That is the first question that confronts us.'

But Kelsiev's sympathies were not attracted in that direction. He really did bring me an article on the Woman Question. It was extremely bad. Kelsiev was angry with me for not publishing it, and he thanked me for this two years later.

He did not want to go back. Work had to be found for him at all costs, and this we set about doing. His theological eccentricities assisted us. We obtained for him the job of correcting the proofs of the Russian edition of the Scriptures published by the London Bible Society, and then handed over to him a heap of papers we had received at various times relating to the Old Believers. Kelsiev enthusiastically undertook the task of arranging and editing them. What he had been groping for and dreaming of lay revealed before him: he glimpsed in the schism a coarsely naïve socialism in an evangelical vestment. This was the best time of Kelsiev's life. He worked passionately, and used to run in to see me in the evening to tell me of some socialistic idea of the Dukhobors or the Molokans, or some communistic doctrine of the Fedoseyevtsy. He was in ecstasies over their wandering in the forests, and found an ideal for his life in wandering among them and becoming the teacher of a socialist Christian schism in Belokrinitsa,[8] or Russia.

And indeed Kelsiev was a vagrant at heart, a vagrant morally and in practice: he was tormented by unstable thoughts, by melancholy. He could not remain in one spot. He had found work, occupation, a livelihood free from want, but he did not find work which would completely absorb his restless temperament; in order to seek it he was ready to go to the ends of the earth or to become a monk, to accept the holy calling without belief in it.

A real Russian, Kelsiev made a new programme of work every month, thought of new schemes and took up a new task without finishing the old one. He worked by bouts, and by bouts did nothing. He grasped things easily, but was at once satisfied and cloyed; he plucked at once all the essence out of a thing, to the last conclusion, sometimes even more than was in it.

The book about the *raskolniki* came off successfully; he published six parts, which were quickly distributed.[9] The government, seeing this, allowed the publication of the information concerning the Old Believers. The same thing happened with the translation of the Bible. The translation from the Hebrew was not successful. Kelsiev tried to perform a *tour de force* and to translate it word for word, regardless of the fact that the grammatical forms of the Semitic tongues do not at all correspond with those of the Slavonic. Nevertheless, the books that were issued were instantly sold, and the Holy Synod, alarmed at the success of the foreign edition, gave its blessing to the publication of the Old Testament in Russian. These back-handed victories were never put down to the credit of our press by anyone.

At the end of 1862 Kelsiev went to Moscow with the object of establishing permanent relations with the *raskolniki*. This journey he ought one day to describe himself. It was incredible, impossible, but it actually took place. The audacity of this trip borders on insanity; its recklessness was almost criminal; but of course I am not going to blame him for that. Incautious chatter abroad might have done a great deal of harm, but that is not the point, and has nothing to do with an appraisal of the journey itself.

On his return to London he undertook, at the suggestion of Trübner, to compile a Russian grammar for English people, and to translate some financial book. He did not complete either of these tasks: his travels had destroyed his remaining *Sitzfleisch*. He was bored by work and sank into hypochondria and depression, but work was needed, for again he had not a penny. Moreover, a new maggot began to gnaw at him. The success of his journey, the courage he had incontestably displayed, the mysterious negotiations, the victory over dangers – all fanned in his breast a flame of vanity which even without this had already been fierce in his heart; in contrary terms to those of Caesar, Don Carlos, and Vadim Passek,[10] Kelsiev would pass his hands through his thick hair and say, shaking his head sadly:

'Not yet thirty, and so much responsibility already taken upon my shoulders!'

From all this it can easily be understood that he would not finish the grammar, but would go away. And go he did. He went to Turkey with the firm intention of getting into even closer contact with the *raskolniki*, forming new ties and if possible remaining there and beginning to preach a Free Church and a communal life. I wrote him a long letter, trying to persuade him not to go, but to stick to his work. But the passion for wandering, the desire to do great deeds and to have a grand destiny, which haunted him, were too strong, and he went.

He and Martyanov disappeared almost at the same time – one, after a series of trials and misfortunes, to bury his dear ones and be lost between Jassy and Galatz, the other to bury himself in penal servitude, to which he was sent by the incredible stupidity of the Tsar and the incredible spite of the vindictive landowning senators.[11]

. . .

[. . .] After making his way to Turkey Kelsiev decided to settle in Tulcea; there he wanted to found a centre for his propaganda among the *raskolniki* and a school for Cossack children and to make the experiment of a communal life, in which profit and loss was to fall equally upon all, and the work, skilled or unskilled, light or heavy, should be shared by all. The cheapness of lodgings and of victuals made the experiment possible. [. . .]

In the summer of 1863 his younger brother Ivan joined him, a fine gifted youth. He had been relegated from Moscow to Perm in connection with the Students' Affair; there he came across a villain of a governor, who oppressed him. After this he was summoned again to Moscow for some depositions; he was threatened with being sent away to somewhere more remote than Perm. He escaped from custody and made his way through Constantinople to Tulcea. His elder brother was extremely glad to see him. He was looking for comrades, and eventually sent for his wife, who was longing to go to him, and had been living under our care at Teddington. [...]

. . .

At the beginning of 1864 two Russian officers, both exiles, Krasnopevtsev and Vasilev,[12] went to Tulcea. At first the little colony set to work amicably. They taught the children and salted cucumbers, patched their clothes and dug in the kitchen garden. Kelsiev's wife cooked the dinner and made clothes for them. Kelsiev was pleased with the beginning, pleased with the Cossacks and with the *raskolniki*, pleased with his comrades and with the Turks.[13]

Kelsiev was still writing us his humorous tales about their settlement, but the black hand of destiny was already menacing the little band of Tulcea community-dwellers. In June 1864, just a year after his arrival, Ivan Kelsiev died of malignant typhus in his brother's arms. He was only three-and-twenty. His death was a fearful blow for his brother, who fell ill himself, but somehow survived. His letters of that period make terrible reading. The spirit which had sustained the recluses drooped, they were overcome by the gloomy tedium; crimes and quarrels began. [...] Krasnopevtsev shot himself. Vasilev went away. Kelsiev, too, could stand it no longer; he took his wife and his children (he had another by then), and without means or aim set off first for Constantinople, then for the princedoms on the Danube. Completely cut off from everyone, cut off for a time even from us, it was then that he broke with the Polish emigrants in Turkey. In vain he sought to earn a crust of bread, with despair he looked at the exhaustion of his poor wife and children. The money we sent him now and then could not be sufficient. 'It happened at times that we had no bread at all,' his wife wrote not long before her death. At last, after long efforts, Kelsiev obtained in Galatz a post as 'overseer of work on the high-roads'. He was oppressed, devoured by boredom. He could not but blame himself for the situation of his family. The ignorance of the barbarous Eastern world outraged him. He pined in it and longed to get away. He had lost his faith in the *raskolniki*; he had lost his faith in Poland; his faith in men, in science, in revolution, was growing more and more unsteady, and it was easy to predict when it too would collapse. He dreamt only of struggling back again into the world at all costs and coming to us, and saw with horror that he could not leave his family. 'If I were alone,' he wrote several times, 'I should set off and follow my nose with a daguerreotype machine or a barrel-organ and, trailing over the world, would appear in Geneva on foot.'

Help was at hand.

Malusha (as they called the elder daughter) went to bed quite well, but woke up ill in the night. By morning she was dead of cholera. A few days later the second child died; the mother was taken to the hospital, and was found to be suffering from acute consumption.

'Do you remember,' she said to him, 'you promised once to tell me, when I was about to die, that it was death? Is this death?'

'It is death, my dear, it is.'

And she smiled once more, sank into unconsciousness and died.

'From the *Moscow News*, 11 June 1867.[14]

'They write to us from Petersburg that the other day the official in charge of the Skulyany Customs House received a letter signed V. Kelsiev warning him that the passenger who would have to present himself at that Customs House with a regular Turkish passport bearing the name of Ivan Zheludkov was none other than himself, Mr Kelsiev, and that, wishing to give himself up to the Russian Government, he requested that he should be arrested and sent to Petersburg.'

CHAPTER IV

THE YOUNGER EMIGRANTS:
THE COMMON FUND

KELSIEV had hardly passed out of our door when fresh people, driven out by the severe cold of 1863, were knocking at it. These came not from the training-schools of the coming revolution but from the devastated stage on which they had already acted rôles. They were seeking shelter from the storm without and seeking nothing within; what they needed was a temporary haven until the weather improved, until a chance presented itself to return to the fray. These men, while still very young, had finished with ideas, with culture; theoretical questions did not interest them, partly because they had not yet arisen among them, partly because what they were concerned with was their application. Though they had been defeated physically, they had given proofs of their courage. They had furled their flag, and their task was to preserve its honour. Hence their dry tone, *cassant*, *raide*, abrupt and rather elevated. Hence their martial, impatient aversion for prolonged deliberation, for criticism, their somewhat elaborate contempt for all intellectual luxuries, among which they put Art in the foreground. What need had they of music? What need of poetry? 'The fatherland is in danger, *aux armes, citoyens!*' In certain cases they were theoretically right, but they did not take into account the complex, intricate process of balancing the ideal with the real, and, I need hardly say, they assumed that their views and theories were the views and theories of the whole of Russia. To blame for this our young pilots of the coming storm would be unjust. It is the common characteristic of youth; a year ago a Frenchman,[1] a follower of Comte, assured me that Catholicism did not exist in France, that it had *complètement perdu le terrain*, and he pointed among others to the medical faculty, to the professors and students who were not merely not Catholics but not even Deists.

'Well, but the part of France,' I observed, 'which neither gives nor hears medical lectures?'

'It, of course, keeps to religion and its rites – but more from habit and ignorance.'

'I can very well believe it, but what will you do with it?'

'What did they do in 1792?'

'Not much: at first the Revolution closed the churches, but afterwards opened them again. Do you remember Augereau's answer to Napoleon when they were celebrating the Concordat? "Do you like the ceremony?" the consul asked as they came out of Notre-Dame. The Jacobin general answered: "Very much. I am only sorry that the two hundred thousand men are not present who went to their graves to abolish such ceremonies!" '

'Ah bah! we have grown wiser, and we shall not open the church doors – or rather we shall not close them at all, but shall turn the temples of superstition into schools.'

'L'infâme sera écrasée,' I wound up, laughing.

'Yes, no doubt of it; that is certain!'

'But that you and I will not see it – that is even more certain.'

It is to this looking at the surrounding world through a prism coloured by personal sympathies that half the revolutionary failures are due. The life of young people, spent in general in a noisy, closed seclusion of a sort, remote from the everyday, whole-sale struggle for personal interests, though it clearly grasps general truths, nearly always comes to grief through a false understanding of their application to the needs of the day.

At first our new visitors cheered us with accounts of the move-ment in Petersburg, of the savage antics of the full-fledged reac-tion, of trials and persecutions, of university and literary parties. Then, when all this had been told with the rapidity with which in such cases men hasten to tell all they know, a pause, a hiatus would follow; our conversations became dull and monotonous.

'Can this,' I thought, 'actually be old age divorcing two genera-tions? Is it the chill induced by years, by weariness, by ordeals?'

Whatever it might be due to, I felt that with the arrival of these new men our horizon was not widened but narrowed. The scope of our conversation was more limited. Sometimes we had nothing to say to one another. They were occupied with the details of their coteries, beyond which nothing interested them. Having once related everything of interest about them, there was nothing to do but to repeat it, and they did repeat it. They took little interest in learning or in public affairs; they even read little, and did not follow the newspapers regularly. Absorbed in memories and expectations, they did not care to step forth into other fields; while we had not air to breathe in that stifling atmosphere. We had been spoiled by different dimensions and were smothered.

Moreover, even if they did know a certain stratum in Petersburg, they did not know Russia at all and, though sincerely desirous of coming into contact with the people, they only approached them bookishly and theoretically.

What we had in common was too general. Advance together, *serve*, as the French say, take action together we might, but it was hard to stand still with arms folded and live together. It was useless even to think of a serious influence on them. A morbid and very unceremonious vanity had long ago taken the bit between its teeth.* Sometimes, it is true, they did ask for a programme, for guidance, but for all their sincerity there was no reality about this. They expected us to formulate their own opinions, and only assented when what we said did not contradict them in the least. They looked upon us as respected veterans, as something past and over, and were naïvely surprised that we were not yet so very much behind themselves.

I have always and in everything feared 'above all sorrows',[2] *mésalliances*; I have always tolerated them, partly through humanity, partly through carelessness, and have always suffered from them.

It was not hard to foresee that our new ties would not last long, that sooner or later they would be broken and that, considering the churlish character of our new friends, this rupture would not come off without disagreeable consequences.

The question upon which our rickety relationship came to grief was just that old question through which acquaintances tacked together with rotten threads usually come apart. I mean money. Knowing absolutely nothing of my resources nor of my sacrifices, they made demands upon me which I did not think it right to satisfy. If I had been able, through all our reverses, without the

---

* Their vanity was not so much great as it was touchy and irritable, and above all, unrestrained in words. They could conceal neither their envy nor a special kind of punctilious insistence on respectful recognition of the position they ascribed to themselves. At the same time they looked down on everything and were perpetually jeering at one another, which was why their friendships never lasted longer than a month.

slightest assistance, to conduct the Russian propaganda for fifteen years, it was only because I had put a careful limit to my other expenses. My new acquaintances considered that all I was doing was not enough, and looked with indignation at a man who gave himself out for a socialist and did not distribute his property in equal shares among people who wanted money without working. Obviously they had not advanced beyond the impractical point of view of Christian charity and voluntary poverty, and mistook that for practical socialism.

The attempts to collect a 'Common Fund' yielded no results of importance. Russians are not fond of giving money to any common cause, unless it includes the building of a church, a banquet, a drinking-party and the approval of the higher authorities.

When the impecuniosity of the exiles was at its height, a rumour circulated among them that I had a sum of money entrusted to me for the purposes of propaganda.

It seemed perfectly right to the young people to relieve me of it.

To make the position clear, I must relate a certain strange incident that occurred in the year 1858.[3] One morning I received a very brief note from an unknown Russian; he wrote to me that he 'urgently needed to see me', and asked me to fix a time.

I happened to be going to London at the time, and so instead of answering I went myself to the Sablonnière Hotel and inquired for him. He was at home. He was a young man who looked like a cadet, shy, very depressed, and with the peculiar rather rough-hewn appearance of the seventh or eighth son of a steppe landowner. Very uncommunicative, he was almost completely silent; it was evident that he had something on his mind, but he could not come to the point of putting it into words.

I went away, inviting him to dinner two or three days later. Before that date I met him in the street.

'May I walk with you?' he asked.

'Of course; there is no danger for me in being seen with you, though there is for you in being seen with me. But London is a big place.'

'I am not afraid' – and then all at once, taking the bit between his teeth, he hurriedly burst out: 'I shall never go back to Russia – no, no, I certainly shall not go back to Russia . . .'

'Upon my word, and you so young?'

'I love Russia – I love it dearly; but there the people . . . I cannot live there. I want to found a colony on completely socialistic principles; I have thought it all over, and now I am going straight there.'

'Straight where?'

'To the Marquesas Islands.'

I looked at him in dumb amazement.

'Yes, yes; it is all settled. I am sailing by the next steamer, and so I am very glad that I have met you today – may I put an indiscreet question to you?'

'As many as you like.'

'Do you make any profit out of your publications?'

'Profit! I am glad to say that now the press pays its way.'

'Well, but what if it should not?'

'I shall make it up.'

'So that no sort of commercial aim enters into your propaganda?'

I laughed heartily.

'Well, but how are you going to pay it off alone? And your propaganda is essential. Please forgive me; I am not asking out of curiosity: when I left Russia for ever, I had the thought in my mind of doing something useful for our country, and I decided . . . well, I only wanted to know first from yourself about finances . . . yes, I decided to leave a small sum of money with you. Should your printing-press need it, or the Russian propaganda generally, then it would be at your disposal.'

Again I had to look at him in amazement.

'Neither the printing-press nor Russian propaganda nor I are in need of money; on the contrary, things are going swimmingly. Why should I take your money? But though I refuse to take it, allow me to thank you from the bottom of my heart for your kind intention.'

'No, sir, it is all decided. I have 50,000 francs. I shall take 30,000 with me to the Islands, and I shall leave 20,000 with you for propaganda.'

'What am I to do with it?'

'Well, if you don't need the money you can give it back to me if I return; but if I don't return within ten years, or if I die – use it for your propaganda efforts. Only,' he added, after a moment's thought, 'do what you like, but . . . but don't give anything to my heirs. Are you free tomorrow morning?'

'Certainly, if you like.'

'Do me the favour of taking me to the bank and to see Roths-child; I know nothing about these things, I can't speak English and I speak French very badly. I want to make haste to get rid of the 20,000 and be off.'

'Very well, I shall accept the money, but on these conditions: I shall give you a receipt.'

'I don't want any receipt.'

'No, but I must give you one, and I shan't take your money without it. Now listen. In the first place,[4] it shall be stated in the receipt that your money is entrusted not to me alone, but to me and to Ogarëv. In the second, since you may get sick of the Marquesas Islands and begin to pine for your native country . . .' (he shook his head). 'How can one know what one does not know? . . . There is no need to specify the object with which you are giving us the capital: we will say that the money is put at the complete disposal of Ogarëv and myself; should we make no other use of it, we shall invest the whole sum for you in securities at five per cent or thereabouts, guaranteed by the English Government. Then I give you my word that we shall not touch your money except in case of extreme necessity for propaganda purposes; you may count upon it in any circumstances, except that of bankruptcy in England.'

'If you insist on taking so much trouble, do so. And tomorrow let us go for the money!'

The following day was an unusually amusing and busy one. It began with the bank and with Rothschild. The money was paid in notes. Bakhmetev at first conceived the guileless intention of changing them into Spanish gold or silver. Rothschild's clerks looked at him in amazement, but when he suddenly, as though half awake, said in very broken Franco-Russian: 'Well, then, lettre crédit Ile Marquise', Kestner, the manager, turned on me an alarmed and anxious look, which said better than any words: 'He is not dangerous, is he?' Besides, never before in Rothschild's bank had anyone asked for a letter of credit to the Marquesas Islands.

We decided to take 30,000 francs in gold and go home; on the way we went into a café. I wrote the receipt; Bakhmetev for his part wrote for me that he put 800 pounds at the complete disposal of myself and Ogarëv; then he went home to get something and I went off to a bookshop to wait for him; a quarter of an hour later he came in, white as a sheet, and announced that of his 30,000 francs, 250, that is ten pounds, were missing. He was utterly over-come. How the loss of 250 francs could so upset a man who had just given away 20,000 without any secure guarantee is another psychological riddle of human nature.

'Haven't you a note too much?' he asked me.

'I haven't the money with me. I gave it to Rothschild, and here is the receipt, precisely 800.' Bakhmetev, who had changed his French notes into sovereigns with no need to do so, scattered 30,000 on Tchorszewski's[5] counter; he counted them and counted them over again; ten pounds were missing, and that was all about it. Seeing his despair, I said to Tchorszewski:

'I'll take that damned ten pounds on myself somehow; here he has done a good deed, and he is punished for it.'

'It is no use grieving and discussing it,' I added to him. 'I propose going to Rothschild's at once.'

We drove there. By now it was after four and the bank was closed. I went in with the embarrassed Bakhmetev. Kestner looked at him and smiled, took a ten-pound note from the table and handed it to me.

'How did it happen?'

'When your friend changed the money he gave me two ten-pound notes instead of two five-pound ones, and at first I did not notice it.'

Bakhmetev looked and looked, and commented:

'How stupid that ten-pound notes and five-pound notes are the same colour; who would notice the difference? You see what a good thing it was that I changed the money into gold.'

His mind was at rest and he came to dine with me; I promised to go and say goodbye to him next day. He was quite ready to start. A little shabby, battered trunk such as cadets or students have, a greatcoat tied up with a strap, and ... and ... 30,000 *francs in gold* wrapped up in a thick pocket-handkerchief, as people tie up a pound of gooseberries or nuts.

This was how the man was setting off for the Marquesas Islands.

'Upon my soul!' I said to him; 'why, you will be robbed and murdered before your ship casts off; you had better put your money in your trunk.'

'It is full.'

'I'll get you a bag.'

'Not on any account.'

And so he went off. During the first days I feared that he would be made away with and that I should incur the suspicion of having sent someone to kill him.

From that day there has been no sight nor sound of him. . . . I put his money in Consols with the firm intention of not touching it except in the case of the printing-press or propaganda being in the utmost straits.

For a long time no one in Russia knew of this; then there were vague rumours, for which we were indebted to two or three of our friends who had given their word to say nothing about it. At last it was learnt that the money really existed and was in my keeping.

This news fell like an apple of temptation, a chronic incitement and ferment. It turned out that everyone needed the money – and I did not give it to them. They could not forgive me for not having lost the whole of my own property, and here I had a deposit given me for the propaganda; and who were 'the propaganda' if not they? The sum quickly grew from modest francs to silver roubles, and was still more tantalising for those who desired to waste it privately for the common cause. They were indignant with Bakhmetev for having entrusted the money to me and not to someone else; the boldest among them declared that it was an error on his part; that he had really meant to give it not to me but to a certain political circle in Petersburg, and that, not knowing how to do this, he had given it to me in London. The audacity of these opinions was the more remarkable that Bakhmetev's surname was as unknown as was his very existence, and that he had not spoken to anyone else of his proposal before his departure, nor had anyone spoken to him since then.

Some needed the money in order to send emissaries; others for establishing centres on the Volga; others still for the publication of a journal. They were dissatisfied with *The Bell*, and did not readily respond to our invitation to work on it.

I absolutely refused to give the money; and let those who demanded it tell me themselves what would have become of it if I had.

'Bakhmetev may return without a farthing,' I said; 'it is not easy to make a fortune by founding a socialist colony in the Marquesas Islands.'

'He is dead for sure.'

'But what if he is alive to spite you?'

'Well, but he gave the money for the propaganda, you know.'

'So far I haven't needed it.'

'But we do.'

'What for precisely?'

'We must send someone to the Volga and someone to Odessa. . . .'

'I don't think that is very necessary.'

'So you don't believe in the indispensability of sending them?'

'I do not.'

'He is growing old and getting miserly,' the most resolute and ferocious said about me in different keys.

'But why mind him? Just take the money from him and have done with it,' the still more resolute and ferocious added, 'and if he resists, we will go for him in the papers and teach him to keep back other people's money.'

I did not give up the money.

They did not go for me in the papers. I was abused in the press much later, but that was over money too. . . .

These more ferocious ones of whom I have spoken were the clumsy and uncouth representatives of the 'New Generation', who may be called the Sobakeviches and Nozdrëvs[6] of Nihilism.

However superfluous it may be to make a reservation, yet I shall do so, knowing the logic and the manners of our opponents. I have not the slightest desire in what I am saying to fling a stone at the younger generation or at Nihilism. Of the latter I have written many times. Our Sobakeviches of Nihilism do not constitute its most powerful expression, but only represent its exaggerated extremes.*

Who would judge of Christianity from the Flagellants of

---

* At that very time in Petersburg and Moscow, and even in Kazan and Kharkov, there were circles being formed among the university youth who devoted themselves in earnest to the study of science, especially among the medical students. They worked honestly and conscientiously but, cut off from active participation in the questions of the day, they were not forced to leave Russia and we scarcely knew anything of them.

Origen or of the Revolution from the Sptember butchers and the *tricoteuses* of Robespierre?

The arrogant lads of whom I am speaking are worth studying, because they are the expression of a temporary type, very definitely marked and very frequently repeated, a transitional form of the sickness of our development from our former stagnation.

For the most part they were lacking in the deportment which is given by breeding, and the staying power which is given by scientific studies. In the first fervour of emancipation they were in a hurry to cast off all the conventional forms and to push away all the rubber fenders which prevent rough collisions. This made difficult the simplest relations with them.

Removing everything to the last rag, our *enfants terribles* proudly appeared as their mothers bore them, and their mothers had not borne them well, not as simple, rather too plump lads but as inheritors of the evil, unhealthy life of our lower classes in Petersburg. Instead of athletic muscles and youthful nakedness, they displayed the melancholy traces of hereditary anaemia, the traces of old sores and of various fetters and collars. There were few among them who had come up from the people. The hall, the barrack-room, the seminary, the petty proprietor's farm survived in their blood and their brains, and lost none of their characteristic features though twisted in an opposite direction. So far as I know, this fact has attracted no serious attention.

On the one hand, the reaction against the old narrow, oppressive world was bound to throw the younger generation into antagonism and negation of their hostile surroundings; it was useless to expect moderation or justice in them. On the contrary, everything was done in defiance, everything was done in resentment. 'You are hypocrites, we shall be cynics; you have been moral in words, we will be wicked in words; you have been polite to your superiors and rude to your inferiors, we shall be rude to everyone; you bow down to those whom you do not respect, we will jostle people without apologising; your feeling of personal dignity consisted in nothing but decorum and external honour, we make it our point of honour to flout every decorum and to scorn every *point d'honneur.*'

But on the other hand, though disowning all the ordinary forms of social life, their character was full of its own hereditary ailments and deformities. Casting off, as we have said, all veils, the most desperate played the dandy in the costume of Gogol's Petukh[7] and did not preserve the pose of the Medici Venus. Their nakedness did not conceal, but revealed, what they were. It revealed that their systematic uncouthness, their rude and insolent talk, had nothing in common with the inoffensive and simple-hearted coarseness of the peasant, but a great deal in common with the manners of the low-class pettifogger, the shop-boy and the flunkey. The people no more considered them as one of themselves than they did a Slavophil in a *murmolka*. To the people these men have remained alien, the lowest stratum of the enemies' camp, skinny young masters, scribblers out of a job, Russians turned Germans.

To be completely free, one must forget one's liberation and that from which one has been liberated, and cast off the habits of the environment out of which one has grown. Until this has been done we cannot help being conscious of the servants' hall, the barrack-room, the government office or the seminary in every gesture they make and every word they utter.

To punch a man on the nose at the first objection he advances – if not with a fist then with a word of abuse – to call Stuart Mill a rascal,[8] forgetting all the service he has done, is not that the same as the Russian master's way of 'punching old Gavrilo in the snout for a crumpled cravat?'[9] In this and similar pranks do you not recognise the policeman, the district officer, the village constable dragging a bailiff by his grey beard? Do you not, in the insolent arrogance of their manners and answers, clearly recognise the insolence of the officers of the days of Nicholas? Do you not see, in men who talk haughtily and disdainfully of Shakespeare and Pushkin, the grandsons of Skalozub,[10] reared in the house of their grandsire who wanted 'to make a Voltaire of his corporal'?[11]

The very leprosy of bribery has survived in high-handed importunity for money, by bias and threats under pretext of common causes, in the feeble impulse towards being fed at the expense of the service and towards avenging a refusal by slander and libel.

All this will be transformed and thrashed out with time. But there is no blinking the fact that a strange soil has been prepared by the Tsar's paternal government and imperial civilisation in our 'kingdom of darkness'. It is a soil on which seedlings of great promise have grown, on the one hand, into worshippers of the

Muravëvs and the Katkovs and, on the other, into the bullies of Nihilism and the impudent Bazarov[12] free-lances.

Our black earth needs a good deal of drainage!

CHAPTER V

M. BAKUNIN AND THE CAUSE
OF POLAND

AT the end of November we received the following letter from Bakunin:

'San Francisco, 15 October 1861.

'Friends, – I have succeeded in escaping from Siberia, and after long wanderings on the Amur, on the shores of the Gulf of Tartary and across Japan, I arrived today in San Francisco.

'Friends, I long to come to you with my whole being, and as soon as I arrive I shall set to work; I shall work with you on the Polish-Slavonic question, which has been my *idée fixe* since 1846 and was in practice my speciality in 1848 and 1849.

'The destruction, the complete destruction, of the Austrian empire will be my last word; I don't say deed: that would be too ambitious; to promote it I am ready to become a drummer-boy or even a scoundrel,[1] and if I should succeed in advancing it by one hair's-breadth I shall be satisfied. And beyond that there appears the glorious, free Slav Federation, the one way out for Russia, the Ukraine, Poland, and the Slavonic peoples generally. . . .'

We had known of his intention to escape from Siberia some months before. By the New Year Bakunin in his own exuberant person was clasped in our arms.

Into our work, into our closed shop of two, a new element had entered, or rather an old element, perhaps a risen shade of the 'forties, and most of all of 1848. Bakunin was just the same; he had grown older in body only, his spirit was as young and enthusiastic as in the days of the all-night arguments with Khomyakov in Moscow. He was just as devoted to one idea, just as capable of being carried away by it, and seeing in everything the fulfilment of his desires and ideals, and even more ready for every experience, every sacrifice, feeling that he had not so much life before him, and that consequently he must make haste and not let slip a single chance. He was fretted by prolonged study, by the weighing of pros and cons and, confident and theoretical as ever, he longed for any action if only it were in the midst of the storms of revolution, in the midst of destruction and danger. Now, too, as in the articles signed 'Jules Elisard', he repeated: 'Die Lust der Zerstörung ist eine schaffende Lust.'[2] The fantasies and ideals with which he was imprisoned in Königstein[3] in 1849 he had preserved, and had carried them complete across Japan and California in 1861. Even his language recalled the finer articles of La Réforme and La vraie République, the striking speeches in La Constituante and at Blanqui's Club. The spirit of the parties of that period, their exclusiveness, their personal sympathies and antipathies, above all their faith in the second coming of the revolution – it was all here.

Strong characters, if not destroyed at once by prison and exile, are preserved by them in an extraordinary way; they come out of them as though out of a faint and go on with what they were about when they lost consciousness. The Decembrists came back from being buried in the snows of Siberia more youthful than the young people who met them, who had been trampled down before they could come to maturity. While two generations of Frenchmen changed several times, turned red and white by turns, advancing with the flood and borne back by the ebb, Barbès and Blanqui remained steady beacons, recalling from behind prison bars and distant foreign lands the old ideals in all their purity.

'The Polish-Slavonic question . . . the destruction of the Austrian empire . . . the glorious free Slav Federation . . .' and all this is to happen straight off, as soon as he arrives in London! And it is written from San Francisco when he has one foot on the ship!

The European reaction did not exist for Bakunin, the bitter years from 1848 to 1858 did not exist for him either; of them he had but a brief, far-away, faint knowledge. He had read them in Siberia, just as he had read at Kaydanov about the Punic Wars and of the fall of the Roman Empire. Like a man who has returned after the plague, he heard who had died, and sighed for them all; but he had not sat by the bedside of the dying, had not hoped that they would be saved, had not followed them to the grave. The events of 1848, on the contrary, were all about him, near to his heart, vivid and in detail; the conversations with Caussidière, the speeches of the Slavs at the Prague Conference,[4] discussions with Arago or Ruge – to Bakunin all these were affairs of yesterday; they were all still ringing in his ears and flashing before his eyes.

In London he first of all set about *revolutionising The Bell*, and in 1862 advanced against us almost all that in 1847 he had advanced against Belinsky. Propaganda was not enough; there ought to be immediate action; centres and committees ought to be organised; to have people closely and remotely associated with us was not enough, we ought to have 'dedicated and half-dedicated brethren', organisations on the spot – a Slavonic organisation, a Polish organisation. Bakunin thought us too moderate, unable to take advantage of the situation of the moment, insufficiently fond of resolute measures. He did not lose heart, however, but was convinced that in a short time he would set us on the right path. While awaiting our conversion Bakunin gathered about him a regular circle of Slavs. Among them there were Czechs, from the writer Fritsch to a musician who was called Naperstok;[5] Serbs who were simply called after their father's names Ioanovic, Danilovic, Petrovic; there were Wallachians who did duty for Slavs, with the everlasting 'esco' at the end of their names; finally, there was a Bulgarian who had been a doctor in the Turkish army, and there were Poles of every diocese – the Bonapartist, the Mieroslawski, the Czartorysczki: democrats without socialist ideas but with a tinge of the officer; socialists, catholics, anarchists, aristocrats and men who were simply soldiers, ready to fight anywhere in North or South America ... and by preference in Poland.

With them Bakunin made up for his nine years' silence and solitude. He argued, lectured, made arrangements, shouted, decided, directed, organised and encouraged all day long, all night long, for days and nights together. In the brief minutes he had free he rushed to his writing-table, cleared a little space from cigarette-ash, and set to work to write five, ten, fifteen letters to Semipalatinsk and Arad, to Belgrade and Tsargrad, to Bessarabia, Moldavia and Belokrinitsa. In the middle of a letter he would fling aside the pen and bring up to date the views of some old-fashioned Dalmatian, then, without finishing his exhortation, snatch up the pen and go on writing. This, however, was made easier for him by the fact that he was writing and talking about one and the same thing. His activity, his laziness, his appetite, and everything else, like his gigantic stature and the everlasting sweat he was in, everything, in fact, was on a superhuman scale, as he was himself; and he was himself a giant with his leonine head and tousled mane.

At fifty he was exactly the same wandering student from the Maroseyka, the same homeless *Bohémien* from the *Rue de Bourgogne*, with no thought for the morrow, careless of money, throwing it away when he had it, borrowing it indiscriminately, right and left when he had not, as simply as children take from their parents, careless of repayment; as simply as he himself would give his last money to anyone, only keeping what he needed for cigarettes and tea. This manner of life did not worry him; he was born to be a great vagrant, a great nomad. If anyone had asked him once and for all what he thought of the right of property, he might have answered as Lalande answered Napoleon about God: 'Sire, in my pursuits I have not come upon any necessity for this right!' There was something childlike, simple and free from malice about him, and this gave him an unusual charm and attracted to him both the weak and the strong, repelling none but the affected *petit bourgeois*. His striking personality, the eccentric and powerful appearance he made everywhere, in a coterie of young people in Moscow, in a lecture-room at Berlin University, among Weitling's Communists and Caussidière's Montagnards, his speeches in Prague, his command at Dresden, his trial, imprisonment, sentence to death, torture in Austria and surrender to Russia – where he vanished behind the fearful walls of the Alexeyevsky ravelin – make of him one of those individualists whom neither the contemporary world nor history can pass by.

That he ever came to marry, I can only put down to the boredom of Siberia. He had piously preserved all the habits and customs of his fatherland, that is of student-life in Moscow: heaps of tobacco lay on his table like stores of forage, cigar-ash covered his papers, together with half-finished glasses of tea; from morning onwards clouds of smoke hung about the room from a regular suite of smokers, who smoked as though they were racing each other, hurriedly blowing it out and drawing it in – as only Russians and Slavs do smoke, in fact. Many a time I enjoyed the amazement, accompanied by a certain horror and perplexity, of the landlady's servant, Grace, when at dead of night she brought boiling water and a fifth basin of sugar into this hotbed of Slav emancipation.

Long after Bakunin left London, tales were told at No. 10 Paddington Green of the way he went on, which upset all the consolidated notions and religiously observed forms and degrees of English middle-class life. Note at the same time that both the maid and the landlady were madly devoted to him.

'Yesterday,' one of his friends told Bakunin, 'So-and-so arrived from Russia; he is a very fine man, formerly an officer.'

'I have heard about him; he is very well spoken of.'

'May I bring him?'

'Certainly; but why bring him, where is he? I'll go and see him. I'll go at once.'

'He seems to be rather a constitutionalist.'

'Perhaps, but . . .'

'But I know he is a chivalrous, fearless and noble man.'

'And trustworthy?'

'He is much respected at Orsett House.'

'Let us go to him.'

'Why? He meant to come to you: that was what we agreed. I'll bring him.'

Bakunin rushes to his writing; he writes and scratches out something, writes it out again, and seals up a packet addressed to Jassy; in his restless expectation he begins walking about the room with a tread which sets the whole house – No. 10 Paddington Green – shaking with his step.

The officer makes his appearance quietly and modestly. Bakunin *le met à l'aise*, talks like a comrade, like a young man, fascinates him, scolds him for his constitutionalism and suddenly asks:

'I am sure you won't refuse to do something for the common cause.'

'Of course not.'

'There is nothing that detains you here?'

'Nothing; I have only just arrived, I. . . .'

'Can you go tomorrow or next day with this letter to Jassy?'

Such a thing had not happened to the officer either at the front in time of war or on the general staff in peace-time. However, accustomed to military obedience, he says, after a pause, in a voice that does not sound quite natural:

'Oh yes!'

'I knew you would. Here is the letter perfectly ready.'

'I am ready to set off at once . . . only . . .' (the officer is embarrassed). 'I had not at all reckoned on such a journey.'

'What? No money? Then say so; that's of no consequence. I'll get it for you from Herzen: you shall pay it back later on. Why, what is it? Only some twenty pounds or so. I'll write to him at once. You will find money at Jassy. From there you can make your way to the Caucasus. We particularly need a trustworthy man there.'

The officer, amazed, dumbfounded, and his companion equally amazed and dumbfounded, take their leave. A little girl whom Bakunin employed on great diplomatic errands flies to me through the rain and sleet with a note. I used to keep chocolate *en losenges* expressly for her benefit, to comfort her for the climate of her native country, and so I give her a big handful and add:

'Tell the tall gentleman that I shall talk it over with him personally.'

The correspondence in fact turned out to be superfluous. Bakunin appeared for dinner, that is an hour later.

'Why twenty pounds for X?'

'Not for him, for the cause; and I say, brother, isn't X a splendid fellow?'

'I have known him for some years. He has stayed in London before.'

'It is such a chance, it would be a sin to let it slip. I am sending him to Jassy, and then he'll have a look round in the Caucasus.'

'To Jassy? And from there to the Caucasus?'

'I see you are going to be funny,' said Bakunin. 'You won't prove anything by jokes.'

'But you know you don't want anything in Jassy.'

'How do you know?'

'I know, in the first place, because nobody does want anything in Jassy; and in the second place, if anything were wanted, you would have been telling me about it incessantly for the last week. You have simply come across a shy young man who wants to prove his devotion, and so you have taken it into your head to send him to Jassy. He wants to see the Exhibition and you will show him Moldo-Wallachia. Come, tell me what for?'

'What inquisitiveness! You never take part in these things with me: what right have you to ask?'

'That is true: in fact, I imagine that it is a secret you will keep from everyone; anyhow, I have not the slightest intention of giving money for couriers to Jassy and Bucharest.'

'But he will pay you back; he will have money.'

'Then let him make a wiser use of it. That's enough; you can send the letter by some Petresco-Manon-Lescaut; and now let's go and eat.'

And Bakunin, laughing himself, and shaking his head, which was always a little too heavy for him, set himself steadily and zealously to the work of eating his dinner, after which he would say each time: 'Now comes the happy moment,' and light a cigarette.

He used to receive everyone, at any time, everywhere. Often he would be still asleep like Onegin, or tossing on his bed, which creaked under him, and two or three Slavs would be in his bedroom smoking with desperate haste; he would get up heavily, souse himself with water, and at the same moment proceed to instruct them; he was never bored, never found them a burden; he could talk without being tired, with the same freshness of mind, to the cleverest or the stupidest man. This lack of discrimination sometimes led to very funny incidents.

Bakunin used to get up late; he could hardly have done otherwise, since he spent the night talking and drinking tea.

One morning some time after ten o'clock he heard someone moving about in his room. His bed stood curtained off in a large alcove.

'Who's there?' shouted Bakunin, waking up.

'A Russian.'

'What is your name?'

'So-and-so.'

'Delighted to see you.'

'Why is it you get up so late and you a democrat?'

Silence: the sounds of splashing water, cascades.

'Mikhail Alexandrovich!'

'Well?'

'I wanted to ask you, were you married in church?'

'Yes.'

'You did wrong. What an example of inconsistency; and here is Turgenev too, having his daughter legally married. You old men ought to set us an example.'

'What nonsense you are talking.'

'But tell me, did you marry for love?'

'What has that to do with you?'

'There was a rumour going about that you married because your bride was rich!'*

'Have you come here to cross-examine me? Go to the devil!'

'Well now, here you are angry, and I really meant no harm. Goodbye. But I shall come and see you again all the same.'

'All right, all right. Only be more sensible next time.'

Meanwhile the Polish storm was drawing nearer and nearer. In the autumn of 1862 Potebnya appeared in London for a few days. Melancholy, pure-hearted, devoted heart and soul to the hurricane, he came to talk to us for himself and his comrades, meaning in any case to go his own way. Poles began to arrive from their country more and more frequently; their language was sharper and more definite. They were moving directly and consciously towards the explosion. I felt with horror that they were going to unavoidable ruin.

'I am mortally sorry for Potebnya and his comrades,' I said to Bakunin, 'and the more so that I doubt whether their aims are the same as those of the Poles.'

'Oh yes they are, yes they are,' Bakunin retorted. 'We can't sit for ever with our arms folded, reflecting; we must take history as it presents itself, or else one will always be too far behind or too far in front.'

Bakunin grew younger; he was in his element: he loved not only the uproar of the revolt and the noise of the club, the market-place and the barricade; he loved the preparatory agitation, the excited and at the same time restrained life, spent among conspiracies, consultations, sleepless nights, conferences, agreements, corrections of ciphers, invisible inks and secret signs. Anyone who has taken part in rehearsals for private theatricals or in preparing a Christmas tree knows that the preparation is one of the best, most exquisite parts of it. But though he was carried away by the preparations of the Christmas tree I had a gnawing at my heart; I was continually arguing with him and reluctantly doing what I did not want to do.

Here I must stop to ask a sorrowful question. How, whence did I come by this readiness to give way, though with a murmur, this weak yielding, though after rebellion and a protest? I had, on the one hand, a conviction that I ought to act in one way, and, on the

_____
* Bakunin took no dowry with his wife.

other, a readiness to act quite differently. This wavering, this dissonance, *dieses Zögernde* has done me infinite harm in my life, and has not even left me with the faint comfort of recognising that my mistake was involuntary, unconscious; I have made blunders *à contre-cœur*; I had all the arguments on the other side before my eyes. I have told in one of my earlier chapters of the part I took in the 13th of June, 1849. That is typical of what I am saying. I did not for one instant believe in the success of the 13th of June; I saw the absurdity of the movement and its impotence, the indifference of the people, the ferocity of the reaction, and the pettiness of the revolutionaries. (I had written about it already, and yet I went out into the square, laughing at the people who went.)

How many misfortunes, how many blows I should have been spared in my life, if at all the crises in it I had had the strength to listen to myself. I have been reproached for being easily carried away; I have been carried away, too, but that is not what matters most. Though I might be committed by my impressionable temper, I pulled myself up at once; thought, reflection and observation almost always gained the day in theory, but not in practice. That is just what is hard to explain: why I let myself be led *nolens volens*. . . .

The reason for my quick compliance was false shame, though sometimes it was the better influences of love, friendship, and indulgence; but did all this overcome my power of reasoning?

After the funeral of Worcell on the 5th of February, 1857, when all the mourners had dispersed to their homes and I, returning to my room, sat down sadly at my writing-table, a melancholy question came into my mind. Had we not lowered into the ground with that just man, and had we not buried with him all our relations with the Polish emigrants?[6]

The gentle character of the old man, which was a conciliating element in the misunderstandings that were constantly arising, had gone for ever, but the misunderstandings remained. Privately, personally, we might love one or another among the Poles and be friendly with them, but there was little common understanding between us in general, and that made our relations strained and conscientiously reserved; we made concessions to one another, that is, weakened ourselves and decreased in each other what was almost the best and strongest in us. It was impossible to come to a common understanding by open talk. We started from different points, and our paths only intersected in our common hatred for the autocracy of Petersburg. The ideal of the Poles was behind them: they strove towards their past, from which they had been cut off by violence and which was the only starting-point from which they could advance again. They had masses of holy relics, while we had empty cradles. In all their actions and in all their poetry there is as much of despair as there is of living faith.

They look for the resurrection of their dead, while we long to bury ours as soon as possible. Our lines of thought, our forms of inspiration are different; our whole genius, our whole constitution has nothing in common with theirs. Our association with them seemed to them alternately a *mésalliance* and a marriage of convenience. On our side there was more sincerity, but not more depth: we were conscious of our indirect guilt, we liked their daring and respected their indomitable protest. What could they like, what could they respect in us? They did violence to themselves in making friends with us; they made an honourable exception for a few Russians.

Locked up as fellow-prisoners in the dark prison-house of the reign of Nicholas, we had more sympathy for each other than knowledge of each other. But as soon as the window was opened a little space, we divined that we had been brought by different paths and that we should disperse in different directions. After the Crimean War we heaved a sigh of relief, and our joy was an offence to them: the new atmosphere in Russia reminded them not of their hopes but of their losses. For us the new times began with presumptuous demands; we rushed forward ready to smash everything; with them it began with requiems and services for the dead. But for a second time the government welded us together. At the sound of firing at priests and children, at crucifixes and women, the sound of firing above the chanting of hymns and prayers, all questions were silenced, all differences were wiped out. With tears and lamentations, I wrote then a series of articles[7] which deeply touched the Poles.

From his deathbed old Adam Czartoryszki sent me by his son a warm word of greeting; a deputation of Poles in Paris presented me with an address signed by four hundred exiles, to which signatures were sent from all parts of the world, even from Polish refugees living in Algiers and in America. It seemed as though in so much we were united; but one step further in, and the difference, the sharp difference, leaped to the eye.

One day Ksawery Branicki, Chojecki and one or two other Poles were sitting with me; they were all on a brief visit to London, and had come to shake hands with me for my articles. The talk fell on the shot fired at Constantine.[8]

'That shot,' I said, 'will do you terrible damage. The government might have made some concessions; now it will yield nothing, and will be twice as savage.'

'But that is just what we want!' Ch. E.[9] observed with heat; 'there could be no worse misfortune for us than concessions. We want a breach, an open conflict.'

'I hope most earnestly that you may not regret it.'

Ch. E. smiled ironically, and no one added a word. That was in the summer of 1861. And a year and a half later Padlewski said the same thing when he was on his way to Poland *through Petersburg*.

The die was cast! . . .

Bakunin believed in the possibility of a rising of the peasants and the army in Russia, and to some extent we believed in it too; and indeed the government itself believed in it, as was shown later on by a series of measures, of officially inspired articles, and of punishments by special decree. That men's minds were working and in a ferment was beyond dispute, and no one saw at the time that the popular excitement would be turned to ferocious patriotism.

Bakunin, not too much given to weighing every circumstance, looked only towards the ultimate goal, and took the second month of pregnancy for the ninth. He carried us away not by arguments but by his hopes. He longed to believe, and he believed, that Zhmud[10] and the Volga, the Don and the Ukraine would rise as one man when they heard of Warsaw; he believed that the Old Believers would take advantage of the Catholic movement to obtain a legal standing for the Schism.

That the league among the officers of the troops stationed in Poland and Lithuania – the league to which Potebnya belonged – was growing and gathering strength was beyond all doubt; but it was very far from possessing the strength which the Poles through design and Bakunin through simplicity ascribed to it.

One day towards the end of September Bakunin came to me, looking particularly preoccupied and somewhat solemn.

'The Warsaw Central Committee,' he said, 'has sent two members to negotiate with us. One of them you know – Padlewski; the other is Giller, a veteran warrior; he took a walk from Poland to the mines in fetters, and as soon as he was back he set to work again.[11] This evening I will bring them to see you, and tomorrow we will meet in my room. We *must define our relations once for all.*'

My answer to the officers was being printed at that time.

'My programme is ready, I will read my letter aloud.'

'I agree with your letter, you know that; but I don't know whether they will altogether like it; in any case, I imagine that it won't be enough for them.'

In the evening Bakunin arrived with three visitors instead of two. I read them my letter. While we were talking and while I was reading, Bakunin sat looking alarmed, as relations are at an examination, or as lawyers are when they tremble lest their client should let something slip out and spoil the whole game of the defence that has been so well arranged, if not strictly in accordance with the truth, anyway for a successful finish.

I saw from their faces that Bakunin had guessed right, and that they were not particularly pleased by what I read them. 'First of all,' observed Giller, 'we shall read the letter to you from the Central Committee.' Milovicz[12] read it; the document, with which readers of *The Bell* are familiar, was written *in Russian*, not quite correctly, but clearly. It has been said that I translated it from the French and altered the sense. That is *not true*. All three spoke Russian well.

The sense of the document was to tell the Russians through us that the provisional Polish Government agreed with us and adopted as its basis for action: '*The recognition of the right of the peasantry to the land tilled by them, and the complete self-determination of every people, the right to determine its own destiny.*'

This manifesto, Milovicz said, bound me to soften the interrogative and hesitating form of my letter. I agreed to some changes, and suggested to them that they might accentuate and define more clearly the idea of the self-determination of provinces; they agreed.

This dispute over words showed that our attitude towards the same questions was not identical.

Next day Bakunin was with me in the morning. He was dissatisfied with me, thought I had been too cold, as though I did not trust them.

'Whatever more do you want? The Poles have never made such concessions. They express themselves in other words which are accepted among them as an article of faith; they can't possibly at the first step, as they hoist the national flag, wound the sensitive popular feeling.'

'I fancy all the same, that they really care very little about the land for the peasants and too much about the provinces.'

'My dear fellow, you will have a document in your hands corrected by you and signed in the presence of all of us; whatever more do you want?'

'I do want something else though!'

'How difficult every step is to you! You are not a practical man at all.'

'Sazonov used to say that before you did.'

Bakunin waved his hand in despair and went off to Ogarëv's room. I looked mournfully after him. I saw that he was in the middle of his revolutionary debauch, and that there would be no bringing him to reason now. With his seven-league boots he was striding over seas and mountains, over years and generations. Beyond the insurrection in Warsaw he was already seeing his 'Glorious and Slav Federation'[13] of which the Poles spoke with something between horror and repulsion; he already saw the red flag of 'Land and Freedom' waving on the Urals and the Volga, in the Ukraine and the Caucasus, possibly on the Winter Palace and the Peter-Paul fortress, and was in haste to smooth away all difficulties somehow, to conceal contradictions, not to fill up the gullies but to fling a skeleton bridge across them.

'There is no liberation without land.'

'You are like a diplomat at the Congress of Vienna,' Bakunin repeated to me with vexation, when we were talking afterwards with the representatives of the provisional Polish Government in his room. 'You keep picking holes in words and expressions. This is not an article for a newspaper, it is not literature.'

'For my part,' observed Giller, 'I am not going to quarrel about words; change them as you like, so long as the main drift remains the same.'

'Bravo, Giller,' cried Bakunin gleefully.

'Well, that fellow,' I thought, 'has come with his horses shod for any season; he will not yield an inch in fact, and that is why he so readily yields in words.'

The manifesto was corrected, the members of the Zhond[14] signed it. I sent it off to the printing-press.

Giller and his companions were fully persuaded that we represented the focus abroad of a whole organisation which depended upon us and would at our command join them or not join them. For them what was essential lay not in words nor in theoretical agreements; they could always tone down their *profession de foi* by interpretations, so that its vivid colours would have altered, faded and vanished.

That the first nuclei of an organisation were being formed in Russia there was no doubt. The first fibrils, the first threads could be discerned with the naked eye; from these threads, these knots, a vast web might be woven, given time and tranquillity. All that was true, but it was not there yet, and every violent shock threatened to ruin the work for a whole generation and to tear asunder the first lacework of the spider's web.

That is just what, after sending the Committee's letter to the press, I said to Giller and his companions, telling them of the prematureness of their rising. Padlewski knew Petersburg too well to be surprised by my words – though he did assure me that the vigour and ramification of the League of Land and Freedom went much further than we imagined; but Giller grew thoughtful. 'You thought,' I said to him, smiling, 'that we were stronger? You were right. We have great power and influence, but that power rests entirely on public opinion, that is, it may evaporate all in a minute; we are strong through the sympathy with us, through our harmony with our own people. There is no organisation to which we could say, "Turn to the right or turn to the left." '

'But, my dear fellow, all the same ...' Bakunin was beginning, walking about the room in excitement.

'Why, is there?' I asked him, and stopped.

'Well, that is as you like to call it; of course, if you go by the external form, it is not at all in the Russian character, but you see....'

'Let me finish; I want to explain to Giller why I have been so insistent about words. If people in Russia do not see on your standard "Land for the Peasants" and "Freedom for the Provinces", then our sympathy *will do you no good at all but will ruin us*; because all our strength rests on their hearts beating in unison with ours. Our hearts may beat more strongly and so be one second ahead of our friends; but they are bound to us by sympathy and not by duty!'

'You will be satisfied with us,' said Giller and Padlewski.

Next day two of them went off to Warsaw, while the third went off to Paris.

The calm before the storm came on. It was a hard, dark time, in which it kept seeming as though the storm would pass over, but it drew nearer and nearer. Then came the *ukaz* 'juggling' with the levying of recruits;[15] this was the last straw; men who were still hesitating to take the final and irrevocable step dashed into the fray. Now even the Whites began to go over to the side of the rebellion.

Padlewski came again; the decree was not withdrawn. Padlewski went off to Poland.

Bakunin was going to Stockholm quite independently of Lapinski's expedition,[16] of which no one thought at the time. Potebnya turned up for a brief moment and vanished after Bakunin. A plenipotentiary from 'Land and Freedom' came from Petersburg *via* Warsaw at the same time as Potebnya; he described with indignation how the Poles who had summoned him to Warsaw had done nothing. He was the first Russian who had seen the beginning of the rebellion; he told us about the murder of the soldiers, about the wounded officer who was a member of the Society. The soldiers thought that this was treachery and began exasperatedly to beat the Poles. Padlewski, who was the chief leader in Kovno, tore his hair, but was afraid to act openly in opposition to his followers.

The plenipotentiary was full of the importance of his mission and invited us to become the *agents* of the League of Land and Freedom. I declined this, to the extreme surprise not only of Bakunin but even of Ogarëv. I said that I did not like this hackneyed French term. The plenipotentiary was treating us as the *Commissaires* of the Convention of 1793 treated the generals in the distant armies. I did not like that either.

'And are there many of you?' I asked him.

'That is hard to say: some hundreds in Petersburg and three thousand in the provinces.'

'Do you believe it?' I asked Ogarëv afterwards. He did not answer. 'Do you believe it?' I asked Bakunin.

'Of course; but,' he added, *'well, if there are not as many now there soon will be!'* and he burst into a roar of laughter.

'That is another matter.'

'The essence of it all is the giving support to feeble beginnings; if they were strong they would not need us,' observed Ogarëv, who was always dissatisfied with my scepticism on these occasions.

'Then they ought to come to us frankly admitting their weakness and asking for friendly help instead of proposing the stupid job of being agents.'

'That is youth,' Bakunin commented, and he went off to Sweden.

And after him Potebnya went off too. With heartfelt sorrow I said goodbye to him. I did not doubt for one second that he was going straight to destruction.[17]

A few days before Bakunin's departure Martyanov came in, paler than usual, gloomier than usual; he sat down in a corner and said nothing. He was pining for Russia and brooding over the thought of returning home. A discussion of the Polish rebellion sprang up. Martyanov listened in silence, then got up, preparing to go, and suddenly stopped in front of me, and said gloomily:

'You must not be angry with me, Olexander Ivanovich; that may be so or it may not, but, anyway, you have done for *The Bell*. What business had you to meddle in Polish affairs? The Poles may be in the right, but their cause is for their gentry, not for you. You have not spared us, God forgive you, Olexander Ivanovich; you will remember what I say. I shall not see it myself; I am going home. There is nothing for me to do here.'

'You are not going to Russia, and *The Bell* is not ruined,' I answered him.

He went out without another word, leaving me heavily weighed down by this second prediction and by a dim consciousness that a blunder had been made.

Martyanov did as he had said; he returned home in the spring of 1863 and went to die in penal servitude, exiled by his 'People's Tsar' for his love for Russia and his trust in him.

Towards the end of 1863 the circulation of *The Bell* dropped from two thousand or two thousand five hundred to five hundred, and never again rose above one thousand copies. The Charlotte Corday from Orlov and the Daniel from the peasants had been right.[18]

Written at Montreux and Lausanne at the end of 1865.

*

NOTES

The ascription of the numbered Notes, drawn from several sources, is as follows:

C.G.: Constance Garnett, translator of *My Past and Thoughts* (6 vols, 1924–7).

H.H.: Humphrey Higgens, who revised Constance Garnett's edition (4 vols, 1968). His edition is referred to in the Notes as, e.g. 'Higgens, I, 236'.

A.S.: *The Collected Works* of A. S. Herzen, edited and published by the Academy of Sciences of the USSR (30 vols, Moscow 1954–65).

A.K.: Aileen Kelly. New notes for the present selection.

Alexander Herzen's own notes, signalled * in the text, appear at the foot of the page.

Cross-references are also made to the translation by J. D. Duff, in The World's Classics edition (Oxford, 1980): *Childhood, Youth and Exile* (Parts I and II of *My Past and Thoughts*): e.g. 'Duff, pp. 1 ff.'

CHAPTER I (pp. 3–10)

1 Vera Artamonovna was Herzen's nanny. See Duff, pp. 3–5. (A.K.)
2 The Senator: Ltv Alekseyevich Yakovlev (1764–1839), Herzen's uncle, his father's middle brother. See Duff, pp. 12–19. (A.K.)
3 Ivan Alekseyevich Yakovlev (1766–1846). See Duff, Pt I passim. (A.K.)
4 Count Aleksey Andreyevich Arakcheyev (1769–1834), the most powerful of Alexander's ministers, dreaded for his cruelty. (A.K.)
5 Count Ernst Johann Biron (1690–1772), Grand Chamberlain, and immensely powerful favourite of the Empress, Anna Ivanovna. (A.K.)

CHAPTER II (pp. 11–23)

1 'The Chemist': Aleksey Aleksandrovich Yakovlev (1795–1868), Herzen's cousin. See Duff, pp. 87–93. (A.K.)
2 Natalya Alexandrovna Zakharin, an illegitimate daughter of The Chemist's father. (H.H.)
3 'I love . . . each other': these words are not in the original of the letter of 1 December 1837 from which Herzen is quoting. (A.S.)
4 Sasha Vyrlin. Sasha is the affectionate diminutive form of both Alexander and Alexandra. (H.H.)
5 Tayan Petrovna Passek, *née* Kuchin (1810–89). Her memoirs throw interesting sidelights on Herzen's narrative. (A.K.)
6 Emilia Aksberg. (A.S.)

7 Skalozub, a character in A. S. Griboyedov's celebrated play, *Woe from Wit*, is the typical coarse, ignorant, blustering military bully. (C.G.)

8 On 9 April 1835. (A.S.)

9 L. V. Passek. (A.S.)

10 Xavier Saintine (1798–1865), a French writer of whose many plays and stories only *Picciola, or the Prisoner's Flower* is still well known. (C.G.)

11 Nikolay Platonovich Ogarëv (1833–77). Herzen's distant cousin and lifelong friend and collaborator. See Duff, pp. 60–6. (A.K.)

12 From Pushkin's *Yevgeny Onegin*, canto VII, stanza 41. (C.G.)

CHAPTER III (pp. 24–41)

1 See the account of Herzen's exile from April 1835 in Duff, pp. 143–271. (A.K.)

2 Karl Ivanovich Sonnenberg, formerly tutor of Herzen's friend Ogarëv. See Duff, pp. 60–3. (A.K.)

3 Half-boots; skull-cap; under-tunic. (H.H.)

4 The reference is probably to Bulgarin, a journalist in close relations with Benckendorf (Chief of the Secret Police). This Bulgarin made many petty personal attacks on Pushkin, who in a well-known poem addresses him by the name Vidocq-Figlyarin. (C.G.) F.-E. Vidocq (1775–1859) was head of the Sûreté, after himself being a criminal. (H.H.)

5 P. P. Medvedev. (A.S.)

6 i.e. 'thy'. She now uses the more familiar form. (H.H.)

7 Matvey Savelevich (c. 1821–43), Herzen's servant from December 1835. (A.K.)

8 Alexander Lavrentevich Vitberg (1787–1855). Architect and artist, friend of Herzen during the latter's Vyatka exile. See Duff, pp. 242–52. (A.K.)

9 Kirill Yakovlevich Tyufyayev (1775–after 1840). Governor in Vyatka from 1834–7. See Duff, Pt II, chs VII and X. (A.K.)

10 See Duff, p. 206. (A.K.)

11 Dmitry Yurevich Shemyaka (1420–53), appanage prince of Zvenigorod. This name came to be used for an unjust, mercenary judge. (A.S.)

12 A portrait of Herzen by A. L. Vitberg, commissioned for Natalya's birthday, 22 October 1836. (A.S.)

13 Natalya had sent the bracelet to Herzen. (A.S.)

CHAPTER IV (pp. 42–51)

1 Herzen and his friend Ogarëv espoused Saint-Simonism in 1833. See Duff, pp. 133–6. (A.K.)

2 Filaret (Vasily Mikhaylovich Drozdov), Metropolitan (1782–1867). See Duff, pp. 107–8. (A.K.)

3 Dmitry Pavlovich Golokhvastov (1796–1849), a cousin of Herzen. Assistant Director of Moscow University from 1831 and Director 1847–9. (A.K.)

4 A reference to the break between Emilia Aksberg and N. M. Satin. (A.S.)

5 M. Yu. Lermontov's poem, 'A Testament'. (A.S.)

6 Yegor Ivanovich Herzen (1803–82). See Herzen's note in Higgens, I, 13. (A.K.)

7 Nikolay Khristoforovich Ketscher (c. 1806–86). Doctor and poet-translator. A friend of Herzen's youth and member of his circles in the 1830s and 1840s. (A.K.)

CHAPTER V (pp. 52–71)

1 9 May 1838: the date of the marriage of Herzen and Natalie Zakharin. (A.S.)

2 See Duff, pp. 166–73 for Herzen's stay in the Krutitsky police-barracks. (A.K.)

3 See Duff, p. 5. (A.K.)

4 The horses in the *troika*. (H.H.)

5 Title of a poem by Ivan Ivanovich Kozlov (1779–1840). (A.K.)

6 Nikolay Ivanovich Astrakov (1809–42), a teacher of mathematics, and Tatyana Alexeyevna (1814–92), his wife, an authoress, and a link between Herzen and his friends in Russia. (A.S.)

7 Published in *The Telescope* in 1836 and known as the 'First Philosophical Letter', this was in fact dedicated to E. D. Panova. In it Chaadayev denounced Russia as stagnant and barbarian because it had developed outside the mainstream of world history as represented by Western Europe. On the enormous impact of Chaadayev's letter on Russia's cultured élite, see above, pp. 174 ff. (A.K.)

8 He is referring ironically to what was said by Sebastian, Minister for Foreign Affairs under Louis-Philippe, in the Chamber of Deputies in 1831, when France was against any military intervention in Poland: 'L'ordre règne à Varsovie.' (H.H.)

CHAPTER VI (pp. 72–4)

1 Date of the birth of the Herzens' first child, Alexander. (A.S.)

CHAPTER I (pp. 77–112)

1 Dates of their meeting in Moscow during Herzen's secret visit, and of their arrival and marriage in Vladimir. (A.S.)

2 From Schiller's poem 'Resignation'. (A.S.)

3 Vadim Vasilevich Passek (1808–42), a member of Herzen's

student circle, and husband of his cousin Tatyana. See Duff, pp. 111–18. (A.K.)

4 In Schiller's play of that name (II, ii). (A.S.)

5 English in the original. (C.G.)

6 Nikolay Diomidovich Oransky, secretary of the Governor of Moscow, was secretary of both Commissions of Enquiry dealing with the affair for which Herzen was arrested in 1834. See Duff, p. 184. (A.K.)

7 Nikolay Vladimirovich Stankevich (1813–40). (A.K.)

8 Mikhail Alexandrovich Bakunin (1814–76). (A.K.)

9 Vissarion Grigorevich Belinsky (1811–48) was the greatest of Russian critics. (A.K.)

10 Mikhail Grigorevich Pavlov (1793–1840), professor of physics, mineralogy and agriculture at Moscow University from 1820. (A.S.)

11 Lorenz Oken (1779–1855), German biologist and philosopher, strongly influenced by Schelling. (A.K.)

12 Timofey Nikolayevich Granovsky (1813–55). (A.K.)

13 Arnold Ruge (1802–80) began his political career with six years' imprisonment in connection with the *Burschenschaft* movement, founded the *Deutsche Jahrbücher* (the journal of the Young Hegelian School), and some ten years later *Die Reform*, a more definitely political paper. From 1849 he lived in England, advocated a universal democratic state, and wrote many books, of which his autobiography is now of most interest. (C.G.)

14 V. P. Botkin lived in the Maroseyka, where Granovsky, Belinsky, and Bakunin stayed with him at various times. Moscow University is in the Mokhovaya. (A.S.)

15 See Herzen's note in Higgens, I, 138.

16 Aleksey Petrovich Yermolov (1772–1861), soldier and diplomat, a hero of the war of 1812. (A.K.)

17 From Pushkin, 'Verses written during a Night of Sleeplessness.' (A.S.)

18 Edward Gans (c. 1798–1839), Hegelian and professor of law. (A.K.)

19 V. I. Filippovich. They met at the end of 1841. (A.S.)

20 Andrey Towjanski (1799–1878) was a Pole, and at one time a member of the Society of Philarets (see below, p. 497, n. 7). He held that there were many Messiahs, of whom Napoleon was one and himself another. (C.G.)

21 His real name was Ganneau (d. 1852). He was a sculptor, and an adventurer well known in Paris between 1820 and 1850. He went in for being a god and called his religion *évadisme* (from Eve and Adam), and himself Mapah (from *mater* and *pater*). He suggested to Dumas that the latter should become his chief disciple. (C.G.)

22 Larisa Dmitriyevna Filippovich. (H.H.)

23 Théroigne de Méricourt, called 'l'Amazone de la liberté', assisted at the taking of the Bastille, and became a popular heroine.

Later on she was publicly whipped by a crowd of women, and lost her reason in consequence of this outrage. (C.G.)

24 In 1762, when she took an active part in the palace revolution which brought Catherine II to the throne. (A.S.)

25 K. G. Carus (1789–1869), a distinguished German physiologist, author of numerous works on anatomy, physiology, and allied subjects. (C.G.)

26 Aesop is said to have been the slave of Xanthos, a philosopher of Samos. (H.H.)

27 Stefan Yavorsky was a famous monk and theologian of the eighteenth century. (C.G.)

28 *Moskvityanin*: monthly journal (1841–56) edited by M. P. Pogodin and S. P. Shevyrëv.

29 Nikolay Alexeyevich Polevoy (1796–1846), author, journalist, historian. (A.S.)

30 *The Tarantas*, a story by Count Vladimir Alexandrovich Sollogub (1814–82), author of comedies and novels satirizing the official class; 'Parasha', an early poem by Turgenev; Mochalov, the great Russian actor, particularly famous for his playing of Hamlet. (C.G.)

31 A reference to the Table of Ranks introduced by Peter the Great, which divided each of the three branches of the state service (civil, military, and court) into fourteen ranks or classes. The intention was to establish a hierarchy based on ability rather than birth. (A.K.)

32 Anna ribbon: the Order of St Anne, one of the eight decorations of Imperial Russia. (A.K.)

33 Peasant cap and long homespun coat. (C.G.)

34 Ivan Vasilevich Kireyevsky (1806–56). (A.K.)

35 Prince Vladimir Fëdorovich Odoyevsky (1804–69), writer and music critic. (A.K.)

36 Ivan Petrovich Sakharov (1807–63), a well-known archaeologist and ethnographer, was a doctor of medicine and lecturer on palaeontology. His discoveries are now regarded somewhat sceptically, but he did much for Russian antiquarian study. (C.G.)

37 Alexander Kazimirovich Meyendorf (1788–1865), a writer on historical and geographical subjects. (C.G.)

38 Ioakinth Bichurin (1777–1853), a monk and at one time an archimandrite, head of the Orthodox Mission to Peking, and later on a translator from the Chinese in the Ministry of Foreign Affairs, was an authority on Chinese language and history. (C.G.)

39 In all probability, A. A. Krayevsky. (A.S.)

40 Vasily Andreyevich Zhukovsky (1783–1852), a poet whose translations had an immense influence on the development of Russian culture. (A.K.)

41 A degree between 'candidate' (i.e. graduate) and doctor. (H.H.)

42 Aleksey Vasilevich Koltsov (1809–42), a poet, best known for his imitations of folksongs. (A.K.)

43 The reference is to the open letter in which Belinsky expressed his passionate indignation at the *Correspondence with Friends* by Gogol. (C.G.)

44 Ogarëv, having left Russia for ever, came to Herzen in London on 9 April 1856. (A.S.)

45 N. P. Sungurov (b. 1805) organized a secret society in Moscow at the end of the 1820s, was arrested in 1831, and sentenced to penal servitude in Siberia, where he died. See Duff, pp. 118–21. (A.K.)

46 Vasily Petrovich Botkin (1811–69), son of a wealthy tea-merchant, a close friend of Belinsky, and a dilettante literary critic and philosopher; Mikhail Nikiforovich Katkov (1818–87), in the 1840s a liberal publicist, later prominent on the Russian Right. (A.K.)

47 The Aksakov brothers: Ivan Sergeyevich (1823–86) and Konstantin Sergeyevich (1817–60); Yuri Fëdorovich Samarin (1819–76); Aleksey Stepanovich Khomyakov (1804–60); the Kireyevsky brothers: Ivan Vasilevich (1806–56) and Pëtr Vasilevich (1808–56). (A.K.)

48 Vasily Ivanovich Krasov (1810–55), a poet, at one time professor of literature at Kiev. His brother Ivan was a teacher of history in the Petersburg secondary schools. (C.G.)

CHAPTER II (pp. 113–39)

1 The Decembrist rising took place on Senate Square. (A.K.)

2 Peter the Great died on 28 January 1725 (A.S.)

3 Alexey Vasilevich Timofeyev (1812–83), a writer of forgotten poems. Nestor Vasilevich Kukolnik (1809–68), was a schoolfellow of Gogol's, and a very popular writer of stories and dramas in the most extreme romantic style. (C.G.)

4 The Moravian Brethren, called after the little town of Herrnhut in Saxony, where they settled in 1722, are a Protestant sect who abjure military service, the taking of oaths, and all distinctions of rank. (C.G.)

5 Count Alexander Khristoforovich Benckendorf (1783–1844), Chief of the Secret Police. (A.K.)

6 Leonty Vasilevich Dubelt (1792–1862), Chief of Staff of the Corps of Gendarmes from 1835 and Director of the Third Division (1839–56). (A.S.)

7 Herzen was summoned to the Third Division on 7 December 1840. The child, Ivan, was born two months later, in February 1841 (A.S.)

8 The lifeguards of Ivan IV. (H.H.)

9 See Duff, p. 166. (A.K.)

10 Paul Louis Courier (1772–1825), a learned and brilliant writer of political pamphlets and letters, who discovered a complete manuscript of Longus's *Daphnis and Chloe*, of which he published a French translation. (C.G.)

11 Not long before his death Benckendorf went over to Catholicism; he died while he was returning from abroad in the steamer *Hercules*. (A.S.)

12 Count Dmitry Nikolayevich Bludov (1785–1864), a Russian statesman (Minister of the Interior, 1832–8). (A.K.)

13 The tyrannical Paul I. (A.K.)

14 Count Pëtr Alexeyevich Pahlen (1745–1826), a Russian statesman, prominent in the reign of Catherine II. (A.K.)

15 Count Alexey Fëdorovich Orlov (1786–1861); from 1844 Chief of Gendarmes and head of the Third Department. (A.S.)

16 See Higgens, IV, 158 ff.

17 Count Mikhail Fëdorovich Orlov (1788–1842), a soldier who had fought in the campaigns against Napoleon; later a member of the Decembrist secret society. On Herzen's acquaintance with Orlov, who attempted to intervene on Herzen's behalf at the time of his arrest, see Duff, pp. 147–50. (A.K.)

18 Prince Mikhail Semënovich Vorontsov (1782–1856), from 1823–44, governor-general of Novorossiya and vice-gerent of the province of Bessarabia. (A.K.)

19 Mikhail Mikhaylovich Speransky (1772–1839), a liberal and an able and trusted minister of Alexander I, was suddenly dismissed and on 17 March 1812 was relegated to Nizhny Novgorod. (H.H.)

20 The biographical details of Bolgovsky, given by Herzen, are not accurate. (A.S.)

21 Perun was the god of sky and of thunder, the chief god of the ancient Slavs. (C.G.)

CHAPTER IV (pp. 147–57)

1 from *Zahme Xenien*, Bk VIII. The second and third lines are omitted. (A.S.)

2 These extracts are inserted here by the author in a slightly altered form. [*Note to the Russian edition of 1921*.]

3 Andrey Stepanovich Kashentsov, valet of Herzen's uncle, the Senator. See Duff, p. 28. (A.K.)

4 Mikhail Semënovich Shchepkin (1788–1863), a famous actor, founder of the realist tradition in Russian dramatic art. (A.K.)

5 Yevgeny Fëdorovich Korsh (1810–97), member of Herzen's circle in the 1840s; editer of *Moskovskie vedomosti* (*Moscow News*). (A.K.)

6 The eponymous Cossack hero of a historical romance by N. V. Gogol, published in the collection *Mirgorod* (1835). (A.K.)

CHAPTER V (pp. 158–72)

1 Written in England. (C.G.)

2 Zurbaran, the Spanish painter of religious subjects. A well-known picture of his is the monk castigating himself before an effigy of the Madonna. (C.G.)

3 Adapted from Addison's *Cato*, V, i. (A.S.)

4 A joking reference to a popular song about a Madrid grisette published by V. P. Botkin in one of his 'Letters from Spain', which contain many observations of the people and the character and beauty of the women. (A.S.)

5 Pëtr Grigorevich Redkin (1808–91), professor of law at Moscow University 1835–48. (A.K.)

6 Puchta, a German professor and authority on Roman law; Savigny, a German university teacher of French origin, and an authority on modern jurisprudence. (C.G.)

7 Count Sergey Grigorevich Strogonov (1794–1882), Director of Moscow University 1835–47. (A.K.)

8 Rotteck, a German university teacher and authority on Roman law. (C.G.)

9 Dmitry Lvovich Kryukov (1809–45), professor of Roman philology and history in Moscow University, member of Herzen's circle of the 1840s. (A.K.)

10 Ludwig Johann Rudolph Agassiz (1807–73), a Swiss natural scientist. (A.K.)

11 From N. P. Ogarëv's poem, 'To Iskander'. (A.S.)

12 The character of Stepan Trofimovich in the earlier chapters of Dostoevsky's novel, *The Possessed* [or *The Devils*] is said to have been modelled on T. N. Granovsky. (C.G.)

13 Gaspard de Coligny (1519–72), a French Protestant leader, killed in the massacre of St. Bartholomew; the Girondins, a political group formed in 1791, were hostile to the Jacobins (Montagnards), and were eliminated by them. (A.K.)

14 Osip Ivanovich Senkovsky (1800–58). Editor of a journal *Biblioteka dlya chteniya* (*The Library of Good Reading*). (A.K.)

15 The friendship between N. P. Ogarëv and T. N. Granovsky was broken off at the end of the 1840s. In particular Granovsky disliked Ogarëv's (second) marriage to N. A. Tuchkov. (A.S.)

16 Granovsky wrote at once on receiving *Sovremennik* (*The Contemporary*) for September 1847, which contains *Doctor Krupov*, a story by Herzen. The letter has survived. (H.H.)

17 Ciceruacchio, a popular leader (his real name was Angelo Brunetti) in Rome, who had great influence from 1847 onwards, supporting the reforms of Pius IX, and was active in bringing about the proclamation of a republic in February 1849. He was captured and shot with his sons in the following July. (C.G.)

18 The Emperor of Austria, Franz Joseph, who ascended the throne in December 1848, at the age of eighteen. (C.G.)

19 Alfred Talandier (1822–90). A French radical who emigrated to England after the 1848 revolution. (A.K.)

20 Originally the lifeguards of Ivan the Terrible; used here of the Secret Police. (H.H.)

21 Heinrich Ritter (1791–1869), philosopher and historian of philosophy. (A.K.)

22 The accusation that V. S. Pechërin had burnt a Protestant Bible was not confirmed at his trial, and he was acquitted. (A.S.)

23 Herzen published in his article 'The Place of Execution' (1857) an annihilating delineation of Nikita Ivanovich Krylov (not the famous writer of fables, who was Ivan Andreyevich), a professor of history at Moscow University, and a state censor 1839–44. (A.S.)

CHAPTER VI (pp. 173–97)

1 From P. Béranger, L'opinion de ces demoiselles. (A.S.)

2 Play by A. S. Griboyedov (1825). (A.K.)

3 See p. 491, n. 17 above.

4 In English. (H.H.)

5 From Pushkin's The Captain. (A.S.)

6 It was not this decoration that Chaadayev received after the battle at Kulm, but the order of St Anne, fourth class. (A.S.)

7 A misquotation from Pushkin's lines 'To a Portrait of Chaadayev'. (A.S.)

8 Count Fëdor Ivanovich Tolstoy (1782–1846), a rich landowner and adventurer, notorious for his violent eccentricities. See Duff, pp. 209–10. (A.K.)

9 Sergey Pavlovich Shipov (1789–1876) had participated in the suppression of the Polish Rising of 1830–1. (A.K.)

10 Stepan Petrovich Shevyrëv (1806–64) was professor of literature at Moscow University, and Mikhail Petrovich Pogodin (1800–75) was professor of history. They were co-editors of the reactionary nationalist periodical Moskvityanin (The Muscovite). (A.K.)

11 Published in 1842. (H.H.)

12 Alexander Ivanovich Turgenev (1785–1846), a distinguished person in his own day, now chiefly remembered for having been a very good friend to Pushkin, was one of the Turgenevs of Simbirsk, and not related to the famous I. S. Turgenev, who has left among his critical articles an obituary notice of this Alexander Ivanovich. (C.G.)

13 German woman writer (1771–1833), who had a literary salon in Berlin which was frequented by Russian writers. (A.S.)

14 The Larins and Lensky are characters in Pushkin's Yevgeny

Onegin. Chatsky is the hero of Griboyedov's *Woe from Wit*, and Famusov is a character in the same play. (C.G.)

15 Sobakevich and Nozdrëv are characters in Gogol's *Dead Souls*. (C.G.)

16 Dmitry Nikolayevich Sverbeyev (1799–1876). (A.K.)

17 Avdotya Petrovna Yelagin (1789–1877), mother of the Kireyevsky brothers. (A.K.)

18 Ilya of Murom, a legendary hero, a defender of his country and righter of wrongs. The tales of him appear to take form in the eleventh and twelfth centuries. (H.H.)

19 See I Corinthians, 1:20. (H.H.)

20 Fëdor Lukich Moroshkin (1804–57), professor of law at Moscow University. (A.S.)

21 The Kireyevskys' mother did not share their views. This is the only explanation I can discover for his being described as 'alone in his own family'. (C.G.)

22 From Lermontov's translation of Goethe's poem, 'Wandrers Nachtlied'. (C.G.)

23 Pavel Ivanovich Pestel (1793–1826). One of the leaders of the Decembrist conspiracy, hanged in 1826. (A.K.)

24 '*Ein guter Mensch in seinem dunklen Drange . . .*' Goethe, *Faust*, Prolog im Himmel. (H.H.)

25 Baron Haxthausen was a learned German who after a visit to Russia at this period wrote an account of the Russian system of land tenure. (C.G.)

26 *The Bell*, sheet 90.

27 Both were authors of a very low order: N. I. Grech (1787–1867), being a trifle more stupid and unscrupulous than F. V. Bulgarin (1789–1859), who was scurrilous in his attacks on Pushkin, and commonly believed to be in the pay of the police. (C.G.)

28 Written in 1855.

29 Arnold Heeren (1760–1842), a professor of history at Göttingen University. (A.S.)

30 Mikhail Trofimovich Kachenovsky (1775–1842), of humble origin and largely self-educated, became editor of the *Vestnik Yevropy* (*European Messenger*), and professor of fine arts, of literature, and later on of history at Moscow University. His sceptical attitude on historical subjects gave offence, and he was superseded in the Chair of History by Pogodin. (C.G.)

31 Johann Georg Adam Forster (1754–94), a German naturalist and writer. He accompanied Cook on his second voyage round the world (1772–4); and in 1793 was sent as a delegate from the town of Mayence to the Convention. (A.S.)

32 The name of the Peace which in 1774 put an end to six years of war between Russia and Turkey. (A.S.)

33 N. M. Yazykov (1803–46), a friend of Pushkin's. (C.G.) Herzen is uniting under one title three poems of Yazykov's: 'To Konstantin Aksakov', 'To our Opponents', and 'To Chaadayev'. Herzen mistakenly took to himself the reference to 'a servant wearing the gorgeous livery of Western culture': the poet himself declares in one of his letters that he had had T. Granovsky in mind. (A.S.)

34 F. F. Vigel, a fiery propagandist of Yazykov's polemical verse. (A.S.)

35 This Glinka, Fëdor Nikolayevich (1786–1880), a poet and one of the founders of the League of Public Welfare, out of which the Decembrist movement developed, was exiled in 1826, but allowed to return later. He was a literary character of the mild and pious type. (C.G.)

36 *The Contemporary* (*Sovremennik*). (A.S.)

37 Printed in Herzen's *The Bell*, sheet 90. (A.S.)

38 From F. Schiller, 'Der Alpenjäger'. (A.S.)

CHAPTER VII (pp. 198–206)

1 See Duff, pp. 107–8. (A.K.)

2 Christian Burmeister (1709–85), a German philosopher, author of the textbook *Logic*, which was translated into Russian. (A.S.)

3 Christian Wolf (1679–1754), a German philosopher and mathematician; teaching founded on his work was given in Russian theological institutions from the end of the eighteenth century. (A.S.)

4 *Raskolniki*, lit. 'schismatics'. (H.H.)

5 The Lycée was transferred in 1844 from Tsarkoye Selo to Petersburg. (A.S.)

6 Nikolay Mikhailovich Satin (1814–73), poet and translator, a member of Herzen's student circle. (A.K.)

7 Stanislav Gabriel Worcell (1799–1857), one of the leaders of the Polish nationalist-liberation movement. (A.S.)

CHAPTER VIII (pp. 207–13)

1 This happened in September 1844, i.e. nearly two years before the death of Herzen's father. (A.S.)

2 See p. 271 above. (H.H.)

3 Benckendorf's report to the Tsar of 7 April 1843 contained the solicitation of S. G. Strogonov, then Director of Moscow University, that Herzen might be permitted, in consequence of his wife's illness, to go to Italy for some months. The report is endorsed in the hand of Nicholas I: '*peregovorim*' – 'Let us talk it over'; and there is a postscript by Benckendorf: '*ne pozvolyayet*' – 'He does not give leave'; the document was countersigned by Dubelt on 9 April 1843. (A.S.)

4 The name of the first stage-post on the way from Petersburg to Moscow. A signpost stood at the crossroads indicating the directions of Moscow, Tsarskoye Selo, Peterhof, and Petersburg. (A.S.)

5 i.e. N. P. Ogarëv, then living on his Penza estate. (A.K.)

6 Quoted from p. 221 above. Chërnaya Gryaz (Black Mire) was the second stage-post on the way from Moscow to Petersburg, where the first change of horses was made; here, by tradition, farewells were arranged for those starting on a journey. (A.S.)

CHAPTER I (pp. 217–20)

1 Written between 1847 and 1852. Various versions of the text were published in French, German, Russian and English in the 1850s. (A.K.)

2 Maria Kasparovna Ern (Reichel), Maria Fëdorovna Korsh, and Luiza Ivanovna Haag, Herzen's mother. (H.H.)

3 Tatyana, the nurse of Herzen's daughter, Natalya (Tata), who accompanied the family as far as the frontier post of Taurogen: see p. 213. (A.S.)

4 From Letters from France and Italy, Letter I. (A.S.)

5 From Pushkin's Yevgeny Onegin, VII, xix. (A.S.)

6 Nikolay Ivanovich Sazonov (1815–62). See pp. 375–90 above.

CHAPTER II (pp. 221–30)

1 In fact on 19 January 1847. (A.S.)

2 Written at the end of 1853.

3 On leaving Russia, Herzen went first to Paris, where he arrived on 20 March 1847. He left for Italy in October 1847 and proceeded to Rome, where the political euphoria of Italy's national awakening was at its most intense. He returned to France after the February Revolution of 1848, arriving on 5 May in Paris, where he remained until June of the following year. (A.K.)

4 This refers to the successful rising in Milan on 18 March 1848 against the Austrian dominion in Lombardy. Charles Albert, King of Piedmont, also declared war on Austria. (A.S.)

5 Apparently Giuseppe Benai (b. 1817). (A.S.)

6 From The Dream, IX, by Byron. English in the original. (C.G.)

7 See Duff, pp. 152–3. (A.K.)

8 Louis-Eugène Cavaignac (1802–1857), the youngest of the three distinguished Frenchmen of that name, was commander-in-chief in 1848, and an unsuccessful candidate for the presidency of the Republic when Louis-Napoleon (afterwards Napoleon III) was elected on 10 December 1848. (C.G.)

9 Louis de Lamoricière (1806–1865), a prominent politician and

general, was exiled in December 1848, and afterwards took command of the Papal troops. (C.G.)

CHAPTER III (pp. 231–49)

1 Count Ksawery Branicki (1812–79), one of the leaders of the Polish aristocratic emigration. (A.K.)

2 Jules Lechevalier (1806–62), who wrote chiefly on economics. (A.S.)

3 'Charles Edmond' (1822–99), a writer and publicist. (A.K.)

4 The July monarchy fell on 24 February 1848. (A.S.)

5 Pierre-Jean David (d'Angers) (1788–1856), a French sculptor (not to be confused with the painter, David). (A.K.)

6 For Herzen's portrait of Worcell, see Higgens, III, 1132–54. (A.K.)

7 A. Mickiewicz was in Russia in 1824–5 to participate in the work of a secret students' society, the Philarets. He met and made friends with a number of Russian writers, including Pushkin. (A.K.)

8 On 22 April 1844. (A.S.)

9 Armand Barbès (1809–70), called the 'Bayard de la démocratie', was a people's representative in 1848, imprisoned in 1849, and set free in 1854. (C.G.)

10 i.e. Louis-Napoleon. (A.S.)

11 In 1848 Adam Mickiewicz had shown himself to be a revolutionary and a democrat; but, like many other workers in the Polish national-liberation movement, he was imbued with Napoleonic illusions, which came out particularly clearly after 10 December 1849, when Napoleon I's nephew, Louis Bonaparte, was elected President of France. Mickiewicz saw in him the continuation of the work of Napoleon I, which had been the work of the Revolution. (A.S.)

12 Ramon de la Sagra (1798–1871), a Spanish economist, took part in the revolutionary movement of 1848 in France, and wrote advocating the views of Proudhon. In 1854 he returned to Spain, and was several times elected a member of the Cortes. He was not seventy in 1848, as Herzen mistakenly assumes, but fifty. (C.G.)

13 Chagrin at the defeat of 1830–1, and the loss of hope in the liberation of Poland bred a mood of mysticism among the Polish emigrants and contributed to the rise of ideas of Messianism. Polish Messianism was the teaching of the peculiar rôle of 'martyred Poland' in the history of peoples, according to which the Polish-people-Messiah was redeeming and liberating all other peoples by its sufferings and its struggle. The representative of this doctrine was Joseph Wronski, a mathematician and philosopher, the author of *Messianism*. The Messianic-mystic mood overcame Mickiewicz too, and induced his spiritual crisis in the 1830s and early 1840s, when he joined the mystic sect of the adventurer

Andrey Towjanski (see above, p. 488, n. 20). (A.S.)

14 A mountain chain in Old Castile, where the French defeated the Spanish in 1808. (C.G.)

15 Louis Narbonne-Lara (1755–1813), a French diplomatist; an adjutant of Napoleon's in 1812. (H.H.)

16 Written in 1856.

17 P. F. O. Rayer was a distinguished French physician and the author of numerous medical works. (C.G.)

18 A character in Gogol's *Dead Souls*. (C.G.)

19 The Jacobins were called *Montagnards* in 1793 because they occupied the highest seats in the Parliament. In 1848–9 the name was given to the supporters of Ledru-Rollin in the Constituent Assembly. (A.S.)

20 Théophile Thorez (1807–69), a French republican of the left and a participator in the Revolution of 1848. (H.H.)

21 Emmanuel Arago (1812–96), the son of the more distinguished F. D. Arago, who was one of the members of the Provisional Government formed after the *coup d'état* of 24 February 1848. The others were Ledru-Rollin, Dupont de l'Eure, Garnier-Pagès, Lamartine, Crémieux, Marrast, Flocon, and Louis Blanc. (C.G.)

22 Jules Bastide (b. 1800), a publicist and politician, was Minister for Foreign Affairs in 1848. He had had an eventful career, and for two years took refuge in England after escaping from prison, where he was thrown for taking part in the riots that followed the funeral of Lamarque in 1832. (C.G.)

23 Nicolas Changarnier (1793–1877), a prominent politician and general, was exiled at the *coup d'état* of 1851, but lived to serve in the Franco-Prussian War of 1870. (C.G.)

24 Hermann Müller-Strübing (1810–93) emigrated to Paris after taking part in the revolutionary events in Berlin in 1848. (A.K.)

25 Alexandre-Auguste Ledru-Rollin (1808–74), member of the French Provisional Government of 1848, and one of the earliest advocates of universal adult suffrage. (C.G.)

26 Auguste-Joseph Guinard (b. 1799), had been one of the first to proclaim the republic in February 1848, and at the head of the 8th Legion had occupied the Hôtel de Ville. (C.G.)

27 Henri-Joseph Forestier (b. 1787), was a painter of merit. He was colonel of the 8th Legion of the National Guard. (C.G.)

28 After the crushing of the demonstration of 13 June 1849, in Paris, and of a series of manifestations in the provinces, the government of Odilon Barrot deprived thirty-three *Montagnards* of their status as deputies, declared them to be enemies of the state and delivered them over for trial. Those who had emigrated were tried *in absentia*. (A.S.)

29 On 15 June 1849. (A.S.)

30 Karl Blind (1826–1907), a writer and revolutionary, was sentenced to eight years' imprisonment for the part he took in the insurrections in South Germany, but was rescued by the mob. He settled in England, where he continued journalistic and propaganda work up to the time of his death. (C.G.)

31 Oran, a province of Algeria in which the French carried on a successful campaign against Abd-el-Kader in 1847. (C.G.)

32 Félix Pyat (1810–89), a journalist, dramatic writer, and communist leader, supported Ledru-Rollin's appeal to the French people in 1849, and on its failure escaped to Switzerland and then to London, where he was a member of the 'European Revolutionary Committee'. He returned to France at the amnesty of 1870, and was in 1871 one of the leaders of the Commune, on the fall of which he again escaped to London. He was condemned to death in his absence, but was again pardoned in 1880. (C.G.)

CHAPTER IV (pp. 250–7)

1 *Epilogue to 1849*: first printed in 1850 in the *émigré* paper *New-York Abend-Zeitung*; incorporated in the Russian edition of *From the Other Shore*, London, 1855. (A.K.)

2 Terence's *Andria*, I, ii. (A.S.)

3 Byron: see p. 224 of this volume. (H.H.)

CHAPTER V (pp. 258–73)

1 Dante, *Inferno*, canto XV, st. 63. (A.S.)

2 Karl Vasilevich Nesselrode (1780–1862), Russian Minister for Foreign Affairs, 1816–56. (A.S.)

3 Mlle Le Normand (1772–1843) was a well-known fortune-teller of the period. (C.G.)

4 Boris Nikolayevich Chicherin (1828–1904), a Right Hegelian professor of law. A member of the circles of Herzen's Moscow friends in the late 1840s, he later developed a conservative view of Russian history which brought him into public conflict with Herzen. See Higgins, II, 624–9, and above p. 438. (A.K.)

CHAPTER VI (pp. 274–93)

1 Lajos Kossuth (1802–94), leader of the national revolutionary movement in Hungary, and of the Hungarian revolutionary forces in 1848–9. (A.K.)

2 In fact four months. (A.S.)

3 Felice Orsini (1819–58). For Herzen's portrait of Orsini, see Higgens, II, 701–4. (A.K.)

4 Pestel was the leader of the Union of the South and Ryleyev of the Union of the North, which combined in the attempt to over-

throw the autocracy and establish constitutional government in Russia on 14 December 1825. (C.G.)

5 Herzen is referring to the Third Division, which watched the conduct of Russians abroad and had prepared for Nicholas I an order urgently ordering him to return to Russia. This summons, and the repressive measures taken against Herzen when he refused to obey it, took place in 1849, not 1850. (A.S.)

6 'Cantonists' were soldiers' sons educated at the government's expense and afterwards sent into the army. (C.G.)

7 James Fazy (1797–1878). A Swiss politician, president of the Canton of Geneva at the time of the political emigration from France in 1849. See the chapter 'James Fazy and the Refugees', Higgens, II, 720–40. (A.K.)

8 Karl Vogt (1817–95), German natural scientist. After participating in the 1848 revolution he emigrated to Switzerland. See Higgens, II, 787–96, for Herzen's description of his relations with the Vogt family. (A.K.)

9 Rossini's *William Tell* was so permeated with ideas of liberation that at the demand of the censorship it was presented on the stage in Austria and Russia under the name of '*Charles the Bold*' with the subject of the libretto altered and dealing with the fifteenth-century Duke of Burgundy who was killed in battle with the Swiss at Murten in the canton of Freiburg. In Russia *William Tell* was put on with a new libretto by R. M. Zotov. (A.S.)

10 Herzen is referring here to the facts that the future Emperor of the French emigrated to Switzerland as a young man, was naturalized at Thurgau and became a Swiss citizen; and that the Tsar Alexander II married Princess Maria of Darmstadt. (A.S.)

11 A reference to Pushkin's *Graf Nulin*, l. 57. (A.S.)

12 The King of Piedmont had earned among the monarchs of Europe the name of 'a rebel', although all his actions were aimed at crushing the revolutionary movement. He communicated his object confidentially to Nicholas I who, however, broke off diplomatic relations with Piedmont, and did not restore them even after the abdication of Charles Albert and the accession of Victor Emmanuel II. This is why there was no Russian ambassador in Piedmont in 1851. (A.S.)

13 Herzen saw Ponsa de la Martino on 12 July 1851. A constitution had been proclaimed in Piedmont on 4 March 1848. (A.S.)

10 See above, p. 486, n. 7.

11 A reference to Griboyedov, *Woe from Wit*, IV, v. (A.S.)

12 Character from Turgenev's novel *Fathers and Children*. (A.K.)

CHAPTER V (pp. 467–83)

1 The word used by Bakunin is *prokhvost*, which is the German *Profoss* (Eng. provost), a military policeman; sometimes an executioner. (H.H.)

2 Bakunin used this phrase for the first time in an article called 'Reaction in Germany: a Frenchman's Account', published over this pseudonym in *Deutscher Jahrbücher* (17–21 October 1842). (A.S.)

3 A fortress in Saxony where political offenders were imprisoned. (A.S.)

4 30 May–12 June 1848. Bakunin adhered to the radical Left. The leading part in the conference was played by the Czech Liberal *bourgeoisie* who put forward an idea for the transformation of the Austrian Empire into a federation of Slav States under the aegis of the Habsburg monarchy. (A.S.)

5 The word means 'thimble' in Russian. (C.G.)

6 On Herzen's relations with Worcell and the Polish emigrants in London, see Higgens, III, 1131–54. (A.K.)

7 'Vivat Polonia', '10 April and the Murders in Warsaw', 'Mater Dolorosa' and others published in *The Bell*. (A.S.)

8 The Grand Duke Constantine Nikolayevich was made viceroy of Poland in 1862. On the day of his arrival in Warsaw, in June of that year, an attempt was made on his life. (A.S.)

9 Charles Edmond was the pseudonym of Chojecki. (A.S.)

10 The country between the lower Neman and the Windau, the inhabitants of which are closely related to the Lithuanians, and from the fourteenth century were included in Lithuania. (C.G.)

11 Agaton Giller was sent to Siberia. When he returned to Poland at the end of the 1850s he joined the right wing of the insurrectionary organisation. (A.S.)

12 Vladimir Milovicz, a Polish revolutionary. (A.K.)

13 'Slava' is the Russian for 'glory'. (C.G.)

14 The Polish Provisional Government. (H.H.)

15 In the autumn of 1862 the tsarist authorities issued an *ukaz* on the levying of recruits in the Kingdom of Poland, which was put into effect according to lists made up beforehand. The authorities tried by this means to put an end to the revolutionary movement in Poland. The conduct of the levy in January 1863, caused the start of the rising. (A.S.)

16 An expedition of about 200 revolutionaries, principally Poles, under the command of a Colonel Lapinski, which set off from

London by steamer for the Lithuanian coast, to reinforce the Polish insurgents. The ship, deserted by its captain and crew, got no further than the Swedish port of Malmö, and the expedition was a fiasco. For Herzen's account of the expedition and Bakunin's part in it, see Higgens, III, 1379 ff. (A.K.)

17 A. A. Potebnya commanded a detachment which participated directly in the Polish rising, and died in March 1863, in the battle at Peschanaya Skala. (A.S.)

18 Herzen is thinking first of his meeting (see above, pp. 442 ff.) with a young Russian woman who announced to him: 'Your friends and supporters will abandon you', and secondly, indicating Martyanov by the name of the prophet Daniel, of his remark about the decrease of the influence of *The Bell* after Herzen's utterances in defence of Poland in revolt. (A.S.)

CHAPTER I (pp. 294–312)

1 This is the section of which Turgenev wrote: 'It is written in tears and blood.' (C.G.) For an objective account of this part of Herzen's life, see E. H. Carr, *The Romantic Exiles* (London, 1933). (H.H.)

2 Higgens, II, 1004–9. (A.K.)

3 E. B. Granovsky. (A.S.)

4 *Letters from France and Italy*, Letter IX. (H.H.)

5 A. V. Koltsov. (A.S.)

6 All this was written to Natalya Alexeyevna Tuchkov, who married Ogarëv in 1850. (C.G.)

7 Natalya (afterwards the wife of Ogarëv) and Yelena (afterwards the wife of Satin, one of the Herzens' circle), with their parents. (C.G.)

8 A great friend of the Herzens', the sister of Yevgeny Korsh. (C.G.)

9 Pavel Vasilevich (1813–87), a literary critic, the author of a brilliant memoir of the Moscow circles of the 1830s and 1840s. (A.K.)

10 Thomas Couture (1815–79), historical painter. The reference is to his picture *Romains de la décadence*. (C.G.)

11 Natalya (Natalie), *née* Zakharin, Herzen's wife. Natalya Tuchkov was first Ogarëv's mistress and then his second wife. She had three children by Herzen: Yelizaveta (Liza), called Herzen, and the twins, Alexey and Yelena, called Ogarëv. (H.H.)

12 From Schiller's poem *Resignation*. (A.S.)

13 Staking myself. (H.H.)

14 The broad purple stripe on the *toga*, worn by Roman senators. (H.H.)

15 See Herzen's 'Ends and Beginnings', Letter 5 in Higgens IV, 1719–23. (A.K.)

CHAPTER II (pp. 313–24)

1 On 23 June 1848, there was a rising of Paris workers which was rigorously put down by General Cavaignac. (A.S.)

2 In the night of 23–4 April 1848, 800 revolutionary emigrants, forming the so-called 'German Democratic Legion', headed by Herwegh, crossed the Rhine with the intention of giving armed assistance to the rising in Baden. This campaign, which Marx openly attacked, ended in complete defeat. The emigrants' battalion gave battle on 27 April to superior forces of Württemburg government troops and, after bitter, prolonged fighting, were compelled to retire. Nearly half the Legion were taken prisoners. The failure of the campaign had a fatal effect on Herwegh's reputation. The German emigrants in Paris accused him, apparently on insufficient foundation, of cowardly conduct on the

battlefield and also of various abuses. The details given by Herzen below about the Baden campaign are drawn from the same unreliable source. (A.S.)

3 A reference to Griboyedov's play *Woe from Wit*, where the ruined proprietor of a theatrical company of serfs bewails the selling off of his 'Amours and Zephyrs' – his actors. (A.S.)

4 Bettina von Arnim (1785-1859). The author of a book well-known in its time: *Goethes Briefwechsel mit einem Kinde*. (H.H.)

5 In 1837 Herwegh stayed for a long time at the house of August Follen at Zürich. The Follen family afforded Herwegh much assistance both moral and material, and helped him to have his first book of verse published. A. Follen was Karl Vogt's grandfather on his mother's side. (H.H.)

6 Ferdinand Flocon (1800-66), was secretary of the Provisional Government of February 1848. (C.G.)

7 Marc Caussidière (1808-61), a French politician and revolutionary. (C.G.)

8 F. K. F. Hecker (1811-81), one of the leaders of the democratic and socialist party of Germany; he attempted on the outbreak of the German revolution to convert the preliminary convention (*Das Vorparlament*) into a permanent assembly. At the head of a band of revolutionaries he invaded Baden, was defeated and fled to Switzerland and from there to America, where he took to farming. In the Civil War he raised a regiment of Germans and commanded a brigade. (C.G.)

9 See Higgens, III, 1086 ff. (H.H.)

10 According to Herwegh's biographers, this statement is not in accordance with the facts. (A.S.)

11 The flat of a sword. (H.H.)

CHAPTER III (pp. 325-30)

1 Ludwig Simon (1810-72), he was a lawyer of Trèves, elected in 1848 to the German National Assembly, as a member for the extreme Left. In July 1849 he escaped to Switzerland, and in his absence was sentenced to death at Trèves. He published in 1855 a book, *Aus dem Exil*. (C.G.)

2 Aleko, the hero of Pushkin's poem 'The Gypsies'. (C.G.)

3 See pp. 266 ff. above. (H.H.)

4 20 November 1850. (A.S.)

CHAPTER IV (pp. 331-9)

1 Jacques Guyot (1810-76), an Italian painter. (A.S.)

2 Serialized in *Notes of the Fatherland* (1845, no. 12; 1846, no. 4)

and separately published in St Petersburg in 1847 (2nd edn. London, 1859). (A.K.)

3 It is clear from a letter of Herwegh's to George Sand that as early as 1842 he knew her *Horace* 'to the last line'. (A.S.)

4 Partly taken from a letter to Haug, written in March 1852. (Herzen.) Ernst Haug took part in the revolution in Vienna in 1848 and in the defence of the Roman republic the following year. (A.K.)

CHAPTER V (pp. 340–3)

1 Vladimir Aristovich Engelson (Weitberg) (1821–57). See above, pp. 391–425. (A.K.)

2 Antonio Mordini (1819–1902) took part in the revolutionary movement in Italy in 1848–9. (A.K.)

CHAPTER VI (pp. 344–54)

1 The title of some verses in *Les rayons et les ombres* by Victor Hugo, who borrowed the phrase from Virgil's Æneid. (A.S.)

2 This should read 'from Geneva'. (A.S.)

3 From a letter, written in Herzen's hand, from Natalya Herzen to Natalya Tuchkov, dated 2–22 March 1852. (A.S.)

4 So literally in the Russian; i.e. 'They are dead'. (H.H.)

5 Marie-Edmond Tessier du Motay, a French chemist; he took part in the 1848 revolution. In 1852 he was tutor to Herzen's son. (A.K.)

6 *Letters from France and Italy.*

7 From *Oceano Nox* by Victor Hugo. (The last line should begin, 'Heurtant de vos fronts morts'. (A.S.)

CHAPTER VII (pp. 355–65)

1 Sir Hudson Lowe (1769–1844), governor of St Helena during Napoleon's exile there. (A.K.) 'Comme nous observions que dans ce climat brûlant nous restions sans ombrage, sans un seul arbre, *On en plantera*, nous a-t-il dit. Quel mot atroce!' Comte de Las Cases: *Le Mémorial de Ste-Hélène* (1951), I, 535. (H.H.)

2 In a letter of 30 June 1852, to M. K. Reichel, Herzen gives the name of this woman, saying that Herwegh 'gets his board from Mme Koch and is a sort of hired favourite'. (A.S.)

3 Ivan Gavrilovich Golovin (1816–90), a Russian *émigré* publicist. For Herzen's relations with him see Higgens, III, 1397–418. (A.K.)

4 Carlo Pisacane (1818–57), see Higgens, II, 695 ff., Agostino Bertani (1812–86); Gaicomo Medici (1819–82), see Higgens II, 704–5. (A.K.) The full text of the letter, dated from Genoa, 23 July 1852, is as follows (translated from the Italian): 'We, the undersigned, being

invited by Mr Herzen (of whose friendship we are proud in consequence of his distinguished merits) to express our opinion concerning his conflict with Mr Herwegh, hereby depose that, in refusing in the present circumstances a duel with Mr Herwegh, Mr Herzen has acted in accordance with our convictions.' The letter is signed by Enrico Cosenz, Carlo Pisacane, Gia(como) Medici, Luigi Mezzacappo, Agostino Bertani, and Camillo Boldoni. Felice Orsini and Antonio Mordini added their signatures to this deposition shortly afterwards. (A.S.)

5 In 1849 with the Italian revolutionary troops under Garibaldi against the French interventionists. (A.S.)

6 Pushkin was killed in a duel by G. C. Dantès, a man who began his career as a page of the Duchesse de Berry, was recommended by her to Nicholas I, and was adopted by the Dutch ambassador to Russia. After killing Pushkin he was expelled from Russia. Later on he became a chamberlain of Napoleon III. (C.G.)

7 The English translation of this letter of Herwegh's, partially preserved in draft form in his notebook, is produced by E. H. Carr, op. cit., pp. 108–9. (A.S.)

8 This passage is puzzling, for 'Charles Edmond' and Chojecki were the same person. (H.H.)

CHAPTER VIII (pp. 366–74)

1 N. Paccelli, a music teacher, the wife of a musician. (A.S.)

2 30 April 1852. (A.S.)

3 From the poem by Pushkin, 'I wander through the noisy streets.' (A.S.)

4 Natalie Tuchkov-Ogarëv is meant. (C.G.)

5 Léonce, Marquis de Vogué (1805–77), a prominent légitimiste. (A.K.)

CHAPTER I (pp. 375–90)

1 Pushkin, 'To a Portrait of Chaadayev' (not quite accurately quoted). (A.S.)

2 Field-Marshal Prince Alexander Alexandrovich Prozorovsky (1732–1809). (A.K.)

3 N. M. Satin. (A.S.)

4 In Pushkin's Yevgeny Onegin. (A.S.)

5 Dmitry Vladimirovich Venevitinov (1805–27), a gifted poet and populariser of German Idealism in Russia. (A.K.)

6 Mikhail Vasilevich Butashevich-Petrashevsky (1821–66), leader of a secret society whose members (who included Dostoevsky) were condemned to death, and led to the scaffold. At the last moment their sentence was commuted to penal servitude in Siberia. (A.K.)

7 See Higgens, III, 1133.

8 Field-Marshal Count Ivan Fëdorovich Paskevich-Erivansky (1782–1856) in 1849 commanded the Russian army which crushed the revolution in Hungary. (A.K.)

9 Polish émigrés in Paris. (A.K.)

10 The reference is to the *Voyage du jeune Anarcharsis en Grèce*, by Jean-Jacques Barthélemy (1779). (C.G.) Anarcharsis is said to have been a Scythian philosopher of the sixth century BC, who visited Athens in the time of Solon. On his return to Scythia he tried to introduce Athenian laws and was killed by his brother, the king. (H.H.)

11 The Riviera di Chiaia, a famous, or notorious, street in Naples. (H.H.)

12 Louis Bouilhet (1822–69) was a great friend of Flaubert, with whom he collaborated. His own works include *Hélène Peyron*, and a very successful drama, *La Conjuration d'Amboise*. (C.G.)

13 See p. 236. (H.H.)

14 Charles Ribeyrolles (1812–1861) a talented writer on *La Réforme*, the organ of the Extreme Left, of which Flocon was editor. (C.G.)

15 See p. 488, n. 21 above. (H.H.)

16 Karl Werder (1806–93), German Hegelian philosopher. (A.K.)

17 Louis Blanc, author of *L'Histoire de Dix Ans*, one of the most widely read books of the epoch. (C.G.)

18 Ernst Haug (A.S.) The real name of Messenhauser was Caesar Wengel, a soldier and writer, who took an active part in the rising of 1848, first at Lemberg and then in Vienna. (C.G.)

19 i.e. Worcell, Sazonov, Golynski (Alexander Vikentevich, a Polish émigré. (H.H., A.K.)

CHAPTER II (pp. 391–425)

1 Engelson had been at the Tsarskoye Selo Lycée. He left it in 1839 without completing the course, but there is no evidence at all that he was expelled. (A.S.)

2 In the Ministry of Foreign Affairs (1844–8). (A.S.)

3 This was published in 1843: the article must in fact have been 'Dilettantism in Science'. (A.S.)

4 Pierre Levassor (1808–70), Paul Grasso (1800–60), French comic actors. (A.K.)

5 Ivan Fëdorovich Gorbunov (1831–95), author and performer of scenes from popular life. (A.K.)

6 Nikolay Gavrilovich Chernyshevsky (1828–89), Russian radical publicist. (A.K.)

7 François Marie Charles Fourier (1772–1837) believed that society as a whole must be organised so as to give full scope to co-operative industry and to the harmonious evolution of human nature. By his scheme each phalanx, of about 1,600 persons, was to inhabit a

phalanstery, or common building. (H.H.)

8 Composed by Col. Ya. I. Rostovtsev and brought into force in 1845. Its spirit determined the whole system of education, and was based on the complete subordination of the personality to the authority of the autocracy and the military-bureaucratic establishment. (A.S.)

9 Natalya Herzen was buried on 3 May 1852. (A.S.)

10 In August 1852. (A.S.)

11 Maria, the illegitimate daughter of C. E. Chojecki. (A.S.)

12 Herzen is referring to an incident which he describes on pp. 364–5. Engelson was the first person he told of his family drama. (A.S.)

13 Haug. (A.S.)

14 Engelson is referring to Chojecki's attitude to the part played by Haug in the conflict between Herzen and Herwegh. (A.S.)

15 In the middle of 1854. (A.S.)

16 Joseph Domengé, tutor to Herzen's son. (H.H.)

17 Quoted in English. (H.H.)

18 The Letter to Alexander was published in *Polyarnaya zvezda* (*The Pole Star*), 1855, Bk I, 11–14. (A.K.) *The Pole Star* was the name of the paper edited by Ryleyev, one of the five Decembrists hanged by Nicholas in 1825. On the anniversary of their execution Herzen brought out the first number of his paper of the same name. (C.G.)

19 *The Pole Star*, 1855, Bk. I.

20 Herzen is confusing a quarrel in March 1855, with one in May of that year which was indeed their final one. (A.S.)

21 Zeno Swentoslawski (b. 1811) took part in the Polish revolutionary movement; emigrated to England. (A.K.)

22 W. J. Linton, a friend of Mazzini and the author of a series of sketches of Italian, French, and Polish exiles, and of Herzen, called *European Republicans*.

23 On 6 May 1855. (A.S.)

24 Baroness Malwida von Meysenbug, authoress of *Memoirs of an Idealist*, was a great friend of Wagner, and also of Nietzsche, whom she cared for at times with motherly kindness. At this date she was living in Herzen's house as the governess of his children, the youngest of whom, Olga, remained in her charge for many years. (C.G.)

25 The reference is to Orsini's attempt to assassinate Napoleon III on 14 January 1858. (A.K.)

26 In English. (H.H.)

27 With this ends that part of *My Past and Thoughts* which was corrected by the author in its final form and published in four volumes. The chapter which follows is now published for the first time, and is that for which, as Herzen himself more than once says, he wrote all the rest. [*Note to the Russian edition—1921.*]

The reference is to 'A Family Drama': see, in this edition, pp. 294–374 ff. (A.K.)

CHAPTER II (pp. 433–45)

1 The 'apogee' of *The Bell* was from 1857–62. (H.H.)

2 Nikolay Onufryevich Sukhozanet (1794–1871), Minister of War 1856–61. (A.K.)

3 Count Vladimir Fëdorovich Adlerberg (1790–1884), Court Chamberlain from 1852. (A.K.)

4 Major-General Aleksey Ivanovich Saburov, brother-in-law of N. M. Satin, a member of the circles of Herzen's youth. (A.K.)

5 From A. S. Griboyedov: *Woe from Wit*, III, xxii. (A.S.)

6 Under this pseudonym, for the purposes of the conspiracy, Herzen concealed the surname of P. L. Pikulin, who left Russia at the beginning of June 1855, and, before arriving at Herzen's in London, spent some time in Vienna (hence the pseudonym). Pikulin had connections with friends of Herzen in Moscow, and brought him a letter from T. N. Granovsky with a postscript by N. Kh. Ketscher. (A.S.)

7 Ludwig Chernetski, a Polish *émigré* who ran the Free Russian Press in London. (A.K.)

8 Nicholas Trübner (1817–84), Herzen's English publisher. (A.K.)

9 For the effect on Herzen of the death of Granovsky, see above, pp. 169 ff. (H.H.)

10 See Higgens I, 73. (A.K.)

11 See Duff, p. 63. (A.K.)

12 In 1859. (A.S.)

13 Yakov Ivanovich Rostovstsev (1803–60), a member of the committees formed to draw up legislation for the emancipation of the serfs. (A.K.)

14 Konstantin Dmitrievich Kavelin (1818–85) a liberal historian who wrote on political and economic questions. (A.K.)

15 The Vehmgerichte were medieval German tribunals which tried capital charges and were greatly dreaded for their severity. (C.G.)

16 See Higgens, II, 624 ff. (A.K.)

17 In the novel of I. I. Lazhechnikov (1792–1860), *The House of Ice*, it is described how Biron's servants, by pouring buckets of water over a disobedient Ukrainian, turned him into a statue of ice. (A.S.)

18 Written in 1864. (H.H.)

19 The era of the orgy of reaction, when part of liberal society turned to nationalism, chauvinism and a state of mind reminiscent of the Black Hundreds, is called by Herzen after two men who personified reaction—Mikhail Katkov and Mikhail Muravëv. (A.S.)

20 In 1853 Prince L. V. Kochubey shot at his steward, I. Saltzmann,

and wounded him; yet not only did he remain unpunished but, by bribing the judges, he managed to get Saltzmann put in prison. Herzen devoted a series of notices in *The Bell* in 1858 and 1859 to the exposure of these abuses, with the result that the case was reviewed and Saltzmann was set free. (A.S.)

21 In *The Bell* for 1 November 1858. (A.S.)

22 In his article *Mikhail Semënovich Shchepkin* (*The Bell*, 1 October 1863). Alexander Mikhailovich Gedeonov (1791–1867) was Director of the Imperial Theatres from 1833–58. (A.K.)

23 Prince Alexander Mikhailovich Gorchakov (1798–1883), Minister of Foreign Affairs from 1856, State Chancellor from 1867. (A.K.)

24 The session was held on 28 January 1861, and was reported in *The Bell* on 1 March. (A.S.)

25 Mikhail Nikolayevich Muravëv (The Hangman) (1796–1866). See Higgens, I, 216–17. (A.K.)

26 See Gogol, *Dead Souls*, Pt I, ch. 4. (H.H.)

27 Count Viktor Nikitich Panin (1801–74), Minister of Justice from 1841–61. (A.K.)

28 Bribes, given at the beginning of the nineteenth century in the form of paper money on which was the signature of A. N. Khovansky, the director of the State Bank. (A.S.)

29 *The Bell* frequently printed revelations of various financial machinations in court and government circles, in which a conspicuously nasty part was played in particular by Mina Ivanovna Burkov, the brazen, covetous favourite of the Court Chamberlain, V. F. Adlerberg. (A.S.)

30 Gogol, *The Government Inspector*, V, viii. (A.S.)

31 P. M. Slivitsky and I. N. Arngoldt were shot by sentence of a court-martial in 1862 for their part in a revolutionary military organisation. A. A. Potebnya twice visited Herzen in London (see above, pp. 474 ff.), participated in the Polish rising and was killed in battle with Russian forces, 4 March 1863. A. A. Krasovsky was sentenced to death in 1862 for spreading proclamations among the soldiers; this was commuted to twelve years' forced labour, which he served at the Alexandrovsky factory at Nerchinsk at the same time as N. G. Chernyshevsky. In 1868 Krasovsky made an unsuccessful attempt to escape, and put an end to his own life. V. A. Obruchev was sentenced to forced labour in 1862 for spreading a proclamation called 'Velikoruss'. All the above were celebrated or commemorated at various times in *The Bell*. (A.S.)

32 Siegmund Serakowski (1827–63), a Polish revolutionary, executed for his participation in the rising in Litva in 1863. (A.K.)

33 Great fires broke out in Petersburg on 28 May 1862, and burned for several days. The tsarist government took advantage of this to carry out a series of repressive, terrorist measures against the

revolutionary camp and endeavoured, by spreading provocative rumours that the fires were the work of students set on by Herzen and N. G. Chernyshevsky, to produce a wave of hatred of the revolutionary young people and their leaders. (A.S.)

34 Nikolay Filippovich Pavlov (1805–64), editor of *Nashe vremya* (*Our Time*), a journal subsidised by the Ministry of the Interior. (A.K.)

CHAPTER III (pp. 446–54)

1 Zygmunt Padlewski (Podlewski) (1835–63), one of the leaders of the Polish rising of 1863. (A.K.)

2 Herzen gives here a short biographical sketch of V. I. Kelsiev, who was temporarily a member of the circle of revolutionary emigrants and became one of the first renegades in the history of the Russian liberation movement. (A.S.)

3 Written when he was imprisoned by the Third Division and addressed to the head of the gendarmes. (A.S.)

4 Pushkin's poems 'To the Calumniators of Russia', 'The Anniversary of Borodino' and, obviously, 'Stanzas', which progressive contemporaries regarded as a departure by the poet from his attitude of liberality. V. G. Belinsky wrote of this in his 'Letter to Gogol' (1847). (A.S.)

5 Sitka I. is off the coast of Alaska; Unalaska is one of the Aleutian Islands. The islands belonged to Russia until 1867, when they were sold to the United States along with Alaska. (A.S.)

6 Utopian socialist group headed by Mikhail Vasilevich Petrashevsky. Its main members, who included Dostoevsky, were arrested in 1849. (A.K.)

7 A vivid description of this part of London in 1845 is given by Engels in his work *The Condition of the Working Class in England*. (A.S.)

8 A district in the Bukovina settled by Russian *raskolniki*. (C.G.)

9 Besides supplying Kelsiev with material for his work on the schism and schismatics, Herzen printed and published them at the Free Russian Press in London, 1860–3. (A.S.)

10 Caesar, Don Carlos, and Vadim Passek. See above, p. 78.

11 Pëtr Alexeyevich Martyanov (1835–65), a former serf who travelled to London to persuade the editors of *The Bell* to give publicity to his grievances against his master. While he was in London his pamphlet 'People and State' was published there, and *The Bell* printed his 'Letter to Alexander II'. Martyanov's views were an odd mixture of hatred of the gentry and nobility and officialdom, and a utopian faith in the 'good' tsar and the possibility that the tsar might summon a *Zemskaya duma*. He spoke against the idea of a Russo-Polish revolutionary alliance, and did

not approve Herzen's attitude to the rising in Poland. Returning voluntarily to Russia, he was arrested on 12 April 1863 and condemned to five years' hard labour and lifelong relegation to Siberia. (H.H., A.K.)

12 Kelsiev had written to Herzen from Tulcea on 23 April 1864, asking Herzen to send him from London some *émigrés*, if there were any, who were decent people, willing to work and live honorably. (A.S.)

13 And this was the awful Tulcea agency with connections with the revolution all over the world, setting fire to Russian villages with money from Mazzini's funds, formidable two years after it had ceased to exist, and even now mentioned in the literature of the detectives and in Katkov's *Police News*.—Herzen means Katkov's *Moscow News*, in which he published an article 'Herzen's Secret Agency at Tulcea' on 2 September 1865. (A.S.)

14 The cutting is pasted into Herzen's MS. (A.S.)

CHAPTER IV (pp. 455–66)

1 G. N. Vyrubov, who had emigrated from Russia in 1864. Herzen was critical of his views and activities, calling him 'Frenchman' and 'doctrinaire' and censuring him for his complete break with his native country. (A.S.)

2 Griboyedov, *Woe from Wit*, I, ii. (H.H.)

3 P. A. Bakhmetev was with Herzen in London in August 1857. (A.S.)

4 Herzen's account corresponds exactly to the contents of a letter of 31 August 1857, from P. A. Bakhmetev to Herzen. After he left London Bakhmetev was not seen in Europe and nothing is known of his further fortunes. His fund remained untouched until 1869. In July of that year, at Ogarëv's request, Herzen gave him half the money in the fund, which was passed on to S. G. Nechayev. After Herzen's death the other half, too, was given by Ogarëv to Nechayev. Herzen's apprehension was realised and Bakhmetev's fund was squandered on adventurist enterprises of Bakunin and Nechayev which were of no use to the Russian revolutionary movement. (A.S.)

5 Stanislas Tchorszewski, a Polish exile who owned a bookshop in London. (A.K.)

6 Two characters in Gogol's *Dead Souls*. (H.H.)

7 A character in Gogol's *Dead Souls*, who was naked when he met Chichikov, the hero of the story. (C.G.)

8 N. V. Sokolov, the economist of *Russkoye slovo* (*The Russian Word*), applied the word 'rascal', in English, to John Stuart Mill in an article in the issue of July 1865. (A.S.)

9 From D. V. Davydov's poem, 'A Contemporary Song'. (A.S.)

10 See above, p. 486, n. 7.

11 A reference to Griboyedov, *Woe from Wit*, IV, v. (A.S.)

12 Character from Turgenev's novel *Fathers and Children*. (A.K.)

CHAPTER V (pp. 467–83)

1 The word used by Bakunin is *prokhvost*, which is the German *Profoss* (Eng. provost), a military policeman; sometimes an executioner. (H.H.)

2 Bakunin used this phrase for the first time in an article called 'Reaction in Germany: a Frenchman's Account', published over this pseudonym in *Deutscher Jahrbücher* (17–21 October 1842). (A.S.)

3 A fortress in Saxony where political offenders were imprisoned. (A.S.)

4 30 May–12 June 1848. Bakunin adhered to the radical Left. The leading part in the conference was played by the Czech Liberal *bourgeoisie* who put forward an idea for the transformation of the Austrian Empire into a federation of Slav States under the aegis of the Habsburg monarchy. (A.S.)

5 The word means 'thimble' in Russian. (C.G.)

6 On Herzen's relations with Worcell and the Polish emigrants in London, see Higgens, III, 1131–54. (A.K.)

7 'Vivat Polonia', '10 April and the Murders in Warsaw', 'Mater Dolorosa' and others published in *The Bell*. (A.S.)

8 The Grand Duke Constantine Nikolayevich was made viceroy of Poland in 1862. On the day of his arrival in Warsaw, in June of that year, an attempt was made on his life. (A.S.)

9 Charles Edmond was the pseudonym of Chojecki. (A.S.)

10 The country between the lower Neman and the Windau, the inhabitants of which are closely related to the Lithuanians, and from the fourteenth century were included in Lithuania. (C.G.)

11 Agaton Giller was sent to Siberia. When he returned to Poland at the end of the 1850s he joined the right wing of the insurrectionary organisation. (A.S.)

12 Vladimir Milovicz, a Polish revolutionary. (A.K.)

13 'Slava' is the Russian for 'glory'. (C.G.)

14 The Polish Provisional Government. (H.H.)

15 In the autumn of 1862 the tsarist authorities issued an *ukaz* on the levying of recruits in the Kingdom of Poland, which was put into effect according to lists made up beforehand. The authorities tried by this means to put an end to the revolutionary movement in Poland. The conduct of the levy in January 1863, caused the start of the rising. (A.S.)

16 An expedition of about 200 revolutionaries, principally Poles, under the command of a Colonel Lapinski, which set off from

London by steamer for the Lithuanian coast, to reinforce the Polish insurgents. The ship, deserted by its captain and crew, got no further than the Swedish port of Malmö, and the expedition was a fiasco. For Herzen's account of the expedition and Bakunin's part in it, see Higgens, III, 1379 ff. (A.K.)

17 A. A. Potebnya commanded a detachment which participated directly in the Polish rising, and died in March 1863, in the battle at Peschanaya Skala. (A.S.)

18 Herzen is thinking first of his meeting (see above, pp. 442 ff.) with a young Russian woman who announced to him: 'Your friends and supporters will abandon you', and secondly, indicating Martyanov by the name of the prophet Daniel, of his remark about the decrease of the influence of *The Bell* after Herzen's utterances in defence of Poland in revolt. (A.S.)